HANDBOOK OF RESEARCH ON FOOD PROCESSING AND PRESERVATION TECHNOLOGIES

Volume 5

Emerging Techniques for Food Processing, Quality, and Safety Assurance

Handbook of Research on Food Processing and Preservation Technologies, 5 volume set:

Volume 1: Nonthermal and Innovative Food Processing Methods

Volume 2: Nonthermal Food Preservation and Novel Processing Strategies

Volume 3: Computer-Aided Food Processing and Quality Evaluation Techniques

Volume 4: Design and Development of Specific Foods, Packaging Systems, and Food Safety

Volume 5: Emerging Techniques for Food Processing, Quality, and Safety Assurance

Innovations in Agricultural and Biological Engineering

HANDBOOK OF RESEARCH ON FOOD PROCESSING AND PRESERVATION TECHNOLOGIES

Volume 5

Emerging Techniques for Food Processing, Quality, and Safety Assurance

Edited by
Monika Sharma, PhD
Megh R. Goyal, PhD, PE
Preeti Birwal, PhD

AAP APPLE ACADEMIC PRESS

First edition published 2022

Apple Academic Press Inc.
1265 Goldenrod Circle, NE,
Palm Bay, FL 32905 USA

4164 Lakeshore Road, Burlington,
ON, L7L 1A4 Canada

CRC Press
6000 Broken Sound Parkway NW,
Suite 300, Boca Raton, FL 33487-2742 USA

2 Park Square, Milton Park,
Abingdon, Oxon, OX14 4RN UK

© 2022 Apple Academic Press, Inc.

Apple Academic Press exclusively co-publishes with CRC Press, an imprint of Taylor & Francis Group, LLC

Library and Archives Canada Cataloguing in Publication

Title: Handbook of research on food processing and preservation technologies / edited by Monika Sharma, PhD, Megh R. Goyal, PhD, PE, Preeti Birwal, PhD.
Names: Sharma, Monika (Food scientist), editor. | Goyal, Megh R., editor. | Birwal, Preeti, editor.
Series: Innovations in agricultural and biological engineering.
Description: First edition. | Series statement: Innovations in agricultural and biological engineering | Includes bibliographical references and indexes. | Contents: Volume 5. Emerging techniques for food processing, quality, and safety assurance.
Identifiers: Canadiana (print) 20210143002 | Canadiana (ebook) 2021014307X | ISBN 9781771889827 (v. 1; hardcover) | ISBN 9781774638514 (v. 1 ; softcover) | ISBN 9781774630037 (v. 2 ; hardcover) | ISBN 9781774638521 (v. 2 ; softcover) | ISBN 9781774630334 (v. 3 ; hardcover) | ISBN 9781774638538 (v. 3 ; softcover) | ISBN 9781774630341 (v. 4 ; hardcover) | ISBN 9781774638545 (v. 4 ; softcover) | ISBN 9781774630358 (v. 5 ; hardcover) | ISBN 9781774638552 (v. 5 ; softcover) | ISBN 9781003153221 (v. 1 ; ebook) | ISBN 9781003161295 (v. 2 ; ebook) | ISBN 9781003184591 (v. 3 ; ebook) | ISBN 9781003184645 (v. 4 ; ebook) | ISBN 9781003184720 (v. 5 ; ebook)
Subjects: LCSH: Food industry and trade. | LCSH: Food—Preservation.
Classification: LCC TP371.2 .H36 2021 | DDC 664/.028—dc23

Library of Congress Cataloging-in-Publication Data

....

CIP data on file with US Library of Congress

....

ISBN: 978-1-77463-036-5 (5-volume set)
ISBN: 978-1-77463-035-8 (hbk)
ISBN: 978-1-77463-855-2 (pbk)
ISBN: 978-1-00318-472-0 (ebk)

ABOUT THE BOOK SERIES: INNOVATIONS IN AGRICULTURAL AND BIOLOGICAL ENGINEERING

Under this book series, Apple Academic Press Inc. is publishing book volumes over a span of 8–10 years in the specialty areas defined by the American Society of Agricultural and Biological Engineers (www.asabe. org). Apple Academic Press Inc. aims to be a principal source of books in agricultural and biological engineering. We welcome book proposals from readers in areas of their expertise.

The mission of this series is to provide knowledge and techniques for agricultural and biological engineers (ABEs). The book series offers high-quality reference and academic content on agricultural and biological engineering (ABE) that is accessible to academicians, researchers, scientists, university faculty and university-level students, and professionals around the world.

Agricultural and biological engineers ensure that the world has the necessities of life, including safe and plentiful food, clean air and water, renewable fuel and energy, safe working conditions, and a healthy environment by employing knowledge and expertise of the sciences, both pure and applied, and engineering principles. Biological engineering applies engineering practices to problems and opportunities presented by living things and the natural environment in agriculture.

ABE embraces a variety of the following specialty areas (www.asabe.org): aquaculture engineering, biological engineering, energy, farm machinery and power engineering, food, and process engineering, forest engineering, information, and electrical technologies, soil, and water conservation engineering, natural resources engineering, nursery, and greenhouse engineering, safety, and health, and structures and environment.

For this book series, we welcome chapters on the following specialty areas (but not limited to):

1. Academia to industry to end-user loop in agricultural engineering.
2. Agricultural mechanization.
3. Aquaculture engineering.
4. Biological engineering in agriculture.
5. Biotechnology applications in agricultural engineering.
6. Energy source engineering.

7. Farm to fork technologies in agriculture.
8. Food and bioprocess engineering.
9. Forest engineering.
10. GPS and remote sensing potential in agricultural engineering.
11. Hill land agriculture.
12. Human factors in engineering.
13. Impact of global warming and climatic change on agriculture economy.
14. Information and electrical technologies.
15. Irrigation and drainage engineering.
16. Micro-irrigation engineering.
17. Milk Engineering.
18. Nanotechnology applications in agricultural engineering.
19. Natural resources engineering.
20. Nursery and greenhouse engineering.
21. Potential of phytochemicals from agricultural and wild plants for human health.
22. Power systems and machinery design.
23. Robot engineering and drones in agriculture.
24. Rural electrification.
25. Sanitary engineering.
26. Simulation and computer modeling.
27. Smart engineering applications in agriculture.
28. Soil and water engineering.
29. Structures and environment engineering.
30. Waste management and recycling.
31. Any other focus areas.

Books published in the Innovations in Agricultural & Biological Engineering Series

- Biological and Chemical Hazards in Food and Food Products: Prevention, Practices, and Management
- Bioremediation and Phytoremediation Technologies in Sustainable Soil Management
  o Volume 1: Fundamental Aspects and Contaminated Sites
  o Volume 2: Microbial Approaches and Recent Trends
  o Volume 3: Inventive Techniques, Research Methods and Case Studies
  o Volume 4: Degradation of Pesticides and Polychlorinated Biphenyls

- Dairy Engineering: Advanced Technologies and Their Applications
- Developing Technologies in Food Science: Status, Applications, and Challenges
- Emerging Technologies in Agricultural Engineering
- Engineering Interventions in Agricultural Processing
- Engineering Interventions in Foods and Plants
- Engineering Practices for Agricultural Production and Water Conservation: An Interdisciplinary Approach
- Engineering Practices for Management of Soil Salinity: Agricultural, Physiological, and Adaptive Approaches
- Engineering Practices for Milk Products: Dairyceuticals, Novel Technologies, and Quality
- Field Practices for Wastewater Use in Agriculture: Future Trends and Use of Biological Systems
- Flood Assessment: Modeling and Parameterization
- Food Engineering: Emerging Issues, Modeling, and Applications
- Food Process Engineering: Emerging Trends in Research and Their Applications
- Food Processing and Preservation Technology: Advances, Methods, and Applications
- Food Technology: Applied Research and Production Techniques
- Handbook of Research on Food Processing and Preservation Technologies:
  - Volume 1: Nonthermal and Innovative Food Processing Methods
  - Volume 2: Nonthermal Food Preservation and Novel Processing Strategies
  - Volume 3: Computer-Aided Food Processing and Quality Evaluation Techniques
  - Volume 4: Design and Development of Specific Foods, Packaging Systems, and Food Safety
  - Volume 5: Emerging Techniques for Food Processing, Quality, and Safety Assurance
- Modeling Methods and Practices in Soil and Water Engineering
- Nanotechnology and Nanomaterial Applications in Food, Health, and Biomedical Sciences
- Nanotechnology Applications in Agricultural and Bioprocess Engineering: Farm to Table
- Nanotechnology Applications in Dairy Science: Packaging, Processing, and Preservation

- Novel Dairy Processing Technologies: Techniques, Management, and Energy Conservation
- Novel Strategies to Improve Shelf-Life and Quality of Foods: Quality, Safety, and Health Aspects
- Processing of Fruits and Vegetables: From Farm to Fork
- Processing Technologies for Milk and Milk Products: Methods, Applications, and Energy Usage
- Scientific and Technical Terms in Bioengineering and Biological Engineering
- Soil and Water Engineering: Principles and Applications of Modeling
- Soil Salinity Management in Agriculture: Technological Advances and Applications
- State-of-the-Art Technologies in Food Science: Human Health, Emerging Issues and Specialty Topics
- Sustainable Biological Systems for Agriculture: Emerging Issues in Nanotechnology, Biofertilizers, Wastewater, and Farm Machines
- Technological Interventions in Dairy Science: Innovative Approaches in Processing, Preservation, and Analysis of Milk Products
- Technological Interventions in Management of Irrigated Agriculture
- Technological Interventions in the Processing of Fruits and Vegetables
- Technological Processes for Marine Foods, from Water to Fork: Bioactive Compounds, Industrial Applications, and Genomics

OTHER BOOKS ON AGRICULTURAL AND BIOLOGICAL ENGINEERING FROM APPLE ACADEMIC PRESS, INC.

Management of Drip/Trickle or Micro Irrigation
Megh R. Goyal, PhD, PE, Senior Editor-in-Chief

Evapotranspiration: Principles and Applications for Water Management
Megh R. Goyal, PhD, PE, and Eric W. Harmsen, Editors

Book Series: Research Advances in Sustainable Micro Irrigation
Senior Editor-in-Chief: Megh R. Goyal, PhD, PE

Volume 1: Sustainable Micro Irrigation: Principles and Practices
Volume 2: Sustainable Practices in Surface and Subsurface Micro Irrigation
Volume 3: Sustainable Micro Irrigation Management for Trees and Vines
Volume 4: Management, Performance, and Applications of Micro Irrigation Systems
Volume 5: Applications of Furrow and Micro Irrigation in Arid and Semi-Arid Regions
Volume 6: Best Management Practices for Drip Irrigated Crops
Volume 7: Closed Circuit Micro Irrigation Design: Theory and Applications
Volume 8: Wastewater Management for Irrigation: Principles and Practices
Volume 9: Water and Fertigation Management in Micro Irrigation
Volume 10: Innovation in Micro Irrigation Technology

Book Series: Innovations and Challenges in Micro Irrigation
Senior Editor-in-Chief: Megh R. Goyal, PhD, PE

Volume 1: Management of Drip/Trickle or Micro Irrigation
Volume 2: Sustainable Micro Irrigation Design Systems for Agricultural Crops
Volume 3: Principles and Management of Clogging in Micro Irrigation
Volume 4: Performance Evaluation of Micro Irrigation Management

Volume 5: Potential Use of Solar Energy and Emerging Technologies in Micro Irrigation

Volume 6: Micro Irrigation Management: Technological Advances and Their Applications

Volume 7: Micro Irrigation Engineering for Horticultural Crops

Volume 8: Micro Irrigation Scheduling and Practices

Volume 9: Engineering Interventions in Sustainable Trickle Irrigation

Volume 10: Management Strategies for Water Use Efficiency and Micro Irrigated Crops

Volume 11: Fertigation Technologies In Micro Irrigation

ABOUT THE EDITORS

Monika Sharma, PhD
Scientist, Dairy Technology Division, Southern Regional Station, ICAR-National Dairy Research Institute, Bengaluru, India

Monika Sharma, PhD, is working as a Scientist, Dairy Technology Division, Southern Regional Station, ICAR-National Dairy Research Institute, Bangalore, India.

She received a BSc (2006) in Food Science and Technology from Delhi University, New Delhi; MSc (2008) in Food Technology from Govind Ballabh Pant University of Agriculture and Technology, Pantnagar; and PhD (2015) in Dairy Technology from ICAR-National Dairy Research Institute (NDRI), Karnal, Haryana, India. She started her career as a Scientist in the Indian Council of Agricultural Research in 2010; as a Scientist in ICAR-Central Institute of Postharvest Engineering and Technology, Ludhiana, Punjab, for more than five years. She is now working as a Scientist at Southern Regional Station, ICAR-National Dairy Research Institute, Bangalore, and is actively involved in teaching and research activities. She has more than ten years of research experience. She has worked in the area of convenience and ready-to-eat foods, functional foods, quality evaluation, composite dairy foods, starch modification and its application in dairy food products, etc. Presently, she is working in the area of functional and indigenous dairy foods.

She has published 25 research papers in peer-reviewed journals, three edited books, four technical bulletins, two technology inventory books, six book chapters, more than 25 popular articles, and more than 20 conference papers. She has successfully guided 6 postgraduate students for their dissertation work. She has worked as Principal Investigator in several research projects and has developed various technologies. She has also conducted entrepreneurship development programs for some of the developed technologies. She has earned several awards, such as an ICAR-JRF award and fellowship (2006–2008), first rank in all India level Agricultural Research Services (2008) examination in the discipline of Food Science and Technology, ICAR-NET (2009), conference awards, institute awards, etc. She is a life member of the Indian Science Congress and the Association of Food Scientists and Technologists (India). Readers may contact her at: sharma. monikaft@gmail.com

Megh R. Goyal, PhD, PE
Retired Professor in Agricultural and Biomedical Engineering, University of Puerto Rico, Mayaguez Campus; Senior Technical Editor-in-Chief, Biomedical Engineering and Agricultural Science, Apple Academic Press, Inc.

Megh R. Goyal, PhD, PE, is a Retired Professor in Agricultural and Biomedical Engineering from the General Engineering Department in the College of Engineering at the University of Puerto Rico–Mayaguez Campus; and Senior Acquisitions Editor and Senior Technical Editor-in-Chief in Agriculture and Biomedical Engineering for Apple Academic Press, Inc. He has worked as a Soil Conservation Inspector and as a Research Assistant at Haryana Agricultural University and Ohio State University.

During his professional career of 52 years, Dr. Goyal has received many prestigious awards and honors. He was the first agricultural engineer to receive the professional license in Agricultural Engineering in 1986 from the College of Engineers and Surveyors of Puerto Rico. In 2005, he was proclaimed as "Father of Irrigation Engineering in Puerto Rico for the Twentieth Century" by the American Society of Agricultural and Biological Engineers (ASABE), Puerto Rico Section, for his pioneering work on micro irrigation, evapotranspiration, agroclimatology, and soil and water engineering. The Water Technology Centre of Tamil Nadu Agricultural University in Coimbatore, India, recognized Dr. Goyal as one of the experts "who rendered meritorious service for the development of micro irrigation sector in India" by bestowing the Award of Outstanding Contribution in Micro Irrigation. This award was presented to Dr. Goyal during the inaugural session of the National Congress on "New Challenges and Advances in Sustainable Micro Irrigation" held at Tamil Nadu Agricultural University.

Dr. Goyal received the Netafim Award for Advancements in Microirrigation: 2018 from the American Society of Agricultural Engineers at the ASABE International Meeting in August 2018. VDGOOD Professional

Association of India awarded Lifetime Achievement Award at 12th Annual Meeting on Engineering, Science and Medicine that was held on 20–21 of November of 2020 in Visakhapatnam, India. A prolific author and editor, he has written more than 200 journal articles and textbooks and has edited over 85 books. He is the editor of three book series published by Apple Academic Press: Innovations in Agricultural & Biological Engineering, Innovations and Challenges in Micro Irrigation, and Research Advances in Sustainable Micro Irrigation. He is also instrumental in the development of the new book series Innovations in Plant Science for Better Health: From Soil to Fork.

Dr. Goyal received his BSc degree in engineering from Punjab Agricultural University, Ludhiana, India; his MSc and PhD degrees from Ohio State University, Columbus; and his Master of Divinity degree from Puerto Rico Evangelical Seminary, Hato Rey, Puerto Rico, USA.

Preeti Birwal, PhD
Scientist (Processing and Food Engineering), Department of Processing and Food Engineering, College of Agricultural Engineering and Technology, Punjab Agricultural University, Ludhiana, Punjab, India

Preeti Birwal, PhD, is working as a Scientist (Processing and Food Engineering) in the Department of Processing and Food Engineering, College of Agricultural Engineering and Technology, Punjab Agricultural University, Ludhiana, Punjab, India. She holds a BSc (2012) in Dairy Technology from ICAR-National Dairy Research Institute (NDRI), Karnal; MSc (2014) in Food Process Engineering and Management from NIFTEM, Haryana; and PhD (Dairy Engineering) from ICAR-NDRI, Bangalore. She is a recipient of MHRD (2008), Nestle India (2009), GATE (2012–2014), and UGC-RGN fellowships (2014–2018).

She is currently working in the area of nonthermal food preservation, fermented beverages, food packaging, and technology of millet-based beer. She has served at Jain Deemed to be University, Bangalore as a member of the board of examiners and placements. She is advising several MTech scholars in food technology. She has participated in several national and international conferences and seminars. She has delivered lectures as a resource person on doubling farmers' income through dairy technology in training sponsored by the Directorate of Extension, Ministry of Agriculture and Farmers Welfare, Government of India. She has named outstanding reviewer of the month by the *Current Research in Nutrition and Food Science Journal*. She has successfully completed AUTOCAD 2D and 3D certification.

She has 18 research papers, one edited book, five book chapters, about 28 popular articles, five conference papers, 56 abstracts, and two editorial opinions to her credit. She has successfully guided five postgraduate students for their dissertation work. She is serving as the external examiner for various Indian state agricultural universities. She is also serving as editor and reviewer of several journals. She is a life member of IDEA. Readers may contact her at: preetibirwal@gmail.com.

CONTENTS

CONTRIBUTORS

Sangita Bansal
Principal Scientist, ICAR-National Bureau of Plant Genetic Resources, IARI Pusa Campus,
New Delhi – 110012, India, Mobile: +91-8448035477, E-mail: Sangita.Bansal@icar.gov.in

Preeti Birwal
Scientist, Punjab Agricultural University, Ferozepur Road, Ludhiana – 141004, Punjab, India,
Mobile: +91 989669633, E-mail: preetibirwal@gmail.com

John David
Professor, SHUATS, Allahabad, Uttar Pradesh, India, Mobile: +91-9453226735,
E-mail: profjohndavid06@gmail.com

Saptashish Deb
Post Graduate Scholar, Department of Food Engineering and Technology,
Sant Longowal Institute of Engineering and Technology, Longowal – 148106, Punjab, India,
Mobile: +91-8787492638, E-mail: sapta121@gmail.com

Gajanan Deshmukh
Senior Research Fellow (RJSTC), College of Dairy Technology, Udgir, Maharashtra, India,
Mobile: +91-8147208662, E-mail: gajanannnn@gmail.com

Megh R. Goyal
Senior Editor-in-Chief (Agriculture and Biomedical Engineering) for AAP Retired Professor in
Agricultural and Biomedical Engineering, University of Puerto Rico-Mayaguez, USA,
Mobile: 001-787-536-0039, E-mail: goyalmegh@gmail.com

Pavan M. Gundu
MTech Scholar, Department of Food Technology, School of Engineering and Technology,
Jain (Deemed-to-be University), Bangalore – 562112, Karnataka, India,
Mobile: +91-9591007045, E-mail: pavanmg777@gmail.com

Mahendra Gunjal
MTech Scholar, Department of Food Technology, Jain (Deemed-to-be) University, Bangalore – 562112,
Karnataka, India, Mobile: +91-965725935, E-mail: mahendragunjal74@gmail.com

C. G. Harshitha
PhD Research Scholar, ICAR-National Dairy Research Institute, Karnal – 132001, Haryana, India,
Mobile: +91-8971923521, E-mail: harshithacg24@gmail.com

Thejus Jacob
Assistant Professor, SHUATS, Allahabad, Uttar Pradesh, India, Mobile: +91-9451595408,
E-mail: thejusjacob88@gmail.com

Abila Krishna
Executive-Quality Assurance Officer, GCMMF, Banas Dairy, Palanpur – 385001, Gujarat, India,
Mobile: +91-9400628353, E-mail: abilakrishna@gmail.com

Arvind Kumar
Assistant Professor, Department of Dairy Science and Food Technology, Institute of Agricultural
Sciences, Banaras Hindu University, Varanasi – 221005, Uttar Pradesh, India,
Mobile: +91-9793583702, E-mail: arvind00000@gmail.com

Leena Kumari
Scientist, ICAR-CIPHET, PAU, Ludhiana, Punjab, India, Mobile: +91-9988695224,
E-mail: leenakumari68@yahoo.co.in

Sharanabasava Kumbar
PhD Research Scholar, ICAR-National Dairy Research Institute, Bangalore – 560030, Karnataka, India,
Mobile: +91-88674222464, E-mail: sharankumbar111@gmail.com

Era V. Malhotra
Scientist, ICAR-National Bureau of Plant Genetic Resources, Pusa Campus, New Delhi – 110012,
India, Mobile: +91-9990088680, E-mail: Era.Vaidya@icar.gov.in

Chaitradeepa G. Mestri
Assistant Professor, College of Horticultural Engineering and Food Technology,
Devihosur University of Horticultural Sciences, Bagalkot – 581110, Karnataka, India,
Mobile: +91 8296563842, E-mail: chaitradeepa869@gmail.com

Sadhna Mishra
PhD Research Scholar, Department of Dairy Science and Food Technology,
Institute of Agricultural Sciences, Banaras Hindu University, Varanasi – 221005, Uttar Pradesh, India,
Mobile: +91-8953236018, E-mail: sadhnamishra2649@gmail.com

Mohona Munshi
Post Graduate Scholar, Department of Food Engineering and Technology,
Sant Longowal Institute of Engineering and Technology, Longowal – 148106, Sangrur, Punjab, India,
Mobile: +91-7717755119, E-mail: mohonamunshi96@gmail.com

Pranali Nikam
PhD Research Scholar, National Dairy Research Institute, Karnal – 132001, Haryana, India,
Mobile: +91-9039509660, E-mail: pranali2801@gmail.com

Shikha Pandhi
PhD Research Scholar, Department of Dairy Science and Food Technology,
Institute of Agricultural Sciences, Banaras Hindu University, Varanasi – 221005, Uttar Pradesh, India,
Mobile: +91-9958382808, E-mail: shikhapandhi94@gmail.com

Nilesh Kumar Pathak
Assistant Professor, Maharaja Agrasen College, University of Delhi, Delhi, India,
Mobile: +91-9953401294, E-mail: nileshpiitd@gmail.com

Ravi Prakash
PhD Research Scholar, Dairy Engineering Section, SRS of ICAR-National Dairy Research Institute,
Adugodi, Bangalore – 560030, Karnataka, India, Mobile: +91-9034966791,
E-mail: rdwivedi.prakash@gmail.com

Syed Insha Rafiq
Young Professional II, ICAR-NDRI Karnal, GT Rd, near Jewels Hotel, Nyaypuri, Karnal – 132001,
Haryana, India, Mobile: 91-8082538113, E-mail: syedinsha12@gmail.com

Syed Mansha Rafiq
Assistant Professor, Department of Food Science and Technology,
National Institute of Food Technology Entrepreneurship and Management (NIFTEM),
Plot No. 97, Sector 56, HSIIDC, Industrial Estate, Kundli – 131028, Haryana, India,
Mobile: 91-9686435525, E-mail: mansharafiq@gmail.com

Dinesh Chandra Rai
Professor and Head, Department of Dairy Science and Food Technology,
Institute of Agricultural Sciences, Banaras Hindu University, Varanasi – 221005, Uttar Pradesh, India,
Mobile: +91-9415256645, E-mail: dcrai@bhu.ac.in

Suvartan Ranvir
Assistant Professor, SHUATS, Allahabad, Uttar Pradesh, India, Mobile: +91-7404940759,
E-mail: suvartanranvir@gmail.com

Menon Rekha Ravindra
Principal Scientist, ICAR-National Dairy Research Institute, Bangalore – 560030, Karnataka, India,
Mobile: 91-8022631422, E-mail: rekhamn@gmail.com

Raman Seth
Principal Scientist and Head, Dairy Chemistry, National Dairy Research Institute, Karnal – 132001,
Haryana, India, Mobile: +91-9416291472, E-mail: ramanseth123@yahoo.co.in

Kanika Sharma
Junior Research Fellow, ICAR-National Bureau of Plant Genetic Resources, IARI Pusa Campus,
New Delhi – 110012, India, Mobile: +91-6239397742, E-mail: kanika2794@gmail.com

Madhusudan Sharma
Post Graduate Scholar, Department of Food Engineering and Technology,
Sant Longowal Institute of Engineering and Technology, Longowal – 148106, Sangrur, Punjab, India,
Mobile: +91-7424830364, E-mail: madysharma95@gmail.com

Monika Sharma
Scientist (Food Science and Technology), Department of Dairy Technology, Southern Regional Station,
ICAR-National Dairy Research Institute, Bangalore – 560030, Karnataka, India,
Mobile: +91-9915018948, E-mail: sharma.monikaft@gmail.com

Nikunj Sharma
MTech Candidate, National Institute of Food Technology Entrepreneurship and Management
(NIFTEM), Plot No. 97, Sector 56, HSIIDC, Industrial Estate, Kundli – 131028, Sonipat, Haryana,
India, Mobile: 91-9205475776, E-mail: nikunjsharma2000@gmail.com

Rajan Sharma
Principal Scientist (Dairy Chemistry), ICAR-National Dairy Research Institute, Karnal – 132001,
Haryana, India, Mobile: +91-9416120181, E-mail: rajansharma21@gmail.com

Humeera Tazeen
Assistant Professor, Department of Food Technology, Jain (Deemed-to-be) University,
Bangalore – 562112, Karnataka, India, Mobile: +91-9655148570, E-mail: humtaz@gmail.com

Neelam Upadhyay
Scientist (Food Technology), ICAR-National Dairy Research Institute, Karnal – 132001, Haryana,
India, Mobile: +91-9255772587, E-mail: neelam1ars@gmail.com

ABBREVIATIONS

2D	two-dimensional
3D	three-dimensional
x	distance below a surface (m)
ABTS	2,2'-azino-bis(3-ethylbenzothiazoline-6-sulfonic acid)
ACA	independent component analysis
AFLPs	amplified fragment length polymorphisms
AgNPs	silver nanoparticles
AMS	automatic milking system
ANN	artificial neural network
ATR	attenuated total reflection
AV	acid value
aw	water activity
BBD	Box-Behnken design
BD	browning determinant
BS	beet sucrose
BSPL	broad spectrum pulsed light
C	carbon
CART	classification and regression tree
CAT	computer-assisted tomography
CCD	charged coupled device
cDNA	complementary deoxyribonucleic acid
CFU	colony-forming unit
CIE	International Commission on Illumination
CL	continuous light
CMOS	complementary metal-oxide semiconductor
C_p	specific heat (J/kg°C)
CPS	cyber-physical system
CR	cylindrical radiators
CT	computed tomography
CV	cross-validation
CW-THz	continuous-wave-terahertz
d	layer depth (m)
DFP	double fluorescence protein
DNA	deoxyribonucleic acid

DOF	degree of freedom
DON	deoxynivalenol
DP	data pre-processing
dpi	dots per inch
DPT	data pre-treatment
DRP	deformation relaxation phenomena
DRS	diffuse reflectance spectroscopy
DSC	differential scanning calorimetric
D_T	decimal reduction time (S)
E	energy (J)
E. coli	*Escherichia coli*
E_0	energy of an incident radiation (J)
ECHO	extraction and classification of homogenous objects
E_d	energy absorbed by a layer of depth d (J)
ELISA	enzyme-linked immunosorbent assay
EM	electromagnetic
EST	expressed sequence tags
EWP	egg white protein
F	energy density or fluence (of a single pulse) (kJ/m^2)
FAO	Food and Agriculture Organization
FDA	fisher's discriminate analysis
FDA	Food and Drug Administration
FFDCA	federal food, drug, and cosmetic act
FGS	fructose: glucose: sucrose
FI	fluorescence imaging
FIR	far-infrared
F_r	power density or fluence rate (kW/m^2)
FSSAI	Food Safety and Standards Authority of India
FT	Fourier transform
FTIR	Fourier transform infrared
F_{tot}	total energy density or fluence (of n pulses) (kJ/m^2)
GC	gas chromatography
GC-MS	gas chromatography-mass spectrometry
gDNA	genomic deoxyribonucleic acid
GE	genetic engineering
GI	gastrointestinal
GM	genetically modified
GMO	genetically modified organisms
GOD	glucose oxidase

GRAS	generally regarded as safe
h	hour
H	hydrogen
HD	high definition
HD	high dimensionality
HDM	hydrodynamic mechanism
HDR	high dynamic range
HFCS	high fructose corn syrup
HHPP	high hydrostatic pressure processing
HI	hypobaric impregnation
HIV	human immunodeficiency viruses
HOG	histogram of oriented gradient
HPLC	high-performance liquid chromatography
HPP	high-pressure processing
HRM	high-resolution melting
HSI	hue, saturation, and intensity
HTST	high-temperature short time
Hz	hertz
IgA	immunoglobulin A
IgE	immunoglobulin E
InGaAs	indium gallium arsenide
IoT	internet of things
IPA	isopropyl alcohol
IPCA	incremental PCA
IR	infrared
ISO	International Organization for Standardization
ISSR	inter simple sequence repeat
KNN	K-nearest neighbor
kV	kilovolt
LAMP	loop-mediated isothermal amplification
LBP	local binary pattern
LC-MS	liquid chromatography-mass spectrometry
LCSM	laser confocal scanning microscopy
LDA	linear discriminant analysis
LED	light-emitting diode
LIFIS	laser-induced FI system
LS-SVM	least square support vector machines
LTLT	low-temperature long time

MALDI-TOF	matrix-assisted laser desorption/ionization-time of flight
MePEO-b-PCL	methoxy poly(ethylene oxide)-block-poly(-caprolactone)
$MgCl_2$	magnesium chloride
MHD	microwave hypobaric drying
MIR	mid-infrared
MLP	multilayer perceptron
MLR	multiple linear regression
MR	magnetic resonance
MRI	magnetic resonance imaging
MS	mass spectrometry
MSC	multiplicative scatter correction
MUFA	monounsaturated fatty acids
MVC	model view controller
n	number of pulses
NDT	non-destructive technique
NEC	necrotizing enterocolitis
ng	nanograms
NIR	near-infrared
NIR-HSI	near infrared-hyperspectral imaging
nm	nanometer
nM	nanomolar
NMR	nuclear magnetic resonance
NPs	nanoparticles
O/W	oil in water
OD	octadecane
OSC	orthogonal signal correction
P	pulse power (kW)
PAS	photoacoustic spectroscopy
PC	principal components
PCA	principal component analysis
PCR	polymerase chain reaction
PCR	principal component regression
PEFs	pulsed electric fields
PFA	prevention of food adulteration
PHB	poly(3-hydroxybutyrate)
PICS	partially inverted cane syrup
PL	pulsed light

PLGA	poly(lactide-co-glycolide)
PLS	partial least square
PLS-DA	partial least squares discriminant analysis
PLSR	partial least squares regression
PLT	pulsed light technology
PME	pectin methylesterase
PNAs	peptide nucleic acids
Poly PLS	poly partial least square
ppm	parts per million
PUFA	polyunsaturated fatty acids
PVDF	polyvinylidene difluoride membrane
QC-PCR	quantitative competition PCR
qPCR	quantitative PCR
RAPD	random amplified polymorphic DNA
RCI	Raman chemical imaging
RFID	radio frequency identification
RFLP	restriction fragment length polymorphism
RGB	red, green, blue
RMSEP	root mean square error of prediction
RNA	ribonucleic acid
ROI	regions of interest
RSM	response surface methodology
RTE	ready-to-eat
SCAR	sequence characterized amplified region
SCARAs	selective compliance articulated robot arms
SCARS	sequence characterized amplified regions
SCoT	start codon targeted polymorphism
SDS	sodium dodecyl sulfate
SDS-PAGE	sodium dodecyl sulfate-polyacrylamide gel electrophoresis
SIMCA	soft independent modeling of class analogy
SLNs	solid lipid nanoparticles
SMUF	simulated milk ultra-filtrate
SNIF-NMR	specific natural isotopic fractionation-nuclear magnetic resonance
SNP	single nucleotide polymorphism
SNV	standard normal variate
SOM	self-organizing maps
SPA	successive projection algorithm

sPCA	sparse PCA
SPFS	surface plasmon-enhanced fluorescent spectroscopy
SPME-GC-MS	solid-phase microextraction followed by gas chromatography-mass spectrometry
SPR	stepped-plate radiator
SSLPs	simple sequence length polymorphisms
SSR	simple sequence repeats
STS	sequence-tagged sites
SURF	speeded up robust features
SVM	support vector machine
SVR	support vector regression
SWIR	short-wave infrared
t	duration of a single pulse (s)
T	temperature (°C)
TFC	total flavonoid content
THz	terahertz
THz-TDS	THz time-domain spectroscopy
TI	thermal imaging
TMV	tobacco mosaic virus
TPC	total phenolic content
UHT	ultra-high temperature
UK	United Kingdom
UV	ultraviolet
UVC	ultraviolet C
UVPL	UV pulsed light
UV-Vis	ultraviolet-visible
v	wave frequency (Hz)
VI	vacuum impregnation
VL	visible light
VNIR	visible and near-infrared
VPL	visible pulsed light
WHO	World Health Organization
WSN	wireless sensor networks
YCbCr	Y Luma component, Cb blue-difference, and Cr red-difference chroma components
ZEA	zearalenone

SYMBOLS

ΔT temperature increase (°C)
α extinction coefficient
γ reflection coefficient
λ wavelength (nm)
ρ specific gravity or density (kg/m^3)

PREFACE

Food preservation has existed since the early times. Food is not only important for relief from hunger, but food with nutrition is also the demand of today's millennium. Food with convenience, good sensorial attributes, longer shelf life is today's challenge. The preservation in the current scenario is done by different methods like heating, cooling, drying, concentration, freezing, and cooling. All these techniques have their own pros and cons. Therefore, now a challenge to save the food from deteriorating its quality in terms of loss of nutrients, texture, sensorial characteristics with longer shelf life has been taken by the food scientists and industrialist. Several new food processing, preservation, and quality assessment technologies have been investigated by researchers to preserve the quality and design of specific nutrient-rich food.

Handbook of Research on Food Processing and Preservation Technologies is a bouquet of various emerging techniques in the food processing sector and quite relevant for the food industry and academic community. Nonthermal techniques (such as high-pressure processing [HPP], pulsed electric field [PEF], pulsed light [PL], ultraviolet [UV], microwave, ohmic heating, electrospinning, nano, and microencapsulation) are a few novel processing techniques that are being investigated thoroughly. The role and application of minimal processing techniques (such as ozone treatment, vacuum drying, osmotic dehydration, dense phase carbon dioxide [DPCD] treatment, and high pressure-assisted freezing) have also been included with a wide range of applications. Literature has reported successful applications on juices, meat, fish, fruits and vegetable slices, food surfaces, purees, milk, and milk products, extraction, drying enhancement, and encapsulation of micro-macro nutrients.

Furthermore, this handbook also covers some computer-aided techniques emerging in the food processing sector, such as robotics, radio frequency identification (RFID), three-dimensional food printing, artificial intelligence, etc. Enough emphasis has also been given on nondestructive quality evaluation techniques (such as image processing, terahertz spectroscopy imaging technique, near-infrared [NIR], Fourier transform infrared [FTIR] spectroscopy technique, etc., for food quality and safety evaluation. A significant role of food properties in the design of specific food and edible films have been elucidated. Thus, this handbook would have significant scope in food processing, preservation, safety, and quality evaluation. The handbook is organized under five volumes.

Volume 5: Emerging Techniques for Food Processing, Quality, and Safety Assurance, discusses various emerging techniques for food preservation, formulation, and non-destructive quality evaluation techniques. This book will serve the professionals working in the area of food science technology, and engineering around the world. The book will also act as a reference book for researchers, students, scholars, industries, universities, and research centers. Each chapter covers major aspects pertaining to principles, design, and applications of various food processing and non-destructive quality evaluation techniques. This book is divided into two parts: Part I: Emerging Technologies for Preservation and Formulation of Foods covers various techniques such as basic approach for Low-temperature-based ultrasonic drying of foods, hypobaric processing of foods, viability of high-pressure technology in food processing industries, application of pulsed electric fields (PEFs) in food preservation, pulsed light technology (PLT) in food processing and preservation, the potential of green nanotechnology in food processing and preservation, and advanced methods of encapsulation. Part II: Non-Destructive Techniques for Food Quality and Safety looks at the basics and detection methods of food authentication, imaging techniques for quality inspection of spices and nuts, FTIR coupled with chemometrics for the evaluation of food quality and safety, and food robots as a tool for quality and safety.

This book volume will serve as a valuable resource of information and excellent reference for researchers, scientists, students, growers, traders, processors, industries, and others for food preservation, processing, and quality aspects of food products.

This book has taken its present shape due to the excellent contribution by the contributing authors, who have been the soul of this compendium. We have mentioned their names in each chapter and also in the list of contributors. We are indeed indebted to them for their knowledge, dedication, and enthusiasm. We expect this book to prove a helpful resource for all the food processing and engineering academicians, food processors, and students.

We also extend our sincere thanks to Apple Academic Press, Inc., for their immense facilitation and suggestions right through this assignment.

We take this opportunity to thank: (1) our families for their motivation, moral support, and blessings in counteracting every obstacle coming in the way; (2) our spouses for their understanding, patience, and encouragement throughout this project; (3) our institutes: PAU, Ludhiana, Punjab; ICAR-NDRI, Karnal; Southern Regional Station, ICAR-NDRI, Bangalore; and UPRM, Mayaguez, for their support during the compilation of this publication.

— Editors

PART I
Emerging Technologies for Preservation and Formulation of Foods

CHAPTER 1

LOW TEMPERATURE BASED ULTRASONIC DRYING OF FOODS

PAVAN M. GUNDU, PREETI BIRWAL, CHAITRADEEPA G. MESTRI, and ABILA KRISHNA

ABSTRACT

Application of power ultrasound in-line with drying and dehydration is highly relevant for the preservation of foods and food products to pursue feasible non-thermal processing. Ultrasound waves boost up heat and mass transfer, resulting in enhanced moisture removal rate during drying due to synergistic effects of sonic waves involving vibration and heating. These cumulative effects of high-intensity ultrasound provide a proficient application for food preservation along with Enhancement in color, texture, vitamins, and anti-oxidant content of food products with better energy efficiency. This chapter discusses and elaborates on the application of high power ultrasonic waves for food drying.

1.1   INTRODUCTION

The process of food preservation by suppressing the growth of microorganisms through the reduction in water activity/availability is one of the oldest preservation methods since prehistoric times [10]. This process is well furnished with specified terminologies in drying and dehydration, as it acts to remove water from food material [31]. Dehydration is one of the methods to minimize the availability of water for microbial multiplication, a degradative reaction, such as non-enzymatic reaction, lipid oxidation, hydrolysis, and enzymatic reaction and thereby delays the spoilage and extends the shelf-life [38].

The drying technique for each particular application critically depends on the type of attachment of the liquid with the solid material, i.e., chemical, mechanical, and physicochemical attachments [9, 59]. Mechanical attachment is associated with the porous materials containing wetting and capillary moisture, whereas adsorption and osmotic moisture is associated with physicochemical attachment. Since ages, these two attachments are treated with conventional drying techniques, such as solar drying, salting, etc., [31].

1.1.1 CONVENTIONAL DRYING

Several techniques, such as solar drying, salting, to modern freeze-drying have been explored based on the principle of demoisturization to extend the shelf-life of food products [67]. Mechanical and thermal treatments are two major and basic conventional techniques for the drying and dehydration of foods. In mechanical drying, water is removed by mechanical forces, and this method is also known as dewatering, which includes pressing, centrifugation, and osmotic dehydration [34]. Dewatering involves the removal of water from the product without any phase change; thus, mechanical or chemical forces are used to draw out the liquid water [103]. The mechanical system removes moisture, which is attached weakly, whereas the thermal drying provides more complete drying, as it is associated with the inclusion of energy in the form of heat to evaporate the moisture content [31].

With these principles of water removal and phase change, air (or convective) drying or freeze-drying (lyophilization) is commonly used in the food industry [76, 79]. Meanwhile, current trends in the search for high nutritional and sensory product traits also have significant implications in the impact on the drying process. Effects on quality attributes [79], such as browning, shrinkage, loss of rehydration capacity, and degradation of thermolabile compounds, and among others will decide the drying efficiency. In addition, several disadvantages have been identified, such as relatively large energy consumption and quality deterioration of the final product in hot air drying, uneven drying, or overheating in microwave drying and high-cost expenditure for freeze and hybrid drying, etc., [44]. Therefore, keeping the product temperature low during dehydration is a key issue to obtain better-quality products.

1.1.2   NONTHERMAL LOW-TEMPERATURE DRYING

The qualitative degradation on conventional thermal drying has opened up the researchers and industries to novel non-thermal low-temperature drying technique, which is the process of dehydration functionalized below the standard room temperature (less than 20 or 25C) to achieve better food quality and to overcome economic losses [39, 91]. Vacuum freeze-drying and lyophilization are examples of low-temperature drying that provides high quality and better preservation of nutritional properties of dried/fresh foods. Although the product deterioration is negligible, yet the freeze-drying is considered to be expensive [31, 39].

Therefore, freeze-drying has been used only for high-quality foodstuffs. Otherwise, drying at low temperature can also be carried out at atmospheric pressure, by blowing low-temperature air around the drying chamber, in which the sample is placed. At atmospheric pressure, however, drying at low temperature is quite slow. Meanwhile, coupling new technologies with convective drying at low temperature is an alternative and feasible means of intensification [74, 91]. However, the application of novel technologies with high thermal effects, such as infrared (IR) radiation or radiofrequency is complicated, as the accurate control of the application is required to avoid overheated spots [9, 10]. This drawback is much less significant in the application of ultrasound, which is mainly based on mechanical effects.

1.1.3   ULTRASONIC DRYING OF FOODS

Ultrasonic drying and dehydration is very propitious as the out-turn of high-intensity ultrasound is highly significant at low temperatures, which decreases the probability of food degradation [21]. Ultrasound is a mechanical waves having a frequency range from 20 kHz to 1 MHz. High-energy (high power, high intensity) ultrasound is distinguished by low-frequency waves (20 to 100 kHz), which are used in drying and minimal processing of foods; and intermediate power/intensity ultrasound is distinguished by medium frequency waves (100 kHz to 1 MHz), and low energy high frequencies (1 to 10 MHz) are used in non-destructive technique (NDT) for the quality assurance of foods [11, 23, 62].

The application of high power, high-intensity ultrasound allows the extraction of liquid content from solids without inducing a change in the liquid phase. In addition, dehydration of heat-sensitive food products by high-intensity ultrasound is one of the best examples for the potential utilization of ultrasonic technique in the food industry [21]. Although advantages of this novel drying technique are protean, reliable, and commercially productive to food industries, yet more research endeavor and innovations are required in the design and development of efficient and sustainable high-intensity ultrasonic systems to reinforce large-scale industrial operations and adoptions [11, 83].

With a conceptualization on qualitative, quantitative, economical, and efficiency aspects of ultrasonic parameters and its attributes over-drying kinetics, this chapter emphasizes on:

1. The phenomenon involved in conventional and modern ultrasound-assisted drying;
2. Principle of ultrasonic waves and its sponge effect influencing convective and diffusive transport of water;
3. Ultrasonic drying equipment and classification of the operational process including ultrasonic pretreatment, airborne application and directly coupled ultrasonic vibration;
4. Drying kinetics and influence of ultrasonic intensity, sonication time, air velocity, temperature, product properties over process efficiency;
5. Water activity, phytochemical, nutritive, and structural quality of the product in ultrasonic drying; and
6. Limitations, current challenges and future scope for large-scale industrial adaption.

1.2 ULTRASONIC DRYING: THEORY AND PRINCIPLES

Utilization of ultrasound is an emerging green technology utilized by the food industries. The ultrasonic waves having frequency >20 kHz is defined as the pressure wave that is oscillating at a frequency above the threshold of human hearing [3, 62]. High energy ultrasonic waves at frequencies between 20 and 500 kHz give intensities greater than one $W \cdot cm^{-2}$, which is disruptive and instigate significant effects on the physical, mechanical, and biochemical properties of foods; therefore, mentioned frequency range is largely used for drying and dehydration operations [23, 62]. On drying due to internal moisture flux, the moisture is transported from the product being dried to its outer surface, and then due to external moisture flux, moisture is released to the outer environment through the boundary layer of the drying medium [10]. Both these internal and external fluxes could be enhanced by ultrasound. Ultrasound, like any acoustic wave, can be transmitted through any substance with elastic properties, such as solids, liquids, semisolids, or gasses [9].

The sound source creates a vibration of the particles, which is transmitted to the adjoining ones. In this way, particles experience periodically repeated cycles (frequency >20 kHz) of compression and expansion [11]. This compression and expansion is cyclic mechanical stress that causes a series of phenomena that are largely dependent on the nature of the system as well as the energy available [3, 38]. A unique sponge compressed and released phenomenon in solid products is associated with every change in the pressure gradient [28]. This mechanical stress, known as the "sponge effect," helps the liquid to flow out of the samples by natural or other microchannels induced by sonic wave propagation [31]. These effects can affect internal resistance to mass transfer too [67]. Meanwhile, high-power ultrasonic waves can induce some agitation that can increase the bulk transport within the fluid due to a reduction in the external resistance to mass transport. In addition, ultrasound interaction at solid-fluid interfaces can induce micro-stirring [77].

Micro-agitation and micro-stirring induced at the immediate vicinity of the solid surface exert a critical significance on the shortening of the diffusion boundary layer thickness [11, 67]. On the other hand, ultrasonic vibration in a liquid may induce inertial cavitation. However, cavitation during ultrasonic air-drying could almost be neglected, as the inception of cavitation needs high-pressure levels of sonic waves during ultrasonic air-drying, thus making it very difficult to reach the threshold level of acoustic energy using current ultrasonic devices [29, 38].

In fact, the application of high-power ultrasonic waves have been used to improve mass transfer kinetics of varied products and processes, such as brining of meat [90], drying of rice [68], carrots [31], onions [17], osmotic dehydration of apples [94] and several extraction processes [52, 75, 87, 104]. Therefore, the use of high-power ultrasonic waves induces a shift in mass transfer kinetics as being a way of accelerating the drying phenomenon [28].

1.2.1 MECHANISM OF WATER TRANSPORT

Application of intense ultrasonic vibrations can enhance the drying, because it increases the mass diffusivity and lowers the viscosity; it creates and heats the bubbles locally and subsequently creates a pressure, which presses out the moisture; because of creating alternating pressure, thus leading to pressure drops so that the water in the capillary channels moves towards the surface [6, 21, 42]. Furthermore, the mechanical effects induced by ultrasound have been classified considering its influence on external (convective) or internal (diffusion) transport during ultrasonic drying. It should be emphasized that during drying, there is an opposite coupled heat and mass transfer [4, 36]. Water moves from the particle to the air, while heat flux is opposite going from air to the particle [40]. Thus, the effects produced by ultrasonic waves jointly affect both transport phenomena in Sections 1.2.1.1 and 1.2.1.2.

1.2.1.1   CONVECTIVE TRANSPORT

During drying while the product surface remains saturated, convective transport is the only significant water transport mechanism controlling the drying rate. This involves water evaporation at the solid surface; thus, water in a liquid state moves from the surface to the air in the gas phase [10]. The importance of convective transport depends on the air turbulence because of its influence on the boundary layer thickness, and resistance to water transport is mainly controlled by this boundary layer thickness [9]. Thus, the only way to increase turbulence and to improve convective transport is in the airflow. In most cases, the higher the airflow and thinner the boundary layer, the faster will be the convective transport. Ultrasound interferes in the airflow because the acoustic waves create an oscillating velocity in the solid-gas interface, further reducing the boundary layer thickness [38].

High-intensity ultrasound can generate acoustic streaming (sound wind), in the solid-gas interface, affecting external resistance to mass transport by enhancing bulk transport [67]. The high-intensity mechanical compressions and expansions produced in the interfaces create micro-streaming, which are formed in the immediate vicinity of the solid surface [3, 11]. Moreover, the pressure variations enhance the evaporation rate, as the water is removed in the negative phase of the pressure cycle, and it does not re-enter in the positive phase [29].

1.2.1.2   DIFFUSION TRANSPORT

In the drying process, diffusion is the critical water transport phenomena occurring in the solid matrix at the molecular level. Ultrasonic waves induce the sponge effect, a rapid series of cyclic compressions and expansions in solid products [28]. This sponge effect induces the removal of fluid from inside of the products. Through intense compressions and expansions, ultrasound creates micro-channels that are better pathways, improving the water removal rate [68].

Micro-channels are the structural effects induced by ultrasound, which facilitate internal diffusion. In addition, if the acoustic energy levels required to induce cavitation could be reached, it would contribute to the removal of the most strongly attached moisture from the solid [31]. These factors are highly relevant for pharmaceutical, chemical, and oil companies to remove traces of solvent and food industries to achieve better non-thermal drying [38].

1.3 DEVICES AND INSTRUMENTATION FOR ULTRASONIC DRYING

A typical instrument for producing and application of high-intensity ultrasound consists of three main components: generator, transducer, and application system. Here, generator acts as a source of energy and generation of ultrasound can be achieved by different methods, namely: Mechanical (aero- and hydrodynamic), thermal (electric discharge), optical (impulse of a high power laser), and with the use of reversible electric and magnetic methods (piezoelectric, electrostriction, magnetostriction) [38, 48]. Among these, only two methods have found practical application during food processing, especially in drying. At the very beginning, ultrasounds were generated by using aerodynamic methods by static and dynamic sirens, etc.

Although Pierre and Jacque Curie discovered piezoelectricity at the end of the 19th century [48, 50], yet this found its practical application only after World War II when high-quality piezoelectric materials were developed, and it became cheap, thus easily available. Piezoelectric transducers generate acoustic waves under the influence of the delivered current or applied magnetic field. The efficiency of these devices, as the ratio of supplied power to generated acoustic power is different, depends on the transducer's construction [48, 83].

Generally, the techniques adopted for generating high-intensity sonic fields convert a kind of energy (from magnetic or electric source) to acoustic energy. The conversion of this energy will be carried out by transducers. Major three types of transducers considered for the processing operations are: magnetostrictive, fluid-driven, and piezoelectric [38, 50]. Magnetostrictive transducer consists of a unique material, which changes in dimension on application of magnetic field. If magnetic field disappears, the unique material helps the transducer to return to its original shape. Fluid-driven transducer generates an acoustic wave when the kinetic energy of fluid makes a mobile part of system vibrate. However, piezoelectric transducers are the most commonly used and adopted for the generation of ultrasound. Piezoelectric transducers generate acoustic energy by changes in size induced by electrical signals in piezoceramic materials [48, 67].

The basic structure of the transducer for drying mainly consists of a piezoelectrically activated vibrator that drives an extensive radiator. The vibrator includes a piezoelectric element that generates the vibration, which is later amplified by a mechanical element. The radiator has an extensive surface area that increases the radiation resistance to vibration. Therefore, the acoustic impedance improves the coupling with the air and the energy transferred into the air. The two types of radiators used for applications in drying processes are: stepped-plate radiator (SPR) and cylindrical radiators (CR) [38, 63].

Circular and rectangular SPRs with an extensive surface area have been used to intensify drying. The radiation pattern is controlled by the stepped profile, so customized patterns can be achieved for specific applications. Although the most commonly used stepped-plate transducer has been the circular one, yet the rectangular plate configuration confers some practical advantages related to its location in processing lines that could be very interesting in industrial applications. Rectangular radiators offer more uniform distribution of radiation between the central and the outer areas. In addition, more uniform vibration involves milder mechanical stress that minimizes damaged by stress. Its main feature in convective drying is in the treatment chamber, where samples are placed between the rectangular radiators and porous material; acting as a sample holder. Suction is applied to help the hot air to remove the extracted water [29, 38].

Controlling the position of the sample holder allows the device to perform airborne or direct-contact application. Meanwhile, for applications in specific volumes, such as a drying chamber, duct, or pipe, CR also has been considered. The basic feature consists of using the walls of the cylindrical chamber as the radiating element; thus, no additional element is necessary to apply ultrasound. In this way, the drying air is forced to go through the ultrasonically activated drying chamber where the samples are placed. In addition, CR were originally designed for fluidized-bed drying, where samples are freely moving in the drying chamber because of the high air velocities applied. The use of sample holder also allows experiments to be carried out in fixed beds by placing samples in trays along the longitudinal axis of the cylinder or distributed randomly [38, 44, 50].

Special high-power transducers are required with adequate standards for better application of high-power sonic waves to different media [63]. For example, some media presents low specific acoustic impedance and a high acoustic absorption as in the fluid media (especially the gasses). Therefore, it is necessary to obtain good impedance matching between the transducer and the medium in order to achieve an efficient energy transmission [31, 38]. In addition, transducers with high power capacity and the extensive radiating area would be needed for large-scale industrial applications.

Most high-intensity commercial transducers adopted in present industrial levels are based on the classical sandwich transducer consisting of piezo-electric element of transduction in a sandwich arrangement with one or more transmission lines called horns or mechanical amplifiers and are formed by extensionally vibrating resonant metallic elements [29, 63, 102]. However, these kinds of transducers have many limitations for industrial applications due to its small radiating area [23, 31, 44]. Many of these problems have been partially solved in the past few years through the development of specific ultrasonic transducers with high-power capacities, good impedance matching the air, large vibration amplitude, and radiating area and optimal control of the radiation pattern [29].

1.4 ULTRASONIC DRYING APPLICATION: TYPES AND CLASSIFICATION

In recent years, many types of transducers and ultrasonic drying instruments have been developed based on convective and diffusive removal of water content from the food material to facilitate low-temperature drying [83]. In addition, there are three main classifications based on the type and mechanism of application, and the operational process involved in ultrasonic drying, namely: ultrasound pretreatment, forced air dehydration assisted by airborne ultrasound, and dehydration by directly coupled ultrasonic vibration.

1.4.1 ULTRASOUND PRETREATMENT

Ultrasound can also be used as a pretreatment on fruits and vegetables and on other foods prior to the drying process. Studies have shown that such pretreatment can result in faster drying, reduction in energy consumption and better product quality [47].

For ultrasound pretreatment, the ultrasound is usually applied either by an ultrasonic bath or by a probe to transmit ultrasound waves into the aqueous media, e.g., distilled water. The acoustic waves induce compression and expansion of the samples (sponge effect), create microscopic channels in the sample, and thus lead to the leakage of intracellular liquid to the surroundings [44]. A study on the application of US as a pretreatment before drying of various foods has shown the reduction in drying time, and higher rehydration properties [45]. Meanwhile, increased water diffusivity on ultrasound pretreatment has caused 11% and 8% reduction in the overall drying time of banana and pineapple, respectively [25, 26].

1.4.2 FORCED-AIR DRYING ASSISTED BY AIRBORNE ULTRASOUND

The application of high-intensity airborne ultrasound for drying and dehydration of food and other products is useful for enhancing the efficacy of the forced-air drying process [29, 31]. Ultrasound application in a conventional hot air drying process can significantly reduce the processing time and energy consumption, without adversely affecting the product quality [89]. Drying experiments with ginseng and other vegetables demonstrated that in comparison with pure convective drying, the airborne ultrasound application had decreased the total power consumption by 20% [6].

The high level of drying intensification by airborne ultrasound can reduce the temperature of the drying agent (air) and enhanced the products final color, texture, and thermosensitive phytochemicals along with the reduction in overall drying time and retention of better product quality and thermophysical characteristics [51].

In airborne ultrasound on the food drying, especially the convective drying, the pressure at gas/liquid interfaces oscillates under high-intensity airborne ultrasound. In negative (compression) phase of the pressure cycle, moisture moves out and it does not re-enter into the sample during the positive (expansion) pressure phase, which accelerates the evaporation rate [44, 51]. In convective drying, the acoustic energy induces an oscillating-velocity effect, which speeds up the drying process. Application of high-power airborne ultrasonic waves produces micro streaming at the interfaces, and the diffusion boundary layer is reduced, thus water diffusion is enhanced [8, 29].

On the other hand, even though high-power airborne ultrasound enhances forced-air dehydration process, but its impact is considered to be limited.

Low penetration of acoustic energy in food materials is the major constraint on drying through the application of airborne ultrasonic radiation due to the mismatch between acoustic impedances [50]. Acoustic impedance is defined as the product of density and sound speed in the material, and the optimum efficacy of energy transfer between two media very much depends on this factor. In addition, optimal levels of ultrasound power, air temperature, velocity, and humidity significantly affect the drying process [53, 54]. Furthermore, a certain level of relative air humidity is necessary for the successful transmission of airborne ultrasound; and above a threshold value of ultrasound power level, the stability of the acoustic field will be decreased [8]. However, this particular application is considered beneficial only to specific drying operation, in which drying material is well matched acoustically to air or to insist in a process, where hot air at moderate temperature is required for product preservation [31, 44].

1.4.3   DEHYDRATION BY DIRECTLY COUPLED ULTRASONIC VIBRATION

Application of high-intensity sonic energy directly coupled to the food products through the stepped-plate transducers is an efficient method for food drying. However, the high attenuation of acoustic energy by a gas medium and the acoustic impedance mismatch between air and the applied system limit the utilization of airborne ultrasound [50, 92]. If the ultrasonic energy is applied directly in contact with the food materials, the problem of energy loss will be solved theoretically [21].

In directly coupled ultrasonic vibration drying system, the advantageous acoustic impedance matching between vibrating transducer plate and the material of food favors the deep infiltration of acoustic energy and increases the effectiveness of drying process [6, 31]. Furthermore, a rapid series of alternative contractions and expansions are created leading to the sponge effect. Alternating stress developed enhances the dehydration by forming microscopic channels for the movement of water. Besides this, high-power ultrasound induces cavitation phenomenon that is also beneficial for the removal of moisture content strongly attached to the food products [50].

Generally, the improvement in the drying rate by contacting ultrasound is better than that of airborne ultrasound. In a study, cylinder slices of carrots were dried at 30°C under a pilot-scale drying system based on the application of direct contact high-intensity ultrasound at low temperatures and it was found that the increase in ultrasound power had increased the drying kinetics and reduced the drying time [21]. Compared to forced-air drying, contact method ultrasound application is not only much faster and consumes low energy; but it is also much effective and powerful so that the final moisture content <1% can also be achieved. The major ascendant of this application is the optimum transfer of energy and plate radiators with the extensive surface area makes it capable of large-scale industrial applications [44, 92].

1.5   DRYING KINETICS AND ULTRASOUND APPLICATION

According to various studies conducted in the past, the application of ultrasound has a positive influence on the drying kinetics of foods. Compared to conventional convective drying, ultrasound pretreatment could increase the drying rate by almost five times with increased effective water diffusivity of 210% in blackberries [88], decreased the total drying time by 28% in banana [60] and increased the effective water diffusivity by 64.3% in pineapple [24], even though at low temperatures. In addition, ultrasound-assisted vacuum drying reduced the drying time of salmon filets by 44.4–67.6% and trout filets by 52.5–75.8% compared to normal vacuum drying and oven drying [31].

Furthermore, application of ultrasound-assisted drying for carrots [14], potatoes [73], and apple samples [89] significantly accelerated the drying process, as the drying time was reduced by 30%, 40%, and 46–57%, respectively; along with increase in effective moisture diffusivity by 40% for carrots. Meanwhile, the ultrasound application can also reduce the energy consumption by almost 10% in the drying process [53, 56]. For example, ultrasound-assisted convective drying of apple samples has reduced the total energy consumption by 42–54% compared to conventional convective drying [89]. On the other hand, few studies also presented a contrary result to the expectation. The ultrasound pretreatment did not influence the drying kinetics of apple, as there was no reduction in the total drying time, and kinetics proceeded in a similar way as without initial sonic pretreatment [65].

The positive or even negative influence of the drying kinetics with the ultrasound pretreatment or assisted drying is due to the comprehensive effect of the drying conditions, food structure, and water gain or loss during the ultrasound application [44]. By applying ultrasound, the cell disruption and microscopic channels might occur, which make the water movement easy by reducing the resistance for water migration and accelerate the drying process. Therefore, the drying kinetics can be enhanced by increasing the ultrasonic parameters, such as ultrasound intensity, sonication time, and amplitude, for which the water loss or gain will be increased [69].

1.6 FACTORS AFFECTING THE EFFICIENCY OF ULTRASONIC DRYING

Novel ultrasonic drying and mass transfers phenomenon can be influenced by the factors, such as ultrasonic power, sonication time, frequency, the amplitude of sonicator probe, mass load density, air velocity [13, 36, 37], temperature [39, 90, 94], and food structure [64, 67]. In fact, the extent of ultrasonic enhancement effect on drying is primarily dependent on these process parameters, which will significantly influence over-drying kinetics and the quality of food materials [44].

1.6.1 ULTRASOUND INTENSITY

The significance and impact of ultrasonic intensity on drying and mass transfer phenomenon have been documented by investigators [43, 61]. Drying acceleration is a linear function of the power of the applied ultrasound [69, 84, 86]. Increasing ultrasound power and decreasing relative air humidity causes acceleration of drying. The drying time was decreased from 235 minutes (pure convective drying) to 185 and 145 minutes at ultrasonic power of 100 W and 200 W, respectively [44].

Arkangelskii et al. [3] reported on the existence of an intensity threshold and they stated that about 140 dB is the threshold for any acoustic evaporation. The significance of ultrasound is noticeable when applied intensity is higher than this critical threshold. Higher intensities than threshold results in Enhancement of ultrasonic effect and no significant effects are noticeable at lower intensities [67]. Furthermore, with the maximum power applied the ultrasound application doubles the drying kinetics [38, 63]. Therefore, an increase in ultrasonic power results in faster drying, because the drying rate is proportional to the ultrasonic intensity.

1.6.2 SONICATION TIME

Sonication time exhibits varying effects on water loss, drying rate, and food quality. In a study involving the application of ultrasound pretreatment on the drying kinetics of banana, the drying rate was increased with the increase in sonication time [5]. In the same manner, papaya lost its water to the osmotic solution and the water loss was increased with increasing sonication time [85]. Similar results were also found in pineapple [16, 24], pomegranate [1], melon [18], guava [50], and strawberry [2]. All these samples presented positive water loss values and the water loss was increased with sonication time.

However, a different result is also been observed in some sonic studies. For example, Nowacka et al. [70] applied ultrasound pretreatment for 10, 20, and 30 minutes in an ultrasonic bath before the convective drying of apples. The drying time was firstly increased with the sonication time and then was decreased, i.e., the apple with 20 minutes of ultrasound pretreatment presented the longest drying time among the three pretreated samples. Meanwhile, as the sonication time was increased, the total color change of food products was also increased significantly [34, 50, 81], while the antioxidant activity was reduced [44, 81, 88].

1.6.3 AIR VELOCITY AND TEMPERATURE

In conventional drying, there is an upper threshold, beyond which air velocity does not influence the drying rate because water removal is completely controlled by the internal diffusion. An increase in the drying air velocity causes a decrease in ultrasound efficiency due to disturbances in the ultrasounds' field pattern [69, 84, 86]. The ultrasonic effect is only significant at low air velocities, thus from approximately 4 m/s, the influence and impact of ultrasound is negligible [69]. The increase in airflow reduces the average sound pressure level from 157 dB to a stationary value of 154 dB for air velocities higher than 8 m/s. The high airflow introduces a high level of noise at ultrasonic frequencies, so that the wind (air) will blur the ultrasonic field [38].

In an airborne ultrasound application, the effect of ultrasound on carrot drying at an air velocity of 3 m/s was almost negligible [29, 30]. During dehydration of persimmons [13], the drying rate was increased at the low range of air velocities (less than 4 m/s) using high-intensity ultrasound (154.3 dB and 21.8 kHz), which influenced the external and internal resistances and enhanced the mass transfer process. These studies clearly indicate the reciprocal effect of the air velocity, so that higher air velocities will reduce the external resistance, but at the same time acoustic field inside the drying chamber is disturbed by high air velocity and ultrasound effects on drying kinetics weakens [13, 36]. However, most industrial convective driers use low air velocities (around 1–2 m/s) because of their large capacities (volumes) [38]. In addition, above a certain power level of ultrasound (80 W), its effectiveness, and drying acceleration decreases regardless of the flow velocity. Therefore, only below that threshold value, both air velocity and ultrasound power accelerate the drying process [8, 69].

The driving force for convective transport during convective drying is the enthalpy of the air mass, which is mainly dependent on air temperature. Thus, water evaporation involves a temperature decrease in the air. When ultrasound is applied, both thermal and mechanical energy is available in the medium to improve drying. However, the increase in temperature [40, 84] and batch mass [14, 37] resulted in a decrease in the efficiency of ultrasound enhancement. Higher the temperature lower is the density of the drying agent (air) and higher will be the attenuation of ultrasounds [69]. In addition, ultrasonic velocity and acoustic impedance are temperature dependent, as ultrasonic propagation is sensitive to temperature.

In a study using airborne ultrasound for drying of carrot slices at 1.3 m/s and different air temperatures, it was observed that the effects of ultrasound on the effective moisture diffusivity and mass transport coefficient were reduced at temperatures above 60°C [37]. Meanwhile, effect of ultrasound on drying kinetics was significant at 60 and 90°C but was undetected at 115°C [30]. This fact could be linked to the proportion of mechanical energy in the medium: the higher the temperature, the lower the ratio of acoustic to thermal energy. Thus, beyond an upper-temperature limit, the acoustic energy is almost negligible compared with the total energy available in the medium. Therefore, drying at high temperature reduces the effectiveness of airborne ultrasonic application and seems to handicap for ultrasonic intensification [38].

1.6.4   MASS LOAD DENSITY

The influence of mass loaded in the drying chamber depends on the available energy per unit of mass. In addition, load densities with a large mass may disrupt the airflow pattern in the drying chamber. At high mass load densities, the differences between experiments with and without ultrasound application were not significant [14]. This fact suggests that when applying 30 kW/m^3, the saturation point should be close to 120 kg/m^3, beyond which the available acoustic energy per unit of mass could be, ignored [38]. On the other hand, an increase in the single batch weight with the same power of ultrasound caused higher energy absorption, which diminished the effectiveness of enhancement [69].

1.6.5 PRODUCT STRUCTURAL CHARACTERISTICS

The propagation of ultrasonic waves is largely dependent on the material through which they travel. The texture of the material being dried could also influence ultrasound effectiveness. An increase in the surface hardness led to a drop in ultrasound efficiency [72–74]. Considering the same acoustic energy level entering the particle, large intercellular spaces, and high-porosity in the soft materials make it more suitable for compression and expansion ultrasonic cycles and facilitate water transfer through the solid mass. However, hard structure and small intercellular spaces with low porosity in the products appear to be more resistant to mechanical stress [22].

Therefore, high acoustic energy levels are required in dense products to achieve the same mechanical stress in the particle. In addition, higher absorption of acoustic energy is also awaited in highly porous materials because of greater gas volume and due to the increase in available internal energy in the particles [38]. Meanwhile, product structure may have an influence on the acoustic energy transfer in the air-solid interface, because the acoustic impedance mismatches with air and product could be affected by structure [6]. For instance, fibrous structure of meat absorbs ultrasound energy and makes it tougher to impart any positive effects. In addition, meat does not contain any air-filled pores, thus the impact of ultrasonication is less effective [67]. All these effects indicate the significance and effects of product structural properties on the effectiveness of ultrasound assisted drying of foods.

1.7 EFFECT OF ULTRASOUND DRYING ON FOOD CONSTITUENTS AND QUALITY PARAMETERS

In the case of food products, quality is a critical parameter, which determines the suitability of a given processing procedure (e.g., drying). Unfortunately, almost all processing procedures somehow influence the quality of products, including physical and chemical ones [6, 23, 38, 69].

1.7.1 EFFECT ON TISSUE AND STRUCTURAL COMPONENTS

Ultrasonic application not only improves mass transport but also involves modifications in the structure of the product being dried. Changes in product structural properties on ultrasound drying arise from both spatial modification in the structure/geometry of the product and changes in the physicochemical properties (such as adhesion force, surface tension, viscosity) [23]. The cells of the internal flesh (endocarp) of eggplant dried using a mild ultrasonic power (18 kW/m^3) maintained their individuality showing a better structure than those dried without ultrasound application [38, 69].

The better structure is linked to shortenings the drying time and the mild mechanical compression and expansion produced by ultrasound. However, when a high ultrasonic power (37 kW/m^3) is applied, a highly compacted tissue appeared, even when the drying time is 70% less than in experiments without ultrasound application [78]. The significant compaction at high acoustic levels is ascribed to the intense compression and expansion produced by ultrasound in the weak structure of eggplant. The same visible and meaningful spatial changes of the product structure is also been seen in microscopic analysis of ultrasound-assisted dried apple [86, 89, 91], and orange peel [41, 71].

Structural alterations in the product surface have been linked to the mechanical effects produced by ultrasound in the interfaces [38]. Unlike mild or low-intensity ultrasound, application of high power ultrasound and increased sonication time results in greater damage to the material structure in comparison to samples dried without ultrasound enhancement, due to increase in porosity, the formation of microchannels, loss in tissue coherence, breakage in cell membrane and cell wall, etc., [69].

1.7.2 EFFECT ON WATER ACTIVITY

High water activity is advantageous for all deteriorative processes, such as oxidation, enzymatic, and non-enzymatic reactions and for the growth of bacteria, fungi, molds, yeasts, etc., [102]. In various studies, ultrasound pre-treatment caused a significant decrease in the water activity of dried apple [53], carrots (significant decrease of 66%) [56], cherry (significant decrease of 18%) [55], cranberry [93], and mushrooms (significant decrease of 19%) [12]. Meanwhile, convective drying of strawberries enhanced with ultrasound had shown a significant decrease of 34% in the water activity compared to the normal convective drying [97]. In addition, the same results has been observed with the drying of green pepper (up to 49%) [99], red beetroot (up to 33%) [100], and raspberry [54], in which water activity was decreased significantly with the assistance of ultrasound.

1.7.3   EFFECT ON COLOR

The color of the products is a very important parameter when judging the product's quality. As reported in the recent studies, there was no influence of ultrasound application on the color of dried products, such as green peas [7] and parsley leaves [96]. Application of ultrasound pretreatment in improving and preservation of color values has also been observed in the drying of carrot [80], seaweed [47], mushroom [12], strawberry [35] and *Agaricus bisporus* slices [46]. However, compared to convective drying, there was decrease in total color change in dried apples [69], potatoes [57], and banana slices [20] and in guava slices [50], after applying assisted ultrasound or ultrasound pretreatment. The differences between the color of ultrasound-assisted and purely convective dried products is imperceptible; and it is also stated that the ultrasound promotes an increase in the lightness of the dried product [72, 74].

The total color change on drying with ultrasonic power of 200 W has been the lowest. However, the color change of apple in ultrasound-assisted drying is higher than that of convective drying [54], due to the atomization of water due to high temperature achieved by the sample's surface during ultrasonic drying. When ultrasonic waves reach the surface of the sample, these will incite atomization by breaking the moisture layer, thus leading to the color change in the products [44].

1.7.4   EFFECT ON PHYTOCHEMICALS AND VITAMINS

Ultrasound-assisted drying helps to preserve phytochemicals and vitamins in the dried products when compared to other conventional methods. This novel drying technology is suitable for preserving antioxidant activity and phenolic compounds because it helps to reduce the drying temperature [6]. The decrease in temperature on the application of acoustic power density of

30.8 W/m^3 has caused a decrease in degradation of polyphenols, flavonoids, and the antioxidant activity in ultrasound-assisted dried apples [86]. Antioxidant activity of carrot samples was improved due to ultrasound application under both convective and microwave-convective drying processes, but sonication had reduced the total polyphenol content [58].

Along with antioxidant activity, total flavonoid content (TFC) was also improved in ultrasound pretreated samples compared to control samples [105]. In addition, total phenolic content (TPC) of the dried garlic slices was increased with increasing ultrasonic intensity (216.8, 902.7, and 1513.5 W/m^2) first and then was dropped [101]. Thus, a modest ultrasound power level can help to retain phenolic compounds, and the loss of antioxidant capacity depends on both drying temperature and ultrasound application.

In a study by Ren et al. [81], the antioxidant activity was reduced with relatively long ultrasound processing time (e.g., 5 minutes). On the other hand, applying ultrasound at a lower temperature of the drying agent resulted with decrease in antioxidant capacity, and at a higher temperature it resulted with an increase in its parametric value; thus, ultrasound enhancement also causes a decrease in total polyphenol and flavonoid content independently of the drying temperature [69].

The ultrasound pretreatment improved the retention of carotenoid content. Total carotenoid content dried by application of ultrasound was increased by an average of 53% compared to the control group [19]. In addition, ultrasound application also improved the carotenoid retention in microwave-convective dried carrots [58]. Furthermore, Allahdad et al. [1] found that the anthocyanin content in ultrasound pretreated pomegranate arils was higher than that of fresh samples. Meanwhile, in relying on the attributes, the application of ultrasound was also found to have no influence on the final content of anthocyanin in dried blueberries [69, 95].

The influence of ultrasound for retention of vitamins could be either positive (mostly in a lower drying temperature) or negative (mostly in higher drying temperature). The high power ultrasound drying allows preserving a high amount of vitamin C. Szadzinska et al. [99] found that the vitamin C retention of green pepper was increased by 55.2% in ultrasound-assisted convective drying. A similar result was observed for ultrasound-assisted dried carrot slices, in which the vitamin C increased by 36.3% and 50.2% at 65 and 70°C, respectively [15]. In addition, vitamins B1, B2, B3, and B6 availability in dried apples was increased with the ultrasound application while the vitamins B5 and E availability has decreased in all the products dried at different conditions [27]. On the other hand, ultrasound also caused significant degradation of the ascorbic acid in dried strawberries [32] and ultrasound-assisted freeze-dried apples [66].

1.7.5   EFFECT ON OTHER RELATED QUALITY PARAMETERS

Enzymes and microorganisms are responsible for many deteriorative processes [102]. Almost all negative biochemical conversions and the activity of pathogenic microorganisms have been affected by a reduction in water activity. In the same instant, the application of ultrasound also leads to the inactivation of enzymes and microbes in the process of drying by its direct and indirect effects [69].

Another important physical parameter describing dried food is its rehydration compliance. According to various studies conducted, there are no significant differences in rehydration ratio between the ultrasound-assisted and pure convective dried products, such as strawberries [32], and freeze-dried red pepper [92]. In addition, the rehydration rate and ratio for potatoes also increased with the application of ultrasound [57]. The same results have been observed in the case of microwave-convective drying of green pepper [98], low-temperature drying of salted cod [74], carrot [33], mushrooms [44] and apple [91], under ultrasound-assisted and pretreatment conditions.

In most applications, ultrasound pretreatment improved the rehydration rate at low temperatures [82]. However, in some exceptional cases by Nowacka et al. [70], rehydration capacity of apple was subjected to ultrasonic treatment for a shorter duration (10 and 20 minutes) and it was found to be lower than the untreated one, due to higher structural modification of samples for lower rehydration rate.

Other relevant physical quality factors are: food texture, hardness, density, and porosity of dried products. Application of ultrasound has caused a significant decrease in the hardness of the cod samples [74], pear [22], cranberry [1], papaya [19], and pomegranate aril [69], under various ultrasound-assisted and pretreatment drying conditions. Both strength and Young's modulus for dried apples rose due to ultrasound application. In addition, ultrasound-assisted dried chips were found to be more brittle and less crispy as compared to the samples dried with pure convective drying [69]. Ultrasound pretreatment also increased the porosity and decreased the density of food products [44]. Meanwhile, ultrasound pretreatment of banana slices significantly reduced the shrinkage by 71.8% [20], and it has been noted that ultrasound application and air temperatures did not significantly affect the products shrinkage.

1.8 LIMITATIONS AND CURRENT CHALLENGES

Facts suggest that the available acoustic energy into the medium seems to be a limitation in the extension of the phenomena associated with ultrasound application due to poor transmission of acoustic energy and low energy available in the medium. The increase in the drying agent temperature also negatively influences the efficiency of the ultrasound enhancement [69]. Other major constraints in the application of ultrasound for assisted drying are:

- Deteriorative influence on product quality by mechanical destruction of the outer skin due to friction and increased sonication time.
- Difficulties in construction and coupling between the transducer and material.
- Difficulties in maintaining hygienic requirements specified for the processing of food products.
- Mechanical effects on ultrasonication may lead to the quality deterioration of food products by the degradation of components, alterations in physical properties, and by the development of off-flavors.
- Product structural properties like fibrous structure in meat that absorbs ultrasound energy and makes it difficult for it to exert any influence.
- The requirement of applying many transducers to attain suitable acoustic power.

In addition, low relative air humidity may also limit the efficient transfer of the ultrasonic waves [38, 69], and the threshold level of power ultrasound, above which effectiveness of ultrasound and drying acceleration decreases regardless of flow velocity. Increasing the duration of ultrasonication at 25 kHz tends to lower the water loss values on assisted drying process [35].

Most of the research studies are limited to the laboratory-scale solutions. Furthermore, the application of ultrasound stays some way from being a feasible and operational technology for industrial preservation and drying due to dearth of information, education, and advance research on full-scale design, scale-up, and on standardization of assisted drying process parameters for food processing and preservation [62].

Operational limitations to generate the needed scale of airborne or assisted ultrasound, and relatively high cost and its energy consumption of such industrial-scale ultrasound equipment can hinder its application in food drying industry. In the case of the drying industry, the lack of an effective technology for the generation of power ultrasound in air has been distinguished as the primary constraint ahead of large-scale application [6]. An effective impact on the material being dried, at an industrial scale, demands the use of a large number of transducers and the development of automated systems, which would raise both investment and operating costs. Therefore, more effort is required in designing ultrasonic transducers with higher power capacities in-line with industrial standards [38].

Ensminger et al. [23] stated that the factors such as scale-up problems, dearth of knowledge, training, and equipment development for ultrasound drying application, and competition from other less expensive methods are major drawback for large-scale applications of high-intensity ultrasonic energy for the process of drying in the food industry.

1.9 FUTURE PERSPECTIVES

Ultrasonic drying is the potential of great commercial importance. Ultrasound waves are non-toxic, safe, and eco-friendly [49]. Application of power ultrasound has a meaningful influence on the kinetics of the drying process. Power ultrasound shortens the time of the drying process by improving heat and mass transfer phenomena [69]. The kinetic improvement also accompanied by an energy savings, and the low energy consumption provides not only cost savings but also a positive environmental impact [38]. In most cases, ultrasound does not cause detrimental effects on food quality attributes, and it enhances retention of fresh flavors and pigments, resulting in products with more appealing flavor, taste, color, and brightness. Moreover, high-intensity ultrasound in amalgamation with other non-thermal methodologies is been noted as an efficacious technique for dehydration of foods [62].

Despite the efforts carried out in these fields, there is still scope for significant improvement to bring ultrasonic technology closer to industrial drying operations. A new transducer technology is required with increased efficiency and power capacity along with improvement in design of radiator to achieve uniform distribution of vibration amplitude in order to compensate the real-time industrial need [31]. In addition, large-scale acceptance of high-intensity ultrasonic processes in the industry depends on unique capabilities offered, utilization of localized reaction zones, and design and operational simplicity [23]. Therefore, optimization of ultrasound-assisted drying in terms of energy consumption and process design on an industrial scale is another area that deserves more research [6].

1.10 SUMMARY

Application of ultrasound through contactless method accelerates the drying kinetics at low temperature and low air velocity ranges, and contact methods are evidently more efficacious especially if static pressure is applied. Moreover, the application of novel ultrasonic technology will significantly influence the food quality. Apart from lowering the water activity, the application of ultrasound also improves the product color and minimizes the loss of nutrients, such as TPC, flavonoid content, antioxidant activity, and vitamin C on drying operation. Even though ultrasonic drying involves efficient power output and lower running cost, yet the process variables can influence the magnitude of the ultrasound effects, and it is necessary to establish the optimum value for each specific application. Therefore, tuning of drying operational parameters and large-scale design of machineries in conjugation with core research on its effect and efficiency and adaptation to multiple food systems play a key role in reaching an efficient ultrasonic drying application to an industrial scale.

KEYWORDS

- **drying kinetics**
- **high-intensity ultrasound**
- **product quality**
- **sonication**
- **sponge effect**
- **ultrasound-assisted drying**
- **ultrasound pretreatment**

REFERENCES

1. Allahdad, Z., Nasiri, M., Varidi, M., & Varidi, M. J., (2019). Effect of sonication on osmotic dehydration and subsequent air-drying of pomegranate arils. *Journal of Food Engineering, 244,* 202–211.
2. Amami, E., Khezami, W., & Mezrigui, S., (2017). Effect of ultrasound-assisted osmotic dehydration pretreatment on the convective drying of strawberry. *Ultrasonics Sonochemistry, 36,* 286–300.

3. Arkhangelskii, M. E., & Stanikov, Y. G., (1973). Diffusion in heterogeneous systems: Part VIII. In: Rozenberg, L. D., (ed.), *Physical Principles of Ultrasonic Technology* (Vol. 2, pp. 277–370). New York – USA: Plenum Press.
4. Aversa, M., Curcio, S., Calabro, V., & Iorio, G., (2007). Analysis of the transport phenomena occurring during food drying process. *Journal of Food Engineering, 78*(3), 922–932.
5. Azoubel, P. M., Baima, M. D. A. M., Da Rocha, A. M., & Oliveira, S. S. B., (2010). Effect of ultrasound on banana drying kinetics. *Journal of Food Engineering, 97*(2), 194–198.
6. Baeghbali, V., Niakousari, M., & Ngadi, M., (2019). Update on applications of power ultrasound in drying food: A review. *Journal of Food Engineering and Technology, 8*(1), 29–38.
7. Bantle, M., & Eikevik, T. M., (2011). Parametric study of high-intensity ultrasound in the atmospheric freeze-drying of peas. *Drying Technology, 29*(10), 1230–1239.
8. Beck, S. M., Sabarez, H., & Gaukel, V., (2014). Enhancement of convective drying by application of airborne ultrasound: Response surface approach. *Ultrasonics Sonochemistry, 21*(6), 2144–2150.
9. Bird, R. B., Stewart, W. E., & Lightfoot, E. N., (2007). *Transport Phenomena* (2nd edn., p. 698). New York, USA: John Wiley & Sons, Inc.
10. Bonaui, C., Dumoulin, E., & Raoult-Wack, A. L., (1996). Food drying and dewatering. *Drying Technology, 14*(9), 2135–2170.
11. Borisov, Y. Y., & Gynkina, N. M., (1973). Acoustic drying: Part IX. In: Rozenberg, L. D., (ed.), *Physical Principles of Ultrasonic Technology* (Vol. 2, pp. 381–470). New York, USA: Plenum Press.
12. Cakmak, R. S., Tekeoglu, O., Bozkır, H., Ergun, A. R., & Baysal, T., (2016). Effects of electrical and sonication pretreatments on the drying rate and quality of mushrooms. *LWT-Food Science and Technology, 69*, 197–202.
13. Carcel, J. A., Garcia-Perez, J. V., Riera, E., & Mulet, A., (2007). Influence of high-intensity ultrasound on drying kinetics of persimmon. *Drying Technology, 25*(1), 185–193.
14. Carcel, J. A., Garcia-Perez, J. V., Riera, E., & Mulet, A., (2011). Improvement of convective drying of carrot by applying power ultrasound: Influence of mass load density. *Drying Technology, 29*(2), 174–182.
15. Chen, Z. G., Guo, X. Y., & Wu, T., (2016). A novel dehydration technique for carrot slices implementing ultrasound and vacuum drying methods. *Ultrasonics Sonochemistry, 30*, 28–34.
16. Correa, J. L. G., Rasia, M. C., Mulet, A., & Carcel, J. A., (2017). Influence of ultrasound application on both the osmotic pretreatment and subsequent convective drying of pineapple (*Ananas comosus*). *Innovative Food Science and Emerging Technologies, 41*, 284–291.
17. Da-Mota, V. M., & Palau, E., (1999). Acoustic drying of onion. *Drying Technology, 17*(4/5), 855–867.
18. Da Silva, G. D., Barros, Z. M. P., & De Medeiros, R. A. B., (2016). Pretreatments for melon drying implementing ultrasound and vacuum. *LWT-Food Science and Technology, 74*, 114–119.
19. Da Silva, J. E. V., & De Melo, L. L., (2018). Influence of ultrasound and vacuum-assisted drying on papaya quality parameters. *LWT-Food Science and Technology, 97*, 317–322.

20. Dehsheikh, F. N., & Dinani, S. T., (2019). Coating pretreatment of banana slices using carboxymethyl cellulose in an ultrasonic system before convective drying. *Ultrasonics Sonochemistry, 52*, 401–413.

21. De La Fuente-Blanco, S., De Sarabia, E. R. F., Acosta-Aparicio, V. M., Blanco-Blanco, A., & Gallego-Juarez, J. A., (2006). Food drying process by power ultrasound. *Ultrasonics, 44*, 523–527.

22. Dujmic, F., Brncic, M., & Karlovic, S., (2013). Ultrasound-assisted infrared drying of pear slices: Textural issues. *Journal of Food Process Engineering, 36*(3), 397–406.

23. Ensminger, D., & Bond, L. J., (2011). Applications of high-intensity ultrasonics: Basic mechanisms and effects. In: Ensminger, D., & Bond, L., (eds.), *Ultrasonics: Fundamentals, Technologies, and Applications* (3rd edn., pp. 459–494). Boca Raton, USA: CRC Press (Taylor & Francis Group).

24. Fernandes, F. A., Gallao, M. I., & Rodrigues, S., (2009). Effect of osmosis and ultrasound on pineapple cell tissue structure during dehydration. *Journal of Food Engineering, 90*(2), 186–190.

25. Fernandes, F. A., Linhares, Jr. F. E., & Rodrigues, S., (2008). Ultrasound as pretreatment for drying of pineapple. *Ultrasonics Sonochemistry, 15*(6), 1049–1054.

26. Fernandes, F. A., & Rodrigues, S., (2007). Ultrasound as pretreatment for drying of fruits: Dehydration of banana. *Journal of Food Engineering, 82*(2), 261–267.

27. Fernandes, F. A., Rodrigues, S., Carcel, J. A., & Garcia-Perez, J. V., (2015). Ultrasound-assisted air-drying of apple (*Malus domestica* L.) and its effects on the vitamin of the dried product. *Food and Bioprocess Technology, 8*(7), 1503–1511.

28. Floros, J. D., & Liang, H., (1994). Acoustically assisted diffusion through membranes and biomaterials. *Food Technology, 48*(12), 79–84.

29. Gallego-Juarez, J. A., (1998). Some applications of airborne power ultrasound to food processing: Chapter 7. In: Povey, M. J. W., & Mason, T. J., (eds.), *Ultrasound in Food Processing* (pp. 127–143). London, UK: Blackie Academic & Professional.

30. Gallego-Juarez, J. A., & Riera, E., (2007). Application of high-power ultrasound for dehydration of vegetables: Processes and devices. *Drying Technology, 25*(11), 1893–1901.

31. Gallego-Juarez, J. A., & Rodriguez-Corral, G., (1999). New high-intensity ultrasonic technology for food dehydration. *Drying Technology, 17*(3), 597–608.

32. Gamboa-Santos, J., Montilla, A., & Soria, A. C., (2014). Impact of power ultrasound on chemical and physicochemical quality indicators of strawberries dried by convection. *Food Chemistry, 161*, 40–46.

33. Gamboa-Santos, J., Soria, A. C., Villamiel, M., & Montilla, A., (2013). Quality parameters in convective dehydrated carrots blanched by ultrasound and conventional treatment. *Food Chemistry, 141*(1), 616–624.

34. Garcia-Noguera, J., Oliveira, F. I., & Gallao, M. I., (2010). Ultrasound-assisted osmotic dehydration of strawberries: Effect of pretreatment time and ultrasonic frequency. *Drying Technology, 28*(2), 294–303.

35. Garcia-Noguera, J., Oliveira, F. I., & Weller, C. L., (2014). Effect of ultrasonic and osmotic dehydration pretreatments on the color of freeze-dried strawberries. *Journal of Food Science and Technology, 51*(9), 2222–2227.

36. Garcia-Perez, J. V., Carcel, J. A., Benedito, J., & Mulet, A., (2007). Power ultrasound mass transfer enhancement in food drying. *Food and Bioproducts Processing, 85*(3), 247–254.

37. Garcia-Perez, J. V., Carcel, J. A., & De La Fuente-Blanco, S., (2006). Ultrasonic drying of foodstuff in a fluidized bed: Parametric study. *Ultrasonics, 44*, 539–543.

38. Garcia-Perez, J. V., Carcel, J. A., & Mulet, A., (2015). Ultrasonic drying for food preservation: Chapter 29. In: Gallego-Juarez, J. A., & Graff, K. F., (eds.), *Power Ultrasonics: Applications of High-Intensity Ultrasound* (pp. 875–910). Cambridge, UK: Woodhead Publishing.

39. Garcia-Perez, J. V., Carcel, J. A., Riera, E., Rossello, C., & Mulet, A., (2012). Intensification of low-temperature drying by using ultrasound. *Drying Technology, 30*(11/12), 1199–1208.

40. Garcia-Perez, J. V., Carcel, J. A., & Simal, S., (2013). Ultrasonic intensification of grape stalk convective drying: Kinetic and energy efficiency. *Drying Technology, 31*(8), 942–950.

41. Garcia-Perez, J. V., Ortuno, C., Puig, A., Carcel, J. A., & Perez-Munuera, I., (2012). Enhancement of water transport and microstructural changes induced by high-intensity ultrasound application on orange peel drying. *Food and Bioprocess Technology, 5*(6), 2256–2265.

42. Greguss, P., (1963). The mechanism and possible applications of drying by ultrasonic irradiation. *Ultrasonics, 1*(2), 83–86.

43. Howkins, S. D., (1969). Diffusion rates and the effect of ultrasound. *Ultrasonics, 7*(2), 129–130.

44. Huang, D., Men, K., Li, D., Wen, T., Gong, Z., Sunden, B., & Wu, Z., (2019). Application of ultrasound technology in the drying of food products. *Ultrasonics Sonochemistry, 63*, 104950.

45. Jambrak, A. R., Mason, T. J., Paniwnyk, L., & Lelas, V., (2008). Accelerated drying of button mushrooms, brussels sprouts and cauliflower by applying power ultrasound and its rehydration properties. *Journal of Food Engineering, 81*(1), 88–97.

46. Jiang, N., Liu, C., Li, D., Zhang, Z., Yu, Z., & Zhou, Y., (2016). Effect of thermosonic pretreatment on drying kinetics and energy consumption of microwave vacuum dried *Agaricus bisporus* slices. *Journal of Food Engineering, 177*, 21–30.

47. Kadam, S. U., Tiwari, B. K., & O'Donnell, C. P., (2015). Effect of ultrasound pretreatment on the drying kinetics of brown seaweed *Ascophyllum nodosum*. *Ultrasonics Sonochemistry, 23*, 302–307.

48. Katzir, S., (2006). The discovery of the piezoelectric effect: Chapter 1. In: Katzir, S., (ed.), *The Beginnings of Piezoelectricity* (pp. 15–64). Dordrecht, The Netherlands: Springer.

49. Kentish, S., & Ashokkumar, M., (2011). The physical and chemical effects of ultrasound: Chapter 1. In: Feng, H., Barbosa-Canovas, G. V., & Weiss, J., (eds.), *Ultrasound Technologies for Food and Bioprocessing* (pp. 1–12). London - UK: Springer.

50. Kek, S. P., Chin, N. L., & Yusof, Y. A., (2013). Direct and indirect power ultrasound-assisted pre-osmotic treatments in convective drying of guava slices. *Food and Bioproducts Processing, 91*(4), 495–506.

51. Khmelev, V. N., Shalunov, A. V., & Barsukov, R. V., (2011). Studies of ultrasonic dehydration efficiency. *Journal of Zhejiang University-Science A, 12*(4), 247–254.

52. Kim, S. M., & Zayas, J. F., (1991). Effects of ultrasound treatment on the properties of chymosin. *Journal of Food Science, 56*(4), 926–930.

53. Kowalski, S. J., & Mierzwa, D., (2015). US-assisted convective drying of biological materials. *Drying Technology, 33*(13), 1601–1613.

54. Kowalski, S. J., Pawlowski, A., Szadzinska, J., Lechtanska, J., & Stasiak, M., (2016). High power airborne ultrasound assist in combined drying of raspberries. *Innovative Food Science and Emerging Technologies, 34*, 225–233.

55. Kowalski, S. J., & Szadzinska, J., (2014). Convective-intermittent drying of cherries preceded by ultrasonic-assisted osmotic dehydration. *Chemical Engineering and Processing: Process Intensification, 82*, 65–70.

56. Kowalski, S. J., Szadzinska, J., & Pawlowski, A., (2015). Ultrasonic-assisted osmotic dehydration of carrot followed by convective drying with continuous and intermittent heating. *Drying Technology, 33*(13), 1570–1580.

57. Kroehnke, J., Musielak, G., & Boratynska, A., (2014). Convective drying of potato assisted by ultrasound. *PhD Interdisciplinary Journal, 1*, 57–65.

58. Kroehnke, J., Szadzinska, J., & Stasiak, M., (2018). Ultrasound-and microwave-assisted convective drying of carrots-process kinetics and product's quality analysis. *Ultrasonics Sonochemistry, 48*, 249–258.

59. Krokida, M. K., Zogzas, N. P., & Maroulis, Z. B., (2001). Mass transfer coefficient in food processing: Compilation of literature data. *International Journal of Food Properties, 4*(3), 373–382.

60. La Fuente, C. I. A., Zabalaga, R. F., & Tadini, C. C., (2017). Combined effects of ultrasound and pulsed-vacuum on air-drying to obtain unripe banana flour. *Innovative Food Science and Emerging Technologies, 44*, 123–130.

61. Lenart, I., & Auslander, D., (1980). The effect of ultrasound on diffusion through membranes. *Ultrasonics, 18*(5), 216–218.

62. Majid, I., Nayik, G. A., & Nanda, V., (2015). Ultrasonication and food technology: A review. *Cogent Food and Agriculture, 1*(1), 1071022.

63. Mason, T. J., (1998). Power ultrasound in food processing-the way forward. Chapter 6. In: Povey, M. J. W., & Mason, T. J., (eds.), *Ultrasound in Food Processing* (pp. 105–126). London, UK: Blackie Academic & Professional.

64. Mavroudis, N. E., Gekas, V., & Sjoholm, I., (1998). Osmotic dehydration of apples: Effects of agitation and raw material characteristics. *Journal of Food Engineering, 35*(2), 191–209.

65. Mierzwa, D., & Kowalski, S. J., (2016). Ultrasound-assisted osmotic dehydration and convective drying of apples: Process kinetics and quality issues. *Chemical and Process Engineering, 37*(3), 383–391.

66. Moreno, C., Brines, C., Mulet, A., Rossello, C., & Carcel, J. A., (2017). Antioxidant potential of atmospheric freeze-dried apples as affected by ultrasound application and sample surface. *Drying Technology, 35*(8), 957–968.

67. Mulet, A., Carcel, J. A., Sanjuan, N., & Bon, J., (2003). New food drying technologies: Use of ultrasound. *Food Science and Technology International, 9*(3), 215–221.

68. Muralidhara, H. S., Ensminger, D., & Putnam, A., (1985). Acoustic dewatering and drying (low and high frequency): Review. *Drying Technology, 3*(4), 529–566.

69. Musielak, G., Mierzwa, D., & Kroehnke, J., (2016). Food drying enhancement by ultrasound: A review. *Trends in Food Science and Technology, 56*, 126–141.

70. Nowacka, M., Wiktor, A., Sledz, M., Jurek, N., & Witrowa-Rajchert, D., (2012). Drying of ultrasound pretreated apple and its selected physical properties. *Journal of Food Engineering, 113*(3), 427–433.

71. Ortuno, C., Perez-Munuera, I., Puig, A., Riera, E., & Garcia-Perez, J. V., (2010). Influence of power ultrasound application on mass transport and microstructure of orange peel during hot air drying. *Physics Procedia, 3*(1), 153–159.
72. Ozuna, C., Alvarez-Arenas, T. G., Riera, E., Carcel, J. A., & Garcia-Perez, J. V., (2014). Influence of material structure on airborne ultrasonic application in drying. *Ultrasonics Sonochemistry, 21*(3), 1235–1243.
73. Ozuna, C., Carcel, J. A., Garcia-Perez, J. V., & Mulet, A., (2011). Improvement of water transport mechanisms during potato drying by applying ultrasound. *Journal of the Science of Food and Agriculture, 91*(14), 2511–2517.
74. Ozuna, C., Carcel, J. A., Walde, P. M., & Garcia-Perez, J. V., (2014). Low-temperature drying of salted cod (*Gadus morhua*) assisted by high power ultrasound: Kinetics and physical properties. *Innovative Food Science and Emerging Technologies, 23*, 146–155.
75. Panchev, I. N. K. N. A., Kirchev, N., & Kratchanov, C., (1988). Improving pectin technology, II: Extraction using ultrasonic treatment. *International Journal of Food Science and Technology, 23*(4), 337–341.
76. Plaza, L., De Ancos, B., & Cano, P. M., (2003). Nutritional and health-related compounds in sprouts and seeds of soybean (*Glycine max*), wheat (*Triticum aestivum* L.) and alfalfa (*Medicago sativa*) treated by a new drying method. *European Food Research and Technology, 216*(2), 138–144.
77. Pugin, B., & Turner, A. T., (1990). Influence of ultrasound on reaction with metals. In: Mason, T. J., (ed.), *Advances in Sonochemistry* (Vol. 1, pp. 81–118). London, UK: JAI Press Inc.
78. Puig, A., Perez-Munuera, I., Carcel, J. A., Hernando, I., & Garcia-Perez, J. V., (2012). Moisture loss kinetics and microstructural changes in eggplant (*Solanum melongena* L.) during conventional and ultrasonically assisted convective drying. *Food and Bioproducts Processing, 90*(4), 624–632.
79. Ratti, C., (2001). Hot air and freeze-drying of high-value foods: A review. *Journal of Food Engineering, 49*(4), 311–319.
80. Rawson, A., Tiwari, B. K., Tuohy, M. G., O'Donnell, C. P., & Brunton, N., (2011). Effect of ultrasound and blanching pretreatments on polyacetylene and carotenoid content of hot air and freeze-dried carrot discs. *Ultrasonics Sonochemistry, 18*(5), 1172–1179.
81. Ren, F., Perussello, C. A., Zhang, Z., Kerry, J. P., & Tiwari, B. K., (2018). Impact of ultrasound and blanching on functional properties of hot-air dried and freeze-dried onions. *LWT-Food Science and Technology, 87*, 102–111.
82. Rice, C., Rojas, M. L., Miano, A. C., Siche, R., Augusto, P. E. D., (2016). Ultrasound pretreatment enhances the carrot drying and rehydration. *Food Research International, 89*, 701–708.
83. Riera, E., Garcia-Perez, J. V., & Acosta, V. M., (2011). Computational study of ultrasound-assisted drying of food materials: Chapter 13. In: Knoerzer, K., Juliano, P., Roupas, P., & Versteeg, C., (eds.), *Innovative Food Processing Technologies: Advances in Multiphysics Simulation* (pp. 265–301). Chichester, West Sussex, UK: John Wiley & Sons, Ltd.
84. Rodriguez, J., Mulet, A., & Bon, J., (2014). Influence of high-intensity ultrasound on drying kinetics in fixed beds of high porosity. *Journal of Food Engineering, 127*, 93–102.

85. Rodrigues, S., Oliveira, F. I., Gallao, M. I., & Fernandes, F. A., (2009). Effect of immersion time in osmosis and ultrasound on papaya cell structure during dehydration. *Drying Technology, 27*(2), 220–225.

86. Rodriguez, O., Santacatalina, J. V., & Simal, S., (2014). Influence of power ultrasound application on drying kinetics of apple and its antioxidant and microstructural properties. *Journal of Food Engineering, 129*, 21–29.

87. Romdhane, M., & Gourdon, C., (2002). Investigation in solid-liquid extraction: Influence of ultrasound. *Chemical Engineering Journal, 87*(1), 11–19.

88. Romero, C. J., & Yepez, B. V., (2015). Ultrasound as pretreatment to convective drying of Andean blackberry (*Rubus glaucus* Benth.). *Ultrasonics Sonochemistry, 22*, 205–210.

89. Sabarez, H. T., Gallego-Juarez, J. A., & Riera, E., (2012). Ultrasonic-assisted convective drying of apple slices. *Drying Technology, 30*(9), 989–997.

90. Sanchez, E. S., Simal, S., Femenia, A., Benedito, J., & Rossello, C., (1999). Influence of ultrasound on mass transport during cheese brining. *European Food Research and Technology, 209*(3/4), 215–219.

91. Santacatalina, J. V., Rodriguez, O., & Simal, S., (2014). Ultrasonically enhanced low-temperature drying of apple: Influence on drying kinetics and antioxidant potential. *Journal of Food Engineering, 138*, 35–44.

92. Schossler, K., Jager, H., & Knorr, D., (2012). Novel contact ultrasound system for the accelerated freeze-drying of vegetables. *Innovative Food Science and Emerging Technologies, 16*, 113–120.

93. Shamaei, S., Emam-Djomeh, Z. A. H. R. A., & Moini, S., (2012). Ultrasound-assisted osmotic dehydration of cranberries: Effect of finish drying methods and ultrasonic frequency on textural properties. *Journal of Texture Studies, 43*(2), 133–141.

94. Simal, S., Benedito, J., Sanchez, E. S., & Rossello, C., (1998). Use of ultrasound to increase mass transport rates during osmotic dehydration. *Journal of Food Engineering, 36*(3), 323–336.

95. Siucinska, K., & Konopacka, D., (2016). The effects of ultrasound on quality and nutritional aspects of dried sour cherries during shelf-life. *LWT-Food Science and Technology, 68*, 168–173.

96. Sledz, M., Wiktor, A., Rybak, K., & Nowacka, M., (2016). The impact of ultrasound and steam blanching pretreatments on the drying kinetics, energy consumption and selected properties of parsley leaves. *Applied Acoustics, 103*, 148–156.

97. Szadzinska, J., Kowalski, S. J., & Stasiak, M., (2016). Microwave and ultrasound enhancement of convective drying of strawberries: Experimental and modeling efficiency. *International Journal of Heat and Mass Transfer, 103*, 1065–1074.

98. Szadzinska, J., Lechtanska, J., & Kowalski, S. J., (2015). Microwave-and infrared-assisted convective drying of green pepper: Quality and energy considerations. *Chemical Engineering and Processing: Process Intensification, 98*, 155–164.

99. Szadzinska, J., Lechtanska, J., Kowalski, S. J., & Stasiak, M., (2017). The effect of high power airborne ultrasound and microwaves on convective drying effectiveness and quality of green pepper. *Ultrasonics Sonochemistry, 34*, 531–539.

100. Szadzinska, J., Mierzwa, D., & Pawłowski, A., (2020). Ultrasound and microwave-assisted intermittent drying of red beetroot. *Drying Technology, 38*(1/2), 93–107.

101. Tao, Y., Zhang, J., Jiang, S., Xu, Y., Show, P. L., Han, Y., Ye, X., & Ye, M., (2018). Contacting ultrasound enhanced hot air convective drying of garlic slices: Mass transfer modeling and quality evaluation. *Journal of Food Engineering, 235*, 79–88.

102. Toledo, R. T., Singh, R. K., & Kong, F., (2007). *Fundamentals of Food Process Engineering* (4ᵗʰ edn., Vol. 297, p. 420). New York, USA: Springer International Publishing AG.
103. Vega-Mercado, H., Gongora-Nieto, M. M., & Barbosa-Canovas, G. V., (2001). Advances in dehydration of foods. *Journal of Food Engineering, 49*(4), 271–289.
104. Vinatoru, M., Toma, M., Radu, O., Filip, P. I., Lazurca, D., & Mason, T. J., (1997). The use of ultrasound for the extraction of bioactive principles from plant materials. *Ultrasonics Sonochemistry, 4*(2), 135–139.
105. Wang, H., Zhao, Q. S., Wang, X. D., Hong, Z. D., & Zhao, B., (2019). Pretreatment of ultrasound combined vacuum enhances the convective drying efficiency and physicochemical properties of okra (*Abelmoschus esculentus*). *LWT-Food Science and Technology, 112*, 11. Article ID: 108201.

CHAPTER 2

HYPOBARIC PROCESSING OF FOOD

SHARANABASAVA KUMBAR and MENON REKHA RAVINDRA

ABSTRACT

Hypobaric/vacuum/sub-baric processing refers to the processing operations carried out under conditions of pressure lower than the normal atmospheric pressure. The manipulation of the pressure to a sub-baric level results in alteration of system performance in terms of its transport phenomena influencing the process rate as well as product quality. Several applications of low-pressure processing are: impregnation, frying, cooking, and drying under hypobaric conditions. This book chapter provides to the reader with an overview of the fundamental aspects of hypobaric processing and systems to generate and maintain the desired vacuum.

2.1 INTRODUCTION

A region subjected to pressure much lower than the normal ambient conditions is said to be under vacuum, and the specification for the level of vacuum is often expressed in the usual barometric specifications of pressure (e.g., mm of Hg, bar, Pa, etc.). Conventional food processing operations under atmospheric conditions have several limitations, such as exposure to a higher temperature leads to the nutrient losses, time-consuming, detrimental effects on the product quality, and the contamination of the product. One of the approaches attempted by the process industry to overcome the afore-stated effects of conventional methods is by exploring the use of vacuum-assisted processing operations, such as frying, cooling, drying, packaging, etc.

This book chapter discusses:

1. The fundamental aspects of hypobaric processing and systems to generate and maintain the desired vacuum;
2. Hypobaric process modeling;
3. The effect of the pressure on process kinetics and thermodynamics; and
4. Various applications of hypobaric processing of foods and food products.

2.2   HYPOBARIC IMPREGNATION (HI) TECHNOLOGY

Hypobaric or vacuum impregnation (VI) was developed for the preparation of high quality and nutritionally rich food products, which use enhanced osmotic dehydration technology. The VI technology helps to rapidly impregnate plant or animal tissue pores with desired external solution producing a product with improved qualities and shelf-life with nominal changes in nutritional and sensory properties.

2.2.1   MECHANISM OF HYPOBARIC IMPREGNATION (HI)

Hypobaric or VI enhances the rate of osmosis in an immersed system due to the co-current effects of two observed phenomena, namely hydrodynamic mechanism (HDM) and deformation relaxation phenomena (DRP). The HDM is the rapid mass transfer when materials with an inherent permeable or porous nature are kept submerged in a liquid phase. The mass transfer is attributed to the in-flow of the extrinsic liquid into the capillary pores of the soaked solids controlled by the volume changes (expansion/compression) of its internal gas phase, due to the cycle of lower pressure (vacuum) and restored atmospheric pressure imposed on the system, ultimately, leading to the impregnation of the porous products.

Under vacuum, the gas occupying the porous interstices of the product structure is partially expelled due to its expansion. In addition, the pressure-induced changes in the surface tension of the liquid within the capillary pores also affect the mass movement. As the pressure cycle switches to its atmospheric level, the internal gas content retained in the pores is compressed and the surrounding liquid phase in the immersion bath occupies the vacated volume in the food pores, thus resulting in osmosis [22].

2.2.2   *HYPOBARIC OR VACUUM IMPREGNATION (VI) SYSTEM*

Hypobaric or vacuum system consists of an impregnation chamber (samples and the external solution), vacuum pump, condenser, pressure regulating valve, system controlling unit (Figure 2.1).

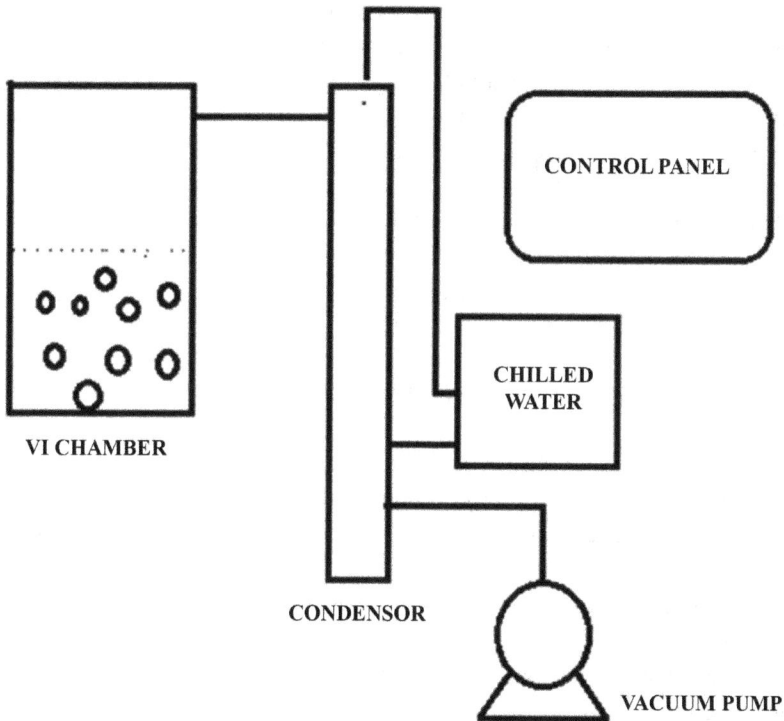

FIGURE 2.1   Schematic diagram of a vacuum impregnation system.

2.3   APPLICATIONS OF HYPOBARIC IMPREGNATION (HI) TECHNOLOGY

2.3.1   *CALORIE DENSIFICATION*

Food materials are generally composed of large quantities of water and air spaces, and these components occupy enormous volumes. The elimination of these calorie-less constituents and replacing or filling the same with nutrients is possible using hypobaric or VI resulting in an increase in the calories available per unit volume. The calorific density of the product is affected by both physical and chemical factors [3].

2.3.2 REDUCTION IN PH

Food materials when conventionally acidified by the blanching and soaking methods suffer as the process is time-consuming and causes loss of bioactive compounds from the product. Several researchers have reported that VI treatment successfully reduced the pH of the food [39].

It is also reported that hypobaric impregnation (HI) increases the rate of pH reduction in vegetables, and it was found to be better than conventional methods [13]. During the process, the rate of diffusion of the hydrogen ions from solution to vegetable tissue increases. This process depends on the vacuum pressure, the porosity of vegetable tissue, mechanical properties, etc.

2.3.3 PRE-FREEZING METHOD

Hypobaric or VI, as a pretreatment for food freezing, is found to enhance the textural quality of frozen ready to consume products such as cut fruits and vegetables. The process also aids in containing the exudation of drip during thawing and reducing the energy consumption during freezing [12, 48]. HI with cryoprotectants like cryostabilizers or hypertonic sugar solution is recommended for reducing freezable water content and decreasing the injury due to ice crystals in frozen materials [34]. The treatment with hypertonic solution leads to osmotic dehydration due to the collective effort of capillary action and pressure difference [22].

2.3.4 FOOD SALTING

HI technology leads to reduction in salting time [8] due to the enhanced salt uptake by the food product under the improved pressure gradient in the capillaries as a result of the HDMs observed during the process. Vacuum-assisted salting of spongy foods also increased the solute yield with faster salting kinetics and uniform distribution of salt in the material. Brine VI has been found to decrease the time of salting in ham for curing and manchego type cheeses while providing a uniform salt allocation in the food.

2.3.5 PRETREATMENT TO DRYING

HI can be applied as a pretreatment method for drying, which results in saving of energy and incorporation of functional/anti-browning/anti-microbial compounds of food. During pre-drying treatment at the first stage, the gas in the capillaries of food escapes from the interior of products. This is followed by the second stage, wherein the system is reverted to atmospheric pressure, which then reduces the internal gas volume in the food matrix. The visco-elastic property of plant tissues takes some time to contract after which the liquid outside the food product will flow into the porous capillary [21].

The combination of HI pretreatment with air-drying resulted in a low water activity product with softer texture compared to the conventional drying process [32]. The HI as a pretreatment also provides better preservation of pigments with narrowed moisture diffusivity of samples [1].

2.3.6 FUNCTIONAL PRODUCTS

The incorporation of probiotics to the matrix of fruit and vegetables using this impregnation technique helps to expand the range of probiotic food produce [24]. Due to the HDM, a greater number of probiotic cells may be incorporated into the food matrix. Vacuum impregnated probiotic-enriched dried samples of apple contained about 10^7 CFU/g of microorganisms. Probiotic-rich apple snacks were formed by impregnating apple cylinders either with two alternate liquid phases rich (10^7–10^8 CFU/ml) in probiotics (apple juice + *S. cerevisiae* and whole milk + *Lactobacillus casei (spp. rhamnosus)* [7].

2.3.7 MODIFIED THERMAL PROPERTIES

HI has been found to improve the thermal properties of fruit and vegetables based products. The composition and structure of food determine the thermal conductivity and diffusion coefficient of the product. Modifying this structure of material through the process prior to the thermal processing may enhance the heat conduction efficiency and improve the product quality. An upswing in the thermal conductivity (15%–24%) in the hypobaric impregnated apples in isotonic sucrose solution with a mild variation in the diffusion rate was observed [33]. It is also opined that the increase in thermal conductivity could be due to the enhanced solute content replacing the displaced gas molecules in the pores [20].

2.3.8   FORTIFIED PRODUCTS

The application of hypobaric or VI for fruits and vegetable fortification with minerals was studied with the primary objective of obtaining nearly 20–25% dietary intake through the samples [21]. Various mathematical models were applied for estimating the desirable amount of minerals in the external medium for the impregnation process. From the predictions of modeling and experimental confirmation, it was established that the process could be used as a successful method to fortify fruit and vegetable pieces with select vitamins, minerals, or desirable bioactive compounds. The calcium distribution in the fortified plant tissues was studied using microanalysis, which indicated that mushroom and eggplant showed calcium in the intercellular spaces while deposition in the xylem was observed for carrots [26].

2.4   ADVANTAGES AND DISADVANTAGES OF HYPOBARIC OR VACUUM IMPREGNATION (VI)

The Hypobaric or VI process has been reviewed in detail [13, 39] and the main advantages listed in comparison to conventional process include rapid completion of the soaking objectives, lower energy requirements, the possibility of carrying out the process under ambient temperature, and recycling of the soak solution for many cycles. However, one of the drawbacks for the poor industrial adoption of this technique includes the possibility of precise control of the operating condition. Another practical and technical issue pointed out [51] relates to the density difference between the porous plant tissue and the impregnation solution resulting in the product floating to the medium surface during hypobaric or VI.

2.5   HYPOBARIC OR VACUUM FRYING

Hypobaric or vacuum frying is an operation where food is fried below atmospheric pressure. The frying process, like conventional frying, renders the fried product acceptable and appetizing to the consumer due to a number of physicochemical changes that occur in the product during the process, including compositional changes, ultimately leading to the characteristic color, appearance, flavor, structure, and texture of the product. The novel process has been successfully applied in the process technology of common sweets and savories. Hypobaric frying technology has been largely applied for plant tissues products (fruits and vegetables), including apple, jackfruit, banana, apricot, green, and gold kiwifruits, carrot, potato, sweet potato, yam, mushroom, and shallots.

2.5.1 VACUUM/HYPOBARIC FRYING SYSTEM

The frying system employed for the hypobaric frying process is similar to the HI system (Figure 2.1), wherein the immersion solution in the tank is replaced with a frying medium (oil). Descriptions of the frying system imply that the primary components of the unit include a frying chamber integrated with a system to reduce the pressure (i.e., a vacuum pump) and a mechanism to condense the evolved vapors, commonly designed as a refrigerated condenser. The frying chamber is designed as a leak-proof airtight vessel integrated with an oil heater and a slotted basket, which has a suitable arrangement to raise and lower it into the heating medium. A vacuum pump is also coupled with the unit to provide the design pressure intended for the frying process.

As with all processing equipments, different design modules for hypobaric frying units, namely, based on the scale of the operation (laboratory, pilot, and industrial-scale) and mode of operation (batch and continuous process) have evolved in practice [2].

The integration of a gas-heated frying chamber with a liquid ring vacuum pump and a water-cooled condenser has been described in the development of a hypobaric process unit [40]. Post-processing, the fried product was surface-deoiled within the frying chamber using centrifugation at 450 rpm. Another approach to the design of the frying chamber includes external heating of the oil followed by pumping of the heated oil into the depressurized frying chamber [45]. The process pressure in the system was maintained using an oil-sealed vacuum pump and a cooling tower type water-cooled condenser.

The application of a liquid ring pump to generate the desired vacuum pressure in the hypobaric frying chamber was successfully attempted for frying of jackfruit chips [14] and the generated vapors were condensed using a cooling tower condenser. Further, the surface oil of the chips was removed by centrifuging under ambient conditions in a remote unit. Use of electric coils within the frying chamber to maintain the temperature of the frying oil was reported during hypobaric frying of potato chips [37]. The system was supported with an in-house centrifuge for de-oiling, oil-seal pump for generating and maintaining the desired vacuum and a refrigerated condenser.

2.5.2   APPLICATIONS OF HYPOBARIC FRYING

Production of a fried potato chip with a significant reduction in the oil content (to the tune of 30%) without compromising the color and crispiness of the product due to hypobaric frying was observed [23]. Similarly, hypobaric frying process is also credited with producing potato chips with nearly 97% reduction in its acrylamide content (an anticipatory carcinogen), in comparison to conventional frying process [25].

The technology has been evaluated for the frying of several plants tissue-based fries, such as carrot crisps, sweet potato, green beans, mango, and blue potato [18, 36]. It is universally accepted that the process reduced oil uptake yielding healthier products with better shelf life due to lower levels of fat oxidation during storage. Further, the lower temperature during processing helped preserve product nutrients in the form of vitamins and active principals for antioxidant activity. The technology has also been applied in the frying of fruit pieces such as pineapple [47] and was found to retain the quality attributes of traditional frying process (pleasant yellow color and good texture), while simultaneously presenting desirable nutritive advantages in the form of lower oil content, enhanced vitamin C retention and antioxidant potential due to phenolic compounds.

High protein products such as fried tofu resulted in a lighter hue and lower fatty acid content when fried using this novel technology [36]. Frying of fish filets using hypobaric process was attempted for gilthead sea bream (*Sparus aurata*), and the product was found qualitatively comparable to the conventionally fried product [2].

The application of this technology for frying indigenous dairy sweetmeats, such as *Gulabjamun* and *Pantoa* was reported [31]. The process resulted in an overall improvement in product quality with significant savings in process time and energy.

2.5.3   ADVANTAGES AND DISADVANTAGES OF HYPOBARIC FRYING

Hypobaric frying technology is being actively mooted as a substitutive technology for conventional frying technology [17]. Among the various advantages listed for this novel technology include, significantly reduced uptake of fat and oil during the process. This couples with the starved oxygen content of the frying environment due to the negative pressure of the process cascades to a positive influence on the product shelf life to reduced potential for the development of oxidative rancidity. The reduced oxygen in the frying environment is also attributed to the improved quality of the frying medium facilitating its reuse over the large number of frying cycles [42, 43] and the reduced microbiological load of the fried produce.

Among the advantages noticed with respect to the physicochemical attributes of the fried product are pleasant color due to reduced browning, and better preservation of nutritive properties of the product. The above advantages are primarily due to the lower temperature of the frying medium possible during hypobaric frying [41]. Another notable benefit noticed in the application of this technology is the significant reduction in processing time that could lead to faster throughput of the frying line [41]. On the flip side, due to the increased accessories required to generate and maintain the desired level of vacuum and the leak proofness desirable for the frying chamber, hypobaric frying units are typically more intensive with regard to its capital investment [27]. However, the cost-benefit ratio is expected to be favorable when technology is adopted for large scale processing.

2.6 HYPOBARIC/VACUUM DRYING

The expulsion and removal of moisture from food products during drying under the influence of a low-pressure environment have been widely documented. The primary advantages of the process spell out due to the improved product quality characteristics as a result of the reduced drying temperature and faster drying rates due to the negative pressure in the drying chamber owing to the sub-atmospheric pressures maintained in the drying chamber. Even though the primary accessory to upgrade a conventionally designed dryer to a hypobaric unit remains an integrated vacuum pump of suitable capacity, the application of an ejector pump as a means to maintain the desired low pressure within the drying chamber was investigated [29, 38].

The analysis identified four major factors influencing the process outcome, namely, operating pressure and temperature, ejector water pressure and make-up air circulated within the dryer. A laboratory vacuum drying process was analyzed using dry coconut press cake as the study material

[30]. The study deduced that the drying rate was determined by the thickness of the product as well as the temperature and pressure in the drying chamber.

The process of hypobaric drying and its effect on the drying rate and retention of principals, including vitamin C and dietary fiber content in coriander leaves was evaluated based on drying experiments using a Box-Behnken design (BBD) [46]. The drying parameters were optimized using response surface methodology (RSM) and drying at 75°C under a pressure of 28 mm Hg was recommended for a dryer loading of 0.63 kg/m^2.

The effect of low temperature during hypobaric drying on preserving phytonutrients in dried sour cherries was assessed [50]. Combinations of drying conditions of temperature (46–74°C) and pressure (17–583 mbar) were studied, and the vacuum assistance during drying was found to have a significant effect on the retention of phenolic compounds as well as color pigments of the dried product.

Integration of the hypobaric drying process with infrared (IR) radiators was also evaluated for its efficacy in drying floral products [6]. Benefits were observed in the integrated process in maintaining the process conditions during drying, while maintaining product quality under efficient energy management.

2.6.1   MICROWAVE HYPOBARIC DRYING (MHD)

Combined application of the two innovative approaches to drying of agricultural produce, viz., microwave, and low pressure (hypobaric atmospheric) have been considered as a potential technology for producing dried foods of superior quality with enhanced process efficiency [9, 11]. This approach is able to concurrently reap the individual advantages of both technologies over traditional drying technologies, resulting in faster drying at significantly low drying temperatures (< 50°C) with proven benefits with regard to the energy use efficiency [19].

Further, the volumetric heating supported using microwaves helped in the dissipation of the energy absorbed by the product along with the spontaneous adjustment of the moisture gradient within. It has been observed that product profile in terms of quality and sensory attributes such as color and flavor was also retained better in the products dried using this combination technology.

The process has been compared to freeze-drying, which is the most reliable drying technique with regard to product quality [4, 28]. It has been estimated that the heating rates achieved by the combination of microwave and hypobaric technologies compared to conventional freeze dryers is at an average of 25 times the latter. The cost economics of this novel combination is, however, difficult to compare. In terms of infrastructural cost, the combination technology is pegged at almost 60% greater than that for commercial lyophilizers, however, due to the process advantages, the operating costs are often cut by a factor of 3–4, projecting a scope for cost recovery in due course.

The application of microwave-hypobaric process for drying of high quality products such as honey was studied [10]. The drying process was completed within 10 min and the temporal analysis using moisture and temperature profiling implied that the process conditions including microwave power, hypobaric pressure, and sample thickness influenced the process parameters.

A laboratory-scale unit for microwave-hypobaric drying was developed and evaluated for the drying kinetics in orange juice-based fruit gel simulant [16]. The process conditions evaluated included pressure of 30–50 mbar and 640–710 W of microwave power rating. Pulsed power delivery for the microwave unit combined with hypobaric drying was employed to dehydrate banana slices and the product quality was deemed superior to the traditionally dried slices [15].

Similarly, a modified domestic microwave oven for the combined hypobaric process was evaluated for the drying of banana slices. The process analysis determined that the initial increase in the product temperature was rapid until the saturation temperature for moisture corresponding to the hypobaric pressure was attained [35]. Thereafter, the temperature gradient depicted a gentle increase till the expulsion of all free moisture present in the slice. The initial efficiency for the drying process was near cent per cent; the sluggish process towards the endpoint of drying caused the efficiency to fall to levels of nearly 35%. However, the hypobaric conditions of the drying environment were found to significantly improve the drying efficiency, especially as the product moisture dropped to lower levels.

Two modes of operation of a combined microwave-hypobaric drying system, namely the continuous and pulsed modes were evaluated using cranberries as study material [49]. For the continuous mode study, microwave drying (250 and 500 W) was attempted in combination with low pressure (at 5.33 and 10.67 kPa), while the pulse mode was operated only at a single power rating, i.e., 250 W for on-off cycles set to 30/60 s-60/90/150 s at both the pressure levels evaluated. The study indicated that the pulsed on-off cycles involving shorter on-longer off timings enhanced the process efficiency. The product dried using the combination technology was evaluated to be better in terms of color and texture compared to the conventionally dried product.

2.7   HYPOBARIC COOKING (SOUS VIDE COOKING)

Hypobaric cooking is often referred to as "Sous vide cooking" in culinary parlance. It is a French word, meaning "under vacuum," that may be broadly described as a process of sealing in food materials along with its juices or cooking sauce/marinade in a heat-stable food-grade vacuum package, followed by immersion cooking under prescribed process conditions. Two popular versions of this special cooking technique have been widely practiced in the food business. The first approach caters to the ready-to-serve category, where the food is prepared using the sous vide technique and then served after the finishing touches over the counter. The second approach is primarily practiced with the assistance of low-temperature preservation as an adjunct technology. The food prepared using standard sous vide technology is then subjected to in-pack rapid chilling or freezing before storage. The frozen sous vide packs are thawed/reheated before finishing and serving on demand [5].

2.7.1   APPLICATIONS OF SOUS VIDE COOKING IN THE FOOD INDUSTRY

Among all the avenues of food processing, sous vide cooking has demonstrated the most promise in its application in the meat cooking industry. The technology has also been adopted in processing of plant-based products, especially cooked vegetables.

The primary reason for meat to be a preferred candidate for sous vide cooking has been owing to the undesirable flavor and texture developed in leaner, younger cuts of meat when cooked conventionally. The locking-in of the juices and flavor components within the packaging in this method of cooking permits chefs to prepare flavorful, succulent, and soft-cooked meat.

The meat cuts for the cooking process are often pre-processed before it is sealed in the sous vide pack. Tough cuts of meat are often subjected to tenderizing process; both mechanical and chemical methods are employed for artificial tenderizing of the meat. The mechanical process involves piercing and disrupting the meat fibers by inserting multiple needles thin or blades into the meat. Another mechanical approach to render the meat tender is by cubing and pounding the meat slabs.

Marination is a common approach for tenderizing meat; the composition of the marinade is often acidic with fermented products such as vinegar, vine, and yogurt being good choices for the marinade. Enzyme rich fruit juices/pulp (with ficin, papain, or bromelain) is preferred for tenderization of the meat. Both injections into as well massaging the marinade on the surface and interstices of the meat are found satisfactory. Use of alcoholic marinades in tenderizing meat for *sous vide* applications is generally not recommended as there is a risk of the sealed pack ballooning during cooking due to its lower vapor pressure.

Another pre-processing practice widely adopted for sous vide meat is curing or brining. Both conventional wet brining as well as equilibrium brining have been reported. In the former, the meat is allowed to stand in salt solution (<10% concentration) for an hour or two, rinsed, and then cooked as per the standard practice. In the equilibration process, the meat is allowed to equilibrate (for a period varying between few hours to couple of days) with the salt content of the brine (around 1% for wet brine, around 2.5% as curing salts). The surface salt is then washed off and cooked using the sous vide technology.

The pre-processed meat/poultry cut is then hypobarically sealed along with its juices, flavorings, etc., in a food-grade pouch; typical pressures used for this method include 10–15 mbar for firm food and 100–120 mbar for liquids. The advantages of this step are many fold. Firstly, it allows the heat to be uniformly distributed throughout the meat while it is being cooked, either by immersion cooking in a precisely controlled water bath or by exposure to steam in a well-designed convective tray oven. Secondly, the reduced oxygen in the headspace of the sealed pack prevents the development of any oxidation products, primarily rancid off-flavors. The airtight seal also checks any scope of contamination with aerial microbes. The last two advantages are more significant for the cook-chill/freeze sous vide packs.

In addition to meat, this promising technique has also found application in the cooking of vegetables [44]. The technique has demonstrated better retention of the phytonutrients present in the product than conventional steaming/boiling processes. Further, it has been noted that sous vide cooking does not disturb the cell walls (primarily composed of polysaccharides such as cellulose) of the cells of the cooked vegetables, and the tenderizing of the vegetables during the cooking process is achieved due to solubilizing the cell wall pectin at temperatures around 85°C. This renders the cell wall-less resistant and leading to the tenderness of the cooked vegetable. Other advantages observed include better retention of the more concentrated and distinct flavor profile of the individual ingredients of the dish cooked in the sous vide bag, as well as the better appeal of the product in terms of color and appearance due to significant reduction in oxidation.

2.8 SUMMARY

A reduced pressure or hypobaric approach to some of the commonly accepted process technologies and unit operations has been found to be advantageous in terms of enhanced product quality and improved process efficiency. Some of the combinations discussed include HI, frying, drying, and cooking. The technical issues related to the generation and maintenance of the desired low pressure are mostly related to the additional costs on process infrastructure. However, the technology has demonstrated great promise and can be economized by adopting on a larger scale of production and widening the spectrum of its application in food processing.

KEYWORDS

- **deformation relaxation phenomena**
- **hydrodynamic mechanism**
- **hypobaric impregnation**
- **kinetics**
- **microwave hypobaric drying**
- **response surface methodology**
- **vacuum**

REFERENCES

1. Alvarez, C. A., Aguerre, R., Gomez, R., & Vidales, S., (1995). Air dehydration of strawberries: Effects of blanching and osmotic pretreatments on the kinetics of moisture transport. *Journal of Food Engineering, 25*(2), 167–178.
2. Andrés-Bello, A., García-Segovia, P., & Martínez-Monzó, J., (2011). Vacuum frying: An alternative to obtain high quality dried products. *Food Engineering Reviews, 3*(2), 63–70.

3. Ashitha, G. N., & Prince, M. V., (2018). Vacuum impregnation: Applications in food industry. *International Journal of Food and Fermentation Technology, 8*(2), 141–151.

4. Attiyate, Y., (1979). Microwave vacuum drying: Industrial applications. *Food Engineering, 51*(2), 78–79.

5. Baldwin, D. E., (2012). Sous vide cooking: A review. *International Journal of Gastronomy and Food Science, 1*(1), 15–30.

6. Bazyma, L. A., Guskov, V. P., & Basteev, A. V., (2006). The investigation of low temperature vacuum drying processes of agricultural materials. *Journal of Food Engineering, 74*(3), 410–415.

7. Betoret, N., Puente, L., & Diaz, M. J., (2003). Development of probiotic-enriched dried fruits by vacuum impregnation. *Journal of Food Engineering, 56*(2/), 273–277.

8. Chiralt, A., & Martínez-Navarrete, N., (2001). Changes in mechanical properties throughout osmotic processes: Cryoprotectant effect. *Journal of Food Engineering, 49*(2/3), 129–135.

9. Zheng-Wei, C., Shi-Ying, X., & Da-Wen, S., (2003). Dehydration of garlic slices by combined microwave-vacuum and air-drying. *Drying Technology, 21*(7), 1173–1184.

10. Zheng-Wei, C., Li-Juan, S., Chen, W., & Da-Wen, S., (2008). Preparation of dry honey by microwave-vacuum drying. *Journal of Food Engineering, 84*(4), 582–590.

11. Zheng-Wei, C., Shi-Ying, X., & Da-Wen, S., (2004). Effect of microwave-vacuum drying on the carotenoids retention of carrot slices and chlorophyll retention of Chinese chive leaves. *Drying Technology, 22*(3), 563–575.

12. Danila, T., & Bertolo, G., (2001). Osmotic pretreatments in fruit processing: Chemical, physical and structural effects. *Journal of Food Engineering, 49*(2/3), 247–253.

13. Derossi, A., Pilli, T. D., & Severini, C., (2010). Reduction in pH of vegetables by vacuum impregnation: A case study on pepper. *Journal of Food Engineering, 99*(1), 9–15.

14. Diamante, L. M., Savage, G. P., & Vanhanen, L., (2012). Optimization of vacuum frying of gold kiwifruit slices: Application of response surface methodology. *International Journal of Food Science and Technology, 47*(3), 518–524.

15. Drouzas, A. E., & Schubert, H., (1996). Microwave application in vacuum drying of fruits. *Journal of Food Engineering, 28*(2), 203–209.

16. Drouzas, A. E., Tsami, E., & Saravacos, G. D., (1999). Microwave/vacuum drying of model fruit gels. *Journal of Food Engineering, 39*(2), 117–122.

17. Dueik, V., & Bouchon, P., (2011). Development of healthy low-fat snacks: Understanding the mechanisms of quality changes during atmospheric and vacuum frying. *Food Reviews International, 27*(4), 408–432.

18. Dueik, V., Robert, P., & Bouchon, P., (2010). Vacuum frying reduces oil uptake and improves the quality parameters of carrot crisps. *Food Chemistry, 119*(3), 1143–1149.

19. Durance, T. D., & Wang, J. H., (2002). Energy consumption, density, and rehydration rate of vacuum microwave and hot-air convection-dehydrated tomatoes. *Journal of Food Science, 67*(6), 2212–2216.

20. Fito, P., & Chiralt, A., (2000). Vacuum impregnation of plant tissues. In: Alzamora, S. M., Tapia, M. S., & Lopez-Malo, A., (eds.), *Minimally Processed Fruits and Vegetables: Fundamental Aspects and Applications* (pp. 189–204). Gaithersburg-USA: Aspen Publication.

21. Fito, P., Chiralt, A., & Betoret, N., (2001). Vacuum impregnation and osmotic dehydration in matrix engineering: Application in functional fresh food development. *Journal of Food Engineering, 49*(2/3), 175–183.

22. Fito, P., & Pastor, R., (1994). Non-diffusional mechanisms occurring during vacuum osmotic dehydration. *Journal of Food Engineering, 21*(4), 513–519.
23. Garayo, J., & Moreira, R. G., (2002). Vacuum frying of potato chips. *Journal of Food Engineering, 55*(2), 181–191.
24. Granato, D., Branco, G. F., & Nazzaro, F., (2010). Functional foods and nondairy probiotic food development: Trends, concepts and products. *Comprehensive Reviews in Food Science and Food Safety, 9*(3), 292–302.
25. Granda, C., Moreira, R. G., & Tichy, S. E., (2004). Reduction of acrylamide formation in potato chips by low-temperature vacuum frying. *Journal of Food Science, 69*(8), E405–E411.
26. Gras, M. L., Vidal, D., Betoret, N., Chiralt, A., & Fito, P., (2003). Calcium fortification of vegetables by vacuum impregnation: Interactions with cellular matrix. *Journal of Food Engineering, 56*(2/3), 279–284.
27. Inprasit, C., (2003). *Scale-Up of Vacuum Fryer for Durian.* Research Report on the Thailand Research Fund (In Thai). doi: https://docplayer.net/amp/38450158-Vacuum-frying-chao-inprasit-ph-d-department-of-food-engineering-kasetsart-university.html.
28. Jacques, T., (1992). *Microwaves: Industrial, Scientific, and Medical Applications* (p. 218). Boston, USA: Artech House.
29. Jaya, S., & Das, H., (2003). Vacuum drying model for mango pulp. *Drying Technology, 21*(7), 1215–1234.
30. Jena, S., & Das, H., (2007). Modeling for vacuum drying characteristics of coconut press-cake. *Journal of Food Engineering, 79*(1), 92–99.
31. Mahesh-Kumar, G., & Ravindra, M. R., (2017). Design of vacuum impregnation chamber for soaking of Gulab jamun in sugar syrup and optimization of wall thickness by finite element analysis (FEA). *International Journal of Environment, Agriculture and Biotechnology, 2*(1), 8–18.
32. Maltini, E., Torreggiani, D., Brovetto, B. R., & Bertolo, G., (1993). Functional properties of reduced moisture fruits as ingredients in food systems. *Food Research International, 26*(6), 413–419.
33. Martínez-Monzó, J., Barat, J. M., González-Martínez, C., Chiralt, A., & Fito, P., (2000). Changes in thermal properties of apple due to vacuum impregnation. *Journal of Food Engineering, 43*(4), 213–218.
34. Martínez-Monzó, J., Martínez-Navarrete, N., Chiralt, A., & Fito, P., (1998). Mechanical and structural changes in apple due to vacuum impregnation with cryoprotectants. *Journal of Food Science, 63*(3), 499–503.
35. Mousa, N., & Farid, M., (2002). Microwave vacuum drying of banana slices. *Drying Technology, 20*(10), 2055–2066.
36. Ophithakorn, T., & Yamsaengsung, R., (2003). Oil absorption during vacuum frying of tofu. In: *PSU-UNS International Conference: Energy and the Environment* (pp. 112, 113). Thailand: Hat Yai Songkhla.
37. Pandey, A., & Moreira, R. G., (2012). Batch vacuum frying system analysis for potato chips. *Journal of Food Process Engineering, 35*(6), 863–873.
38. Pipatpong, W., & Sumpun, C., (2014). Performance evaluation of a water ejection type in vacuum drying system. *Energy Procedia, 52*, 588–597.
39. Radziejewska-Kubzdela, E., Biegańska-Marecik, R., & Kidoń, M., (2014). Applicability of vacuum impregnation to modify physicochemical, sensory, and nutritive

characteristics of plant origin products: A review. *International Journal of Molecular Sciences, 15*(9), 16577–16610.

40. Ram, Y., Ariyapuchai, T., & Prasertsit, K., (2011). Effects of vacuum frying on structural changes of bananas. *Journal of Food Engineering, 106*(4), 298–305.
41. Rosana, M. G., Silva, P. F. D., & Gomes, C., (2009). The effect of de-oiling mechanism on the production of high quality vacuum fried potato chips. *Journal of Food Engineering, 92*(3), 297–304.
42. Shyi-Liang, S., Lung-Bin, H., & Hwang, L. S., (1998). Effect of vacuum frying on the oxidative stability of oils. *Journal of the American Oil Chemists' Society, 75*(10), 1393–1398.
43. Shyi-Liang, S., & Hwang, L. S., (2001). Effects of processing conditions on the quality of vacuum fried apple chips. *Food Research International, 34*(2/3), 133–142.
44. Stea, T. H., Johansson, M., Jägerstad, M., & Frølich, W., (2007). Retention of folates in cooked, stored, and reheated peas, broccoli and potatoes for use in modern large-scale service systems. *Food Chemistry, 101*(3), 1095–1107.
45. Steagsinee, S., (2011). Edible coating and post-frying centrifuge step effect on quality of vacuum-fried banana chips. *Journal of Food Engineering, 107*(3/4), 319–325.
46. Thirugnanasambandham, K., & Sivakumar, V., (2016). Enhancement of shelf-life of *Coriandrum sativum* leaves using vacuum drying process: Modeling and optimization. *Journal of the Saudi Society of Agricultural Sciences, 15*(2), 195–201.
47. Tinoco, P., Rosalba, M., & Perez, A., (2008). Effect of vacuum frying on main physicochemical and nutritional quality parameters of pineapple chips. *Journal of the Science of Food and Agriculture, 88*(6), 945–953.
48. Xie, J., & Zhao, Y., (2003). Improvement of physicochemical and nutritional qualities of frozen marionberry by vacuum impregnation pretreatment with cryoprotectants and minerals. *The Journal of Horticultural Science and Biotechnology, 78*(2), 248–253.
49. Yongsawatdigul, J., & Gunasekaran, S., (1996). Microwave-vacuum drying of cranberries. Part II: Quality evaluation. *Journal of Food Processing and Preservation, 20*(2), 145–156.
50. Zdravko, S., Tepić, A., Vidović, S., Jokić, S., & Malbaša, R., (2013). Optimization of frozen sour cherries vacuum drying process. *Food Chemistry, 136*(1), 55–63.
51. Zhao, Y., & Xie, J., (2004). Practical applications of vacuum impregnation in fruit and vegetable processing. *Trends in Food Science and Technology, 15*(9), 434–451.

CHAPTER 3

VIABILITY OF HIGH-PRESSURE TECHNOLOGY IN THE FOOD INDUSTRY

MOHONA MUNSHI, MADHUSUDAN SHARMA, and SAPTASHISH DEB

ABSTRACT

High-pressure processing (HPP) has become the most suitable non-thermal technology for processing dairy products, fish, meat products, fruits, vegetables, and beverages. The HPP is based on the use of the high pressure (300–600 MPa) through a pressure transmitting medium (mostly water) to inactive the enzymes and microorganisms activities without spoiling the nutritional quality of the product. HPP also avoids loss of various volatile compounds. This chapter gives a clear idea on different applications of HPP and the effects of HPP on the nutritive standard and safety of the processed products.

3.1 INTRODUCTION

Food processing transforms the raw food into a consumable product and extends the shelf-life of the processed products without compromising the taste and nutritional values. Food processing industries are gaining significant attention because the scarcity of food is increasing day by day because of exponential growth of population, losses due to improper processing techniques, and inadequate storage facilities. When compared between developed and developing countries, developing countries are facing more food scarcity issues. Apart from food scarcity, developing countries are also facing unhealthy, non-nutritional, and foodborne disease-related issues.

One of the main reasons behind these issues is the selection of traditional food processing methods, such as thermal treatment, enzymatic treatment, dehydration, changes in pH, use of chemical preservatives, etc. Though these techniques can increase the shelf-life of the product and fulfill the consumer demand, yet they can also deteriorate the nutritional value of the final processed product. In addition, these techniques may induce some property changes in the processed product. To overcome these issues and to protect the food products from deterioration by the spoilage microbes without compromising the nutritive value of the processed product, it has become necessary to introduce some novel food processing techniques.

The main objective to adopt novel food processing techniques is to kill the food pathogens and harmful microbes without compromising the chemical components (protein, vitamins, flavonoids, bioactive compounds, etc.), and physical components (color, texture, structure, etc.), and also to improve the shelf-life. Among these novel processing techniques, high-pressure processing (HPP) technology has revealed significant possibility as a substitute for conventional processing techniques [58]. HPP has the capability to maintain the freshness, nutritional quality and taste of the final product almost similar to the raw product [48].

HPP technique is based on Le Chatelier's principle and Pascal's law. Le Chatelier's principle affirms that if a system is in equilibrium condition, then the responses of the system will always minimize the disturbance [62]. Similarly, in HPP, the pressure reduces the volume and at the same time opposite reaction increases the volume, which ultimately minimizes the disturbance. Similarly, Pascal's law states that pressure applied in a closed liquid is uniformly distributed throughout the liquid, which is also applicable in HPP. Therefore, HPP is independent of holding time and mass of the produce.

HPP equipment is normally of two types: batch type and semi-continuous type. Batch type is suitable for solid food products and semi-continuous type is used for liquid or slurry type food products [73]. Pressure vessel with the ability to sustain high pressure up to 1000 MPa (Figure 3.1) is used in HPP equipment to hold the product; furthermore, the vessel with product is immersed in a liquid.

Pressure transmission takes place through the liquid. Water, silicone oil, ethanol, glycol, and sodium benzoate are normally utilized as a medium for pressure transmission purpose. Most of the HPP equipments for industrial use are batch systems (Figure 3.2), where the application of pressure takes place in the product through a pressure transmitting medium enclosed in a chamber or vessel.

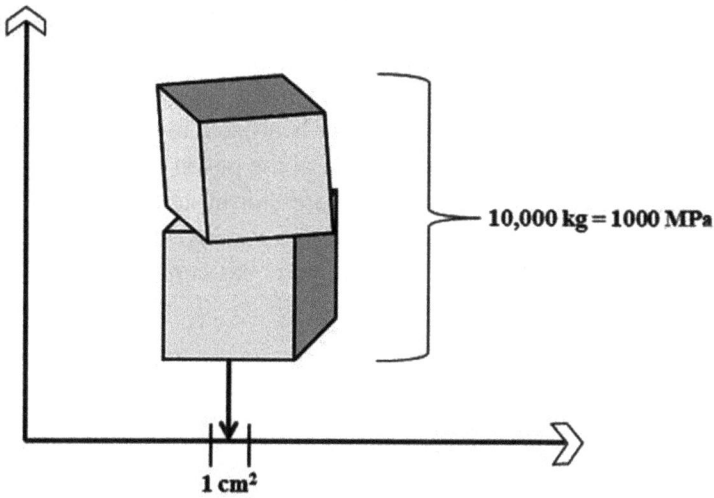

FIGURE 3.1 The 1000 MPa pressure is equivalent to 10,000 kg of weight on an area of one cm².

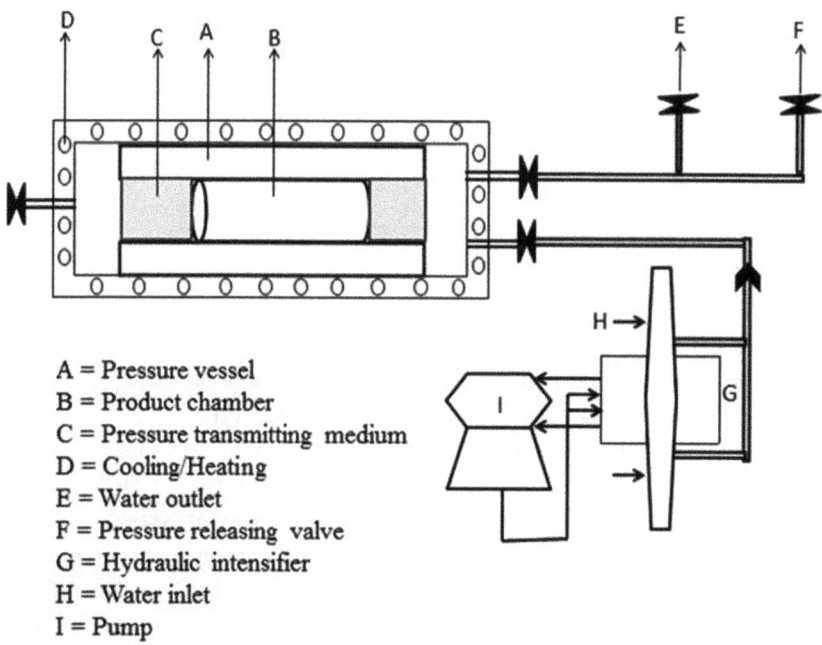

A = Pressure vessel
B = Product chamber
C = Pressure transmitting medium
D = Cooling/Heating
E = Water outlet
F = Pressure releasing valve
G = Hydraulic intensifier
H = Water inlet
I = Pump

FIGURE 3.2 Schematic view of batch type HPP.

Piston or creation of vacuum normally is adopted to apply pressure. Pressure generation using piston is again of two types: direct compression system and an indirect compression system [63]. A piston having a small diameter at one end and a large-diameter on another end is used in direct compression system. Pump with low-pressure capacity is utilized to drive the large diameter end. Whereas the small diameter end of the piston is utilized to generate the pressure (Figure 3.3). In indirect compression system, a pressure medium with high pressure is siphoned into a sealed vessel from a tank utilizing a high-pressure intensifier till the pressure level has come to the desired level (Figure 3.4).

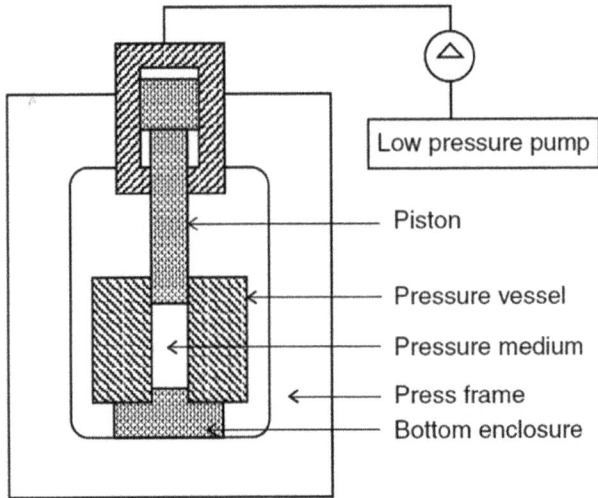

FIGURE 3.3   Direct compression technique used in HPP to increase the pressure.

Once the pressure reaches to the desired level, the application of pressure is stopped to save energy. Then pressure is maintained at the required level until the holding time is over. After the completion of the process, the machine releases the pressure, then the chamber opens up to smack down the product, and the similar process gets repeated. One full cycle (Figure 3.5) consists of: loading the product inside the vessel or chamber, application of pressure till the pressure reaches up to the desired level, maintaining the pressure till the holding time get over then depressurization and unloading of the product is called one cycle; and the time required for the whole cycle is called 'cycle time.' Output depends on cycle time, which ultimately determines the production rate of any industry. Therefore, to increase the production rate, product-holding time is reduced to minimize the cycle time. While selecting an HPP system, the main thing to be taken into consideration is working pressure. Along with increase in an inappropriate working pressure, the number of failures and appropriate working pressure can improve the life of the machine.

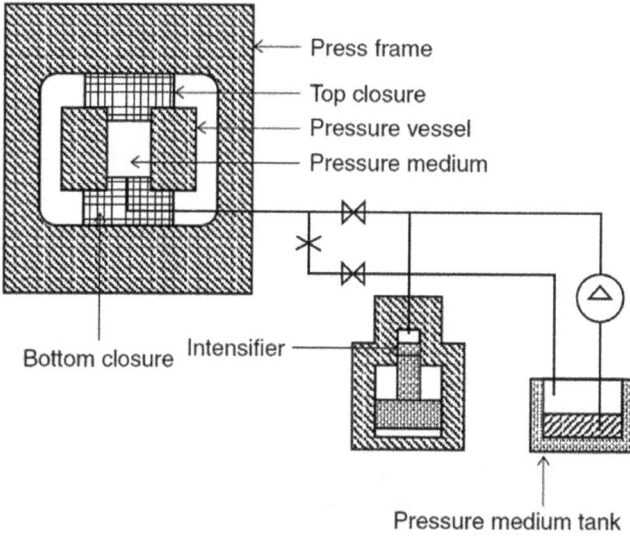

FIGURE 3.4   Indirect compression technique used in HPP to boost the pressure.

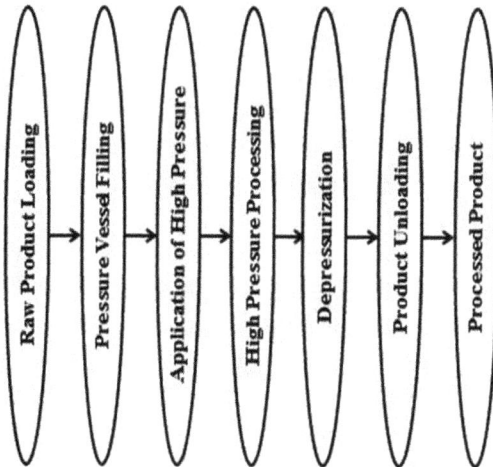

FIGURE 3.5   Process cycle of batch type HPP machine.

Apart from restoring the quality, taste, appearance, nutritional efficacy of the final product, HPP requires lower energy than the traditional thermal processing method due to no requirement of heating and subsequent cooling of the product, which makes it more eco-friendly. HPP is conveniently used for the packaged product; hence it reduces the chances of the secondary contamination. Compared to thermal treatment, there are very fewer changes observed in the final product using HPP technology. This technology can provide a high level of safety and food quality with huge benefits to the manufacturers and the buyers.

The prime focus of this chapter is to elaborate and explain:

1. The potential applications of HPP technology in various food-processing industries;
2. Its effect on chemical, physical, and microbial components of the food products; and
3. Future scope and areas need to develop in HPP also discussed.

3.2 HIGH-PRESSURE PROCESSING (HPP): BRIEF HISTORY

Application of pressure in food processing started in the late nineteenth century. Starch conversion to glucose and acid inversion of sucrose are some of the earliest studies. However, the application of pressure at >100 MPa initiated a new era of HPP technique. The first inactivation of bacteria using HPP technique was done in raw milk by Bert Hite in 1899. Hite showed that application of high pressure expanded the time frame of realistic usability of milk by 96 h at 680 MPa and souring also got delayed when treated for 1 h at 200 MPa [31]. Scientist Larson in 1918 confirmed that high pressure could inactivate microbial growth. Research on engineering characteristics of HPP started at the beginning of the 20th century. Phase change, thermal conductivity, compressibility tests were some of the engineering studies conducted during 1909–1923 [4–6].

Conversion from chlorobenzene to phenol was the first study of HPP technology on chemical reactions conducted in 1920 [7]. The 20th century is considered as the golden era of HPP for experiments and studies. Whereas due to high cost and technical challenges, HPP took some time to come into picture at commercial level the end of 20th century. The first commercial level utilization of HPP started in Japan in 1990, and today, almost all developed countries are using this novel technology.

3.3 RAMIFICATION OF HPP ON PHYSICOCHEMICAL AND MICROBIOLOGICAL COMPOSITION OF FOOD

3.3.1 CHEMICAL COMPOSITION

Plant and animals are major sources of human food, and the main chemical components of food coming from plants and animals are water, protein, fat, starch, vitamins, and minerals. The contents of these components vary from food to food and product to product. For example, water content varies from 15 to 90% depending on the kind of food item. Protein and fat available in food is mainly crude protein and crude fat. Crude protein is basically $N \times 6.25$, where N is the assumed 16% nitrogen content present in most of the plant and animal products which does not contain any non-protein nitrogen; and 6.25 is a conversion factor for nitrogen to protein.

Crude lipid is a mixture of various types of triacylglycerols. Starch is also one of the main nutrients. Digestive starch from both plant and animal products is converted to glucose and provides energy to the human body. Whereas, vitamins and minerals are present in minor quantities compared to major elements but play a vital role in human body growth and immunity.

Desirable quantity of these components in different plant and animal-based products (such as protein in meat and egg products, fat in fishes, the starch in millet and potato, and vitamins and minerals in vegetables and fruits) can be controlled by the food manufacturing industries through adopting proper methods of processing and storage. For example, in India, a huge quantity of crops, vegetables, fruits, fish, and meat products get wasted due to inadequate processing facilities. Conventional and old processing techniques have led to various quality changes in processed products due to the presence of pathogenic bacteria, enzymes, and oxygen. Nutritional qualities get deteriorated because of thermal, chemical, and high or low pH treatments [28].

Moisture loss due to evaporation, nutrient loss, and denaturation of protein due to heating, oxidation, leaching out of vital vitamins and minerals during blanching and cooking are the major drawbacks or disadvantages of conventional processing techniques. To overcome these issues and to save the quality of major food components, novel modern processing technologies have been introduced.

Understanding the impact of HPP on the chemical properties of the final products compared to the raw material will provide the information needed for the acceptability of HPP technique. The HPP technique has various effects on biochemicals present in food components. These effects have been studied widely, and the studies are ongoing. Many of these effects have been put together for a better understanding and to get the idea about the consequences of HPP on these macro- and micronutrients of foods.

3.3.1.1   WATER

Water is one of the major parts of any food item. Water quantity available in any food product provides the data for maintaining the standard of the product. The presence of water in food components can be determined by water activity (a_w), which is generally nearly equal 1.0 in most of the foods to be processed. High pressure influences the compression of water in any food item, which results in reducing the distance among the molecules of water due to the confined space [72]. Freezing properties of water are altered during the HPP, which generates the chances for the food storage below 0°C without the crystallization of ice and fast thawing of frozen foods [59]. Water ionization can also be affected by high pressure. Generally, water molecules in the presence of ionized compounds attach themselves to the ions present on the compounds, which ultimately produce a compact molecular structure. Acidic or basic compounds of biomolecules can be ionized by volume reduction and can be enhanced by HPP on the product [33].

3.3.1.2   PROTEIN

Presence of folded polypeptide chains is basically a three-dimensional structure of proteins. The delicate balance of the multidimensional structure of proteins, the porous regions in the structure and its interaction within the peptide chains and the solvent attributes to the functionality of the proteins and its activity in the food and multicellular organisms [3].

The magnitude of enforced pressure, type, and concentration of protein determines the functionality and structural transformation of protein during HPP treatment. High pressure acts as a catalyst to disturb this delicate balance and protein interaction. Pressure can have both reversible and irreversible effects on proteins. At the lower range of pressure (<400 MPa), the changes can be reversible, and the number of hydrogen bonds gets increased; while at the higher range of pressure (>400 MPa), the changes are irreversible and reduce these hydrogen bonds.

Therefore, only weak bonds (such as hydrogen bonds, ionic bonds, hydrophobic bonds) are disrupted, and only the secondary structure of protein gets affected by high pressure. Whereas during thermal treatments, covalent bonds (such as peptide disulfide bonds) and non-covalent bonds are disrupted [66]. Research has shown that there have been very few structural changes occurring in protein when denaturation takes place by the application of pressure; and along with that, oligomeric, monomeric, and aggregated species are formed when the protein get pressurized without the application of high temperature and any artificial chemicals.

Functional properties of food get affected due to structural deviations because of the application of high pressure. Various functional properties of proteins (such as solubility, agglomeration, precipitation, gelation, and emulsification) depend on the structure of proteins. Variations in these properties can provide various applications in the food industry, such as meat tenderization, reduction of allergic effects, increased binding of ligands, effects on solubility of various proteins, etc. For example, gelation helps to stabilize and establishing milk gel, the effect on myosin resulting the meat structure, etc. A common change due to the high pressure is the activation or inactivation of enzyme, which occurs because of the decrease in volume of the food matrix. The inactivation rate of enzymes increases by the combination of pressure with moderate temperature. It has been found that gelation occurred in rabbit myosin, when pressurized at 140–280 MPa pressure for 30 min at room temperature.

Codfish muscle when processed at 400 MPa pressure for 20 min at room temperature, produces harder gels than the cooked muscle. The gels formed by pressurization were translucent in appearance compared to the heat-treated (90°C for 15 min) muscle, which was opaque in color [14]. Heating along with pressurization induced less thermal denaturation and improved the deactivation of enzymes and meat tenderization. HPP partially preserved the protein from denaturation; and secondly, the increase in proteolytic activity causes myofibrillar protein breakdown and formation of various molecular fragments, which help in water-binding capacity.

The HPP has the capability of adjustment in the intermolecular and the intra-molecular forces, to influence the textural and structural behavior proteins, which lead to modification of the proteins and produce a new product. HPP treatment and its consequences on proteins can easily define the importance and role of HPP in food industries. All the changes in protein structure and properties due to the high-pressure application are of great value towards the safety, quality, and consumer acceptance of the high-pressure treated foods.

3.3.1.3 LIPIDS

Lipids are composed of fats and oils that are rich sources of energy. The molecular structures of lipids have a greater portion of triacylglycerols, which are three fatty acid esters of glycerol. This molecular structure is greatly responsible for the characteristics of lipids [54].

Oxidation rate and the crystallization rate of triacylglycerols are functional properties of lipids. The rate of crystallization depends upon chemical composition, temperature, emulsion as well as the phase of compounds. HPP can augment the crystallization rate through change in phase transition temperature [8]. Amount of applied pressure and holding time basically influence the consequences of high pressure on crystallization rate. High pressure can have both effects of increasing or decreasing the rate of crystallization that depends on the product to be processed and the parameters for the processing.

Apart from lipid crystallization, lipid oxidation is also one of the common drawbacks in food industries. Mainly the meat and seafood processing industries are facing this problem, where oxidation of lipid is one of the major reasons behind spoilage of food product specially cooked product during storage. This phenomenon occurs in foods due to the presence of biological and inorganic catalysts, such as enzymes, and metal ions. In addition, thermal treatment at high temperature breaks covalent carbon(C)-carbon(C) and carbons (C)-hydrogen (H) bonds and produce number of lipid alkyl radicals that lead to the promotion of oxidation [67]. Therefore, controlling of enzymes, removal of metal ions and controlling the temperature are the possible way to overcome this issue.

There are basically two types of oxidation in food products, such as metal oxidation and hydroperoxide decomposition. Peroxides and hydroperoxides are primary products of oxidation; and epoxides, aldehydes, and ketones are the secondary oxidation products [63]. These products deteriorate the flavor and structural properties of food. Along with lipid, cholesterol also gets oxidized by heating and enzymatic activity. Oxidation of cholesterol produces oxysterols. Oxysterols accumulation in the human body causing different age-related diseases and oxysterols are mainly present in milk, eggs, meat, and fish products [37].

Research studies reveal that HPP is an emerging new technique to limit the lipid degradation by deactivating the enzymes and elimination of metal ions by enhancing sensorial characteristics compared to heat treatments [66].

Various research studies show that there is no effect on lipid oxidation when pressure is applied within the range of 400–500 MPa; however, pressure above 500 MPa starts to induce changes. At higher pressure (600 to 800 MPa), peroxides, and thiobarbituric acid values (AV) start to increase. It has been observed that peroxides and thiobarbituric acid got increased in pork meat fat at 800 MPa pressure for 20 min [13]. Whereas, in washed muscle fibers and minced pork processed at 300 to 800 MPa for 20 min, the thiobarbituric acid to determine the lipid oxidation was 20% lower [49]. This problem could also be shortened by the addition of 0.02% of citric acid. High pressure more than 400–500 MPa increases the development of peroxides and thiobarbituric acid alongwith lipids aldehydes such as nonanal, pentanal, hexanal and heptanal. These products are accountable for the development of unpleasant odor in final processed food products [23].

Beef, chicken, ham, and yak are other meat products, where oxidation of fat has been observed while processing at high pressure. Whereas, the exceptional case has been observed in turkey meat, where oxidation took place even at a lower pressure (<400 MPa) while processing for a longer duration (>30 min). However, pressure range within 400–500 MPa with appropriate holding time is found totally safe and there is no reported evidence regarding radical formation at this pressure range. Therefore, this could be considered as a critical range and safe range of pressure to overcome the lipid oxidation problem and to produce safer high-pressure processed food.

3.3.1.4   STARCH

Starch provides 70–80% of calories to our body and is an outstanding source of energy. Starch is normally crystalline in nature and not soluble in water. When heat is applied with an adequate amount of water, then starch gets gelatinized, and the crystalline and molecular order of starch starts to change [70]. Once the crystalline and molecular structure of starch is changed, then starch granules start to absorb water leading to the fragmentation of starch granules. Rheological properties of starch get highly affected by the gelatinization process.

Pressure can be used for the gelatinization of starch; and research shows that gelatinization due to the application of pressure is better than thermally induced gelatinization. Pressure-induced gelatinization is done in two steps [54]: (a) the hydration of starch granules, which results in swelling of these granules and disruption of crystalline regions of starch; and (b) makes the crystalline structure accessible to move water.

In the pressure treated starch gelatinization, the structure of starch granules remains the same while heat-treated starch gelatinization distorts the starch granules. The covalent bonds are not affected by high pressure (600 MPa); hence the structural changes are minimally followed by minimum effects on rheological properties. It has been reported that at higher pressure (600 MPa) with the presence of excess water, starch granules continue to maintain their granular shape, which is not true during the heat treatment. Similarly, in comparison with thermally treated starch, high pressure (550 MPa) treated starch shows lower viscosity. Therefore, to get the required product texture and structure, the knowledge for high pressure-driven starch gelatinization is very crucial and vital for the utilization of HPP technology in starchy food products [77].

3.3.1.5   VITAMINS AND MINERALS

Fresh fruits and leafy vegetables are abundant sources of varieties of micronutrients. These essential micronutrients for a healthy body are: vitamins, tocopherols, phenolic acids, carotenoids, flavonoids, etc. They effectively reduce the chances of carcinogenic diseases, cardiovascular diseases and other biological complications [50]. Fruits and vegetables with a high content of vitamins and minerals are always considered as a healthy diet. However, it is very difficult to conserve these highly sensitive micronutrients after processing with traditional thermal and chemical processing methods. However, thermal and chemical methods conserve the food product and improve the storage life of the final processed products, but these methods are highly detrimental to micronutrients.

To avoid the detrimental effects from the traditional methods, HPP technology is utilized. HPP shows the capability to conserve the quality and various properties of the processed product similar to the raw product, homogeneous processing ability due to equal pressure distribution throughout the product during processing, energy-saving ability, and eco-friendly potentiality compared to thermal treatment.

HPP apply multiple pulses of pressure and can control the temperature within 105°C and processing takes place for a small duration leading to minimum losses of essential micronutrients [47]. When compared to available research studies on the effect of HPP on the microbial and enzymatic activity of processed items, the effect of HPP on vitamins and minerals of processed products are insignificant. However, available research shows that HPP at moderate temperature on various fruits and vegetable products (tomato puree, kiwifruit puree, strawberry puree, orange juice, apple juice, lemon juice, etc.), are highly efficient and there are slight changes in their nutritional components (flavonoids, carotenoids, vitamins: A, B1, B2, C, E, etc.). However, if pressure or temperature or both reach to the maximum level, then there could be a chance of degradation of vitamins and minerals.

3.3.2 PHYSICAL COMPOSITION

Physical factors of any food product are color, texture, yield, and sensory quality; and these factors are also responsible to determine the quality of any given food. Consumer acceptance of any food highly depends on these factors. HPP provides various effects on these parameters and can either enhance or degrade the acceptance of food product for consumers.

3.3.2.1 COLOR

Color is a significant parameter while considering the consumer acceptance of any food item. The traditional value of color in a food product is well known. Changes in color during processing come under structural changes, which highly influence the suitability of the product. Traditional thermal and chemical processing treatments have a high effect on the color of any processed product, which may reduce the acceptability of the processed product. The color of a food product depends on various components such as (a) for fruits and vegetables, it depends on natural pigments, such as chlorophyll, anthocyanins, carotenoids, etc.; (b) meat color depends on myoglobin content.

It has been found that natural pigments present in fruits and vegetables do not get affected during moderate pressure treatment, and there are no changes in color. However, during high pressure and heat treatment browning results in color degradation [53]. At the same time, some compounds related to color in fruits and vegetables (such as anthocyanins) were found to be unstable during storage when using HPP treatment because of the presence of ascorbic acid and incomplete enzyme inactivation [53]. Along with these color changes, there is also browning and condensation with phenolic compounds [10]. Therefore, adjustment of pressure and temperature for the selected product for processing is the main criteria that need to be followed while using HPP technology for fruits and vegetables.

During meat processing, pressure application in some meats results in denaturation of protein, which can further change the functional properties of the food products and the change in color. This change in color mainly depends on product, such as raw meat and poultry products because of the change in myoglobin and iron oxidation due to high-pressure treatment [51].

HPP treatment for already cooked meat products does not show these effects because of the denaturation of protein during the cooking process. Whereas, application of HPP for already cooked meat products enhances the color or appearance. It has been observed that high pressure processed raw meat products cannot be put directly in the market shelves like fresh meat products due to the change in color and further cooking is required before the sale. However, in the case of fish products, this effect is not observed, therefore fish products can undergo pressure treatment without any consideration of color changes.

3.3.2.2   TEXTURE

The texture is also a vital physical parameter in terms of consumer acceptance due to the sensory quality. HPP treatment induces both permanent and temporary effects on the texture of the processed food. Food product with high moisture content has least effect, which is not seen to a significant degree, which can degrade its value. Whereas, in the case of gaseous foods and foods with air pockets, the effect is permanent due to the shrinkage of the food product by the removal of gas or air pockets, along with that shape distortion has also been observed. Research shows that for the food product with no air pockets, high pressure does not result in any permanent change or minimum texture changes have been observed. However, some food products need some changes in the texture, such as cheese during processing, and HPP treatment can be used to get those desirable changes in texture, and it also can help to accelerate the development of functional properties of the final product [57].

HPP treatment also induces the gelation process. Pressure-induced gels are soft, smooth, and uniform in texture than the heat-treated gels. Therefore, for processing of fish, gelation is an important requirement of HPP over the thermal treatment. The pressure-treated gelation does not affect the color of the product, which can be observed in the thermal treatment.

3.3.2.3   YIELD

Yield is the final product developed with respect to the amount of ingredients used. Yield losses drastically affect the economy of the food manufacturer. Yield of some food products was affected by HPP, which causes the ultimate effect on the cost of the product. Yield loss due to HPP mainly depends on the product selected for processing and the intensity of the applied pressure. However, in most cases, HPP gives more yield than the thermal treatment. It has been reported that weight loss of thermally treated sausages was higher than the high pressure treated sausages [51]. Similar advantages have also been observed during the processing of cheese and oyster using HPP.

3.3.2.4   SENSORY QUALITY

Maximum flavoring compounds available in food (mainly fruits and vegetables) are volatile in nature and small in size compared to other nutrients. These compounds are highly affected by thermal treatment and can spoil the taste, flavor, color, and other aspects of sensory properties. It has been reported that there is no specific change on the taste and aroma of vegetables and fruits due to HPP treatment, which can extend the shelf-life of the product while maintaining the taste, flavor, color, and other aspects of sensory properties.

It has been observed that during sensory evaluation, panel members provided with high pressure processed food and fresh food were not able to differentiate among the sensory properties in both cases [35]. Though HPP can alter the various enzymatic and chemical reactions, yet the direct effect of HPP has not been observed while processing of vegetables and fruits at room temperature. It is always advised to keep the high pressure processed fruits and vegetables at low temperature during storage.

3.3.3   MICROBIOLOGICAL COMPOSITION

Microorganisms play a key role in the quality of food products. Most of these microorganisms have been causing undesirable changes in the food quality, such as flavor, decrease in nutritional value, etc.

Microorganisms, such as *L. bulgaricus*, *S. cerevisiae*, *L. thermophilus*, *Acetobacter*, etc., are helpful during the food processing especially during fermentation, which provides aroma and taste to the product. On the contrary, there are microorganisms, such as *C. botulinum, Salmonella, E. coli* O157:H7, *Listeria monocytogenes* that cause food spoilage and are ultimately harmful for consumer's health. The HPP and ultrasonication, along with other processing techniques can decrease the microbial contamination and improve the shelf-life and safety.

HPP not only reduces contamination but also retains the nutritional value of the product. On the contrary, conventional thermal processing techniques of cooking, blanching, the addition of other chemicals and methods of preservation (salting, addition of sugar, etc.), have become a threat to food quality.

3.3.3.1 BACTERIA

The bacteria causing food poisoning are *Campylobacter, S. aureus, Clostridium, and E. coli.* The growth phase of bacteria is dependent on pressure resistance [32]. Heat resistant bacterias are also pressure-resistant, such as *Salmonella typhimurium, Vibrio parahaemolyticus, L. monocytogenes*, which can survive during pressurization. Salmonella found in pork loin and *P. fluorescens* in 6% fat of ovine milk undergo inactivation with HPP [19].

Various media used for platings, such as soy and nutrient agar was applied, which showed *E. coli* had injury only at 182 MPa. *Yersinia enterolytica* is one of the most harmful food poisoning bacteria and it was reduced to 5 log cycles at 275 MPa for 15 min in the phosphate buffer. Similarly, *L. monocytogenes* require 375 MPa and *salmonella* at 450 MPa, *E coli* O157:H7 require 700 MPa to get inactivated. *Lactobacillus plantarum* in its exponential phase was resistant to pressure and *Faecalis* can grow at 200 MPa, which produces a large amount of ammonia, which was neutralized by high pressure; the enzyme present in bacterial cell got deactivated and the cells no longer were able to multiply. Studies have shown that Gram-negative bacteria are less resistant to heat and pressure than Gram-positive bacteria. For example *L.* and *S. aureus* are heat-resistant. These can also withstand for 15 min at 750 MPa. Therefore, in accordance to microbiological safety, the pressure treatment should be adjusted according to the target organism to be eliminated [69].

Human milk is effective against the pathogenic bacteria, such as *E. coli, L., Monocytogenes, S. aureus.* Human milk resistance to bacteria cannot be compared with bovine milk, which gets affected by different pathogenic microorganisms and are not resistant to deterioration. Study on bovine milk was carried out by HPP induced injury in the growth of the cells at all temperatures while processing at a pressure of 400 MPa. The injured cells could recover in milk because of tailing, which occurs in the potential phase of cell growth at 43°C. Therefore, tailing is caused by pressure treatment reducing the population of HPP treatment. Further research is needed to find out the growth temperature and why temperature is causing tailing [29]. Effect of HPP in bacteria still needs attention.

3.3.3.2 BACTERIAL SPORES

Spores are made up of central spore-forming cells by the process called sporulation (a type of cell division). It is covered by protective layers, the outermost layers are proteinaceous called exosporium, and the core is composed of various vegetative bacterial cells, such as DNA, ribosomes, etc. They have very unique characters because of which they are resistant to both chemical and physical agents. Therefore, it is a challenge to remove spores of bacteria during food processing. Under HPP, it has been found that spores are resistant up to 1200 MPa at room temperature. Therefore, improvement in HPP conditions was required with other treatment for the spores to get inactivated. The most common spore-forming bacteria are *bacillus* and *clostridium*. Various studies have been carried out regarding the spore formers, so that germination and growth can be inhibited with pH <4.5 [81].

It has been found that spores of *C. sporogenes* are less pressure-resistant than the spores of *B. coagulans* and others; therefore, high temperature was required to eliminate spores from food products. Elimination of the spores took pressure holding time of 5 min and it was possible at 1400 MPa/54°C and 800 MPa/75°C in meat broth. The strains of *C. sporogenes* were completely destroyed.

B. cereus, B. licheniformis and B. stearothermophilus were 4×10 cfu spores/mol and these were destroyed by double-pulse treatment (200 MPa/ min) followed by (900 MPa/min) for 3 min at 70°C. Therefore, the spores can be easily inactivated at low pH [19]. Some spores are heat sensitive and pressure-sensitive, which do not get destroyed even at high pressure. For example, *B. megatarium* was not inactivated in a treatment for 40 min at 1000 MPa. Therefore, an oscillatory pressure treatment may generally be required for complete inhibition of the bacterial spores. HPP treatment between 500 and 600 MPa withholding time can reduce the spores by a factor of $>10^8$ [69].

Pressure and temperature treatment are also preferred. Mild heat pre-treatment will lead to the germination of spores and subsequent pressure at >500 MPa will lead to inactivation. Spores *B. cereus* and *subtilis* can be easily inactivated by processing at 400 MPa. The 4 log cycle reduction was observed in *B. coagulans* by pressure-heat treatment. Thus, inhibiting bacterial spores by high-pressure treatment only is very difficult; therefore, it should be either combined with double pulse pressure treatment for shelf-stable foods.

The oscillatory treatment is best suitable for inactivation: first pressure treatment is given for spore germination and the second time treatment at high pressure would inactivate the germination of spores. The highly resistant spores of *C. sporogenes* were reduced to 3 log cycles by this treatment. HPP can thus reduce the bacterial spore count in food products.

3.3.3.3 YEASTS AND MOLDS

Yeasts (unicellular) and molds (such as mushroom) producing hyphae are subdivisions of the fungi family. Yeasts are not pathogenic microbes but play an important role in the spoilage of microorganisms. Yeast producing spores required higher temperature for their inactivation. Example of yeasts, such as *Paecilomyces* spp., *Euroticum, Byssochlamys* spp., etc., require 800 MPa pressure at 70°C with inoculum of <10⁶/ml [19]. Fungi, such as *Fusarium graminarum* release mycotoxin named deoxynivalenol (DON) and zearalenone (ZEA). These two cause irritation in the gastrointestinal (GI) tract, immature births, infertility, etc., and these are also carcinogenic. The DON and ZEA can only be inactivated completely at 550 MPa at 45°C for 20 min. Other fungal spores, which decrease the palatability and nutritional value, make it rancid causing severe health problems, such as cancer, hepatic, and neurological problems, etc. Fungal spore can be completely inactivated in peptone water at 380 MPa at 60°C for 30 min [39].

Olives containing mold species (such as *Penicillium* spp., *Cladosporium* spp., *Aspergillus* spp.) were studied; and after HHP application mold flora was reduced on an average of 90% level at 25°C, which concluded that HHP suppressed the mold flora in olives [74]. Fungal species (such as *S. cerevisiae, H. uvarum, P. anomala,* and *R. stolonifer*) generally cause spoilage of raw fruits and vegetable products. The conidia, which are hydrophilic in nature and are easily wettable, can be inactivated after pressure of 600 MPa for 1 min. These spores are very toxic for human health on consumption.

Even at 600 MPa, *Eupenicilium* was eliminated at 10^7 Cfu/ml within 10 min. A high pressure can also inactivate *Byssochlamys* spp.

The baroprotective effect (increase in solute concentration) along with HPP was studied on yeast cells (*S. cerevisiae*) and fungal spores (*P. expansum, R. stolonifer*) and these were formulated in sucrose citric phosphate buffer, which was subjected to 600 MPa pressure varying with times. Among these, the mold spores showed the strongest resistance to HPP. However, with an increase in the concentration of sucrose solution, yeasts showed a gradual resistance. It was concluded that during any commercial use of HHP in fruit preparations, only the efficient design of the process should be taken into account [26].

3.3.3.4 OTHER MICROORGANISMS

Microbes also include viruses, parasites that have a minor role in degrading the food quality. TMV (tobacco mosaic virus) was found to get inactivated at 920 MPa. However, human immunodeficiency viruses (HIV) were found to be more sensitive to HPP than the TMV. HIV viruses when treated with 400–600 MPa at 10 min were reduced by 10^5 log cycles. On the other hand, bacteriophages remained unaffected at 300–400 MPa. Herpex virus was inactivated when treated for 10 min at 400 MPa. Parasites (*Giardia lamblia* and *Cryptosporidium*) still lack information about pressure resistance. *Trichinella spiralis* got inactivated at 200 MPa for 10 min. *Cyclospora* are comparatively safe when compared with other bacterias. All of this microbe inactivation still needs further research studies [19].

A list of harmful microorganisms and their sources, along with the required HPP treatment to stop their activity for safe and healthy food products, is given in Table 3.1.

TABLE 3.1 HPP Treatment Required for the Deactivation of Various Harmful Microorganisms for Safe and Healthy Foods

Harmful Microorganism	HPP Treatment	Source(s)	References
Acinetobacter spp.	300 MPa	Minced mackerel	[19]
Bacillus cereus	500 MPa, 30 min	Milk (infant food)	[81]
Clostridium sporogens PA3679	680 MPa, 60 min	Meat broth	[69]
E. coli O157:H7	680 MPa	Milk, poultry	[32]
Fusarium graminearum	550 MPa, 20 min	Maize	[39]

TABLE 3.1   *(Continued)*

Harmful Microorganism	HPP Treatment	Source(s)	References
Listeria monocytogenes	345 MPa, 20 min	Whole milk, orange juice	[29]
Moraxella spp.	200 MPa	Minced mackerel	[19]
Pseudomonas fluorescence	300 MPa	Ovine milk	[19]
Saccharomyces cerevisiae	350–500 MPa	Jams	[69]
Salmonella enteritidis	450 MPa	Poultry, meat	[32]
Salmonella typhimurium	350 MPa	Meat, poultry, egg yolk	[32]
Staphylococcus aureus	700 MPa	Milk, poultry	[32]
Streptococcus faecalis	400 MPa	Minced muscle of albacore tuna	[69]
Trichinella spiralis	200 MPa, 10 min	Muscle tissue	[19]
Vibrio parahaemolyticus	173 MPa,10 min	Seafood (oysters)	[9]

3.4   APPLICATIONS OF HPP IN THE FOOD INDUSTRY

Consumers with health consciousness always look for less-processed, additive-free, and fresh food products. Today, social media and mass communication has significantly increased the awareness regarding healthy foods and human well-being and the side effects of processed foods and food products [78].

Research and development teams from food industries, academia are continuously trying to fulfill the consumer demand by developing different non-thermal technologies, such as pulsed electric field, irradiation, HPP, etc. The applications of HPP have become a reality in the USA, Japan, Spain, etc., [14]. Though HPP treatment was first used in the dairy industry for the improvement of milk stability followed by fruit and vegetable products [40], yet presently food-processing industries mostly use this technology for fruits and vegetables, meats, and seafoods, and dairy products [80].

In addition, HPP technology is highly effective in producing baby foods as it ensures an increase in the level of food safety and security without adding any additives. The applications of HPP technology in different food processing sectors for maintaining the quality and to improve the storage life of the processed product are discussed in this section.

3.4.1 FRUITS AND VEGETABLE INDUSTRY

Fruits and vegetables have vital micronutrients, minerals, dietary fibers, antioxidants, vitamins, etc., [11]. Preservation of these micronutrients is very crucial and important. During processing, it is difficult to kill the microbes and to stop the enzymatic activity to increase the storage life without degrading the freshness of the product through thermal processing techniques. The application of conventional processing techniques in the fruits and vegetable industry reduces the stability of the product and increases the chances of spoilage. Apart from the inactivation of enzymes and microorganisms, traditional processing techniques also use additives to maintain food quality and flavors. These are the main reasons that non-thermal processing techniques are gaining enormous attention among researchers and industry personals while processing fruits and vegetables [20, 34].

HPP technology is highly advisable for fruits and vegetables as it ensures food security and nutritional quality and flavor and aroma [15]. In comparison with conventional thermal processing techniques, HPP technology can also be executed at room temperature minimizes energy utilization, and the food is packed; therefore, it does not directly come in contact with the pressure device [34, 64]. HPP effectively conserves the nutritional values and sensory attributes as it has the least effects on molecular compounds and covalent bonds, such as flavor, color of fruits and vegetable products [53].

Studies reveal that HPP treatment has the ability to control the changes in color that take place in green vegetables at moderate and high temperature. HPP effect on leafy green vegetables makes it more intense. For example, HHP was used in green beans at 500 MPa for 1 min at room temperature [42]. However, the increase of temperature caused the change in color from green to olive green in the case of green beans and basils after high-pressure treatment of 1000 MPa/75°C/80 sec and 860 MPa/75°C/80 sec, respectively [43].

Apart from chlorophyll, carotenoids and anthocyanins are also important to maintain the red, yellow, orange, and blue color of fruits and vegetables, and they remain stable during high-pressure treatment. However, anthocyanins are exceptions and sometimes become unstable during storage due to reactions caused by enzymes, which are incompletely inactivated during HPP [24], and it is also due to the ascorbic acid effect on anthocyanins [41]. Apart from color pigment, browning also plays a vital role in the color change of high pressure processed vegetables and fruits, especially at the time of storage.

HPP treatment changes the pectinase activity, permeability in the cell, cell metabolites, firmness, and rheological properties of fruits and vegetables responsible for textural changes. At the time of processing, different temperature and pressure combination can change the pectinase to create a new texture, which is not possible with thermal processing. High-pressure treatment disturbs the cell permeability, which gives a firm texture with the soaked appearance and the later intercellular spaces never get filled with gasses; and because of that, the newly formed texture never get changed [53]. The firmness of celery, carrot, pear, orange, and pineapple does not change during HPP. The change in rheological properties of fruits and vegetable after HPP treatment totally depends on the type of the product [1].

The fresh flavor of fruits and vegetables does not change due to HPP, because the small molecular compounds and their structures are not affected [53]. However, some negative changes have been observed in terms of flavor of HP processed fruits and vegetable products due to enhancement of enzymatic and chemical reactions, changes in hexanal content [52] and oxidation of free fatty acids (such as linoleic acid and linolenic acid). Whereas, modified flavor in strawberry puree has also been reported [44].

Most of the developed countries have already accepted high pressure processed fruits and vegetable products [34]. Earlier, these products were only sold in natural food stores, but now these products are available in mainstream retail outlets. However, care must be taken during fruit and vegetable processing using HPP as incomplete inactivation of microbes and enzymes may lead to quality changes during storage. At the same time, proper packaging method is needed to reduce the browning and the other detrimental effects.

3.4.2   MEAT AND SEAFOOD INDUSTRY

Meat and seafood products are highly deteriorative due to the spoilage caused by the pathogenic microorganisms. HPP technology can give the meat and seafood industry a promising potential to produce advanced, contemporary, secure, and ready to serve products.

High-pressure technology was initially applied to check the effect on microorganisms in meat products, and research shows that meat products processed using HPP technology achieved higher microbiological stability by the changes in structure and texture of the processed product, due to change in meat proteolysis, muscle enzymes and myofibril proteins [64].

Inactivation of the target microorganisms due to HPP simultaneously resulted in introducing new packaging technique, which increased the shelf-life of the processed product.

USA and Canada already approved HPP technology for meat industries to inactivate microorganisms, such as *Salmonella* spp. and *Listeria monocytogenes* in pork and poultry items. HPP not only kills these microorganisms during processing, but also prohibits their growth during the storage period. In conventional technique, salt is mixed with cooked meat products to improve texture, water-binding capacity and taste but the combined application of salt (1.5–3%) and high pressure give high quality improvement [17]. Further, it has been found that meat product cooked at high pressure (500 MPa) at 70°C can increase the sheer force of the final product and causes structural changes. However, comparison between only cooked, cooked with the application of high pressure and cooked, high pressure and application of salt showed that cooked, high pressure and the salt combination gave fully disrupted structure [17].

The combination of cooking, high pressure, and addition of salt can improve the solubility of protein, water-holding capacity and provide a firm structure. However, a higher concentration of salt is not appreciated by the consumers because of taste and health issues. In the case of meat sausages, still this technique is acceptable. However, in the case of whole muscle meat products, it is totally unacceptable. To overcome this issue while producing whole muscle meat products, initially high pressure is applied to the meat product to have different gelling properties when compared to thermal treatment; and further to reduce the salt concentration, potato starch and carrot fiber is used as an alternative.

Starch and fiber has the ability to reduce the salt concentration from 3% to 1.2%. Fiber and starch have more impact on textural properties than salt and can also improve the water holding capacity. Apart from starch and fiber, ß-Glucan (basically a polysaccharide with high water binding capacity) also has the potential to reduce the salt concentration [12]. With the combination of these techniques, HPP technology can increase the storage life of the processed meat products by inactivating the microbes and enzymes without deteriorating the palatability and nutrients of the final product. High pressure processed meat product showed the average storage life of 20–25 days at 4°C of storage temperature [9], whereas under chilled storage condition, it can be extended up to 120 days.

Overall high-pressure technique showed the effective results in quality control and shelf-life extension of meat products, without compromising the nutritional and sensory quality. At the same time, its low-temperature eco-friendly HPP technique significantly overcomes the issues related to other conventional processing techniques. At present, commercially HPP technology is being applied to different ready-to-eat (RTE) meat products, such as cooked sliced ham, pre-cooked beef and chicken, chorizo, salami, mortadella, sausages, bacon, whole-sliced, and dried meat.

Similarly, in the seafood industry, oysters (shellfish) are the most cultivated seafood all over the world, which is prone to microorganism's infections. Mostly *Vibrio parahaemolyticus* and *V. vulnificus* were successfully killed by HPP without degrading the sensory attributes. Recently HPP has been recognized to be a valid process to reduce the pathogens Vibrio bacteria and to increase the shelf-life. In oysters, the adductor muscle is denatured by HPP, which helps in sucking of the meat out from the shell instead of manual sucking. However, there were changes in body color and other descriptive characteristics.

Recent advances of HPP have been used for lobsters and salmons but still work remains to conclude that HPP is beneficial for the improvement of the quality of the fish. However, at a pressure of 250 to 500 MPa, microorganisms, and product yield were reduced. HPP also helped in giving texture effect in uncooked suromis by making protein substrates accessible to transglutaminase [9].

In albacore muscle when pressurized at less than 22°C for 93 days and stored at–20°C, the high pressure denatured the gel proteins, which resulted in improvement of the texture. HHP also improved lipid stabilization and the color [71]. HPP in seafoods can extend the keeping quality along with improving the physical profile of fishes.

3.4.3   DAIRY INDUSTRY

Application of thermal treatments, such as pasteurization, ultra-high temperature (UHT) treatment are common practices in the dairy industry to extend the shelf-life of the dairy and milk products for around 7 to 20 days [45]. Though high-temperature treatment increases the keeping quality of the product, yet it simultaneously reduces the nutritional values of the processed milk product due to the destruction of various heat-sensitive micronutrients. To solve this issue, dairy industries are trying to adopt non-thermal techniques, such as HPP, cold processing, etc.

HPP deactivates almost all the pathogenic bacteria and harmful microbes present in milk at 400–600 MPa [55]. The survival rate of pathogenic bacteria and harmful microbes is determined by the rate of pressure, holding time, applied temperature, and the growth rate of microorganisms [27]. Whereas, pressure from 200 to 400 MPa required for the microbial death [60]. The temperature during thermal processing and pressure during non-thermal processing can alone inactivate pathogenic bacteria, but it has been observed that a combination of these two can dramatically improve the inactivation rate of spoilage microorganisms, thus yielding safer processed milk products. HPP not only preserve the dairy products, but it also improves the physicochemical, sensory, and rheological qualities of the final treated product.

HPP is basically an amalgamation of high pressure and heat; and while processing milk it acts as sterilization to inactivate microorganisms without compromising the desired nutritional quality [21]. Presently HPP is mainly applied in fluid milk processing, yogurt, cheeses, ice cream and cream and butter processing industries to improve the standard and to reduce the microbial load on the final product.

Normally fluid milk treated at 680 MPa at room temperature for 10 min gives 5–6 microbial log cycle reduction. However, pressure and temperature combination during fluid milk processing are decided based on storage time. The shelf-life of raw milk was 25 days when treated at 350 MPa pressure and 0°C. Whereas, at the same pressure, if the temperature was increased to 5°C, shelf-life got reduced from 25 days to 18 days; and at 10°C it reduced to 12 days, respectively. Apart from the pressure-temperature combination, the combination of various anti-microbial peptides (such as lacticin, lactoferricin, lactoferrin) with pressure are found highly effective and efficient to improve the standard and safety of the final product [46].

High pressure processed yogurt is becoming highly popular, where milk passes through high-pressure treatment [76] before fermentation; and low fat and creamy yogurt without the addition of external polysaccharides are produced. High-pressure application in yogurt acts as an alternative to food additives, which gives aroma, mouthfeel, and taste without adding any artificial additives [68]. Likewise, high pressure treated cheeses and ice creams are also dominating conventionally processed cheeses and ice creams. High pressure treated cheeses contain high moisture, zero salt and amino acids compared to raw and conventionally processed cheeses [75].

High pressure processed ice creams are low-fat ice creams, having a slower melting rate with higher sensory properties due to the formation of protein gels during processing [36]. In case of cream and butter processing using high-pressure technology, whipping properties of butter and cream get upgraded and serum loss gets reduced because of better crystallization of milk fat [18]. 100 MPa pressure and 8–9°C temperature have been found to be optimum for HPP of cream and butter. Research showed that HPP treatment can significantly enhance the ripening quality of dairy cream used for making the butter.

Although HPP has greater efficiency to control the dairy industry compare to thermal treatment, yet HPP treatment increases the free fatty acid level, reduces the casein micelle size, denatures the whey protein of milk [45]; and combined application of high pressure and temperature resulted in some cooked flavor in the final product, which deteriorates the sensory quality [79]. However, future research and development can make this novel technology highly useful for dairy industry to produce minimally processed, microbial free and longer shelf-life dairy products.

3.4.4   INFANT FOODS INDUSTRY

Infant foods are specially proposed to replace human milk when the mother is not able to breastfeed her baby. Infant foods are also coming in recent frames with the increase in the number of working mothers. Earlier infant foods included modified cow milk with the drawbacks of inadequate nutrition and improper development of infants. With recent technological improvements and advanced formulations, infant foods have achieved similarity with human milk [38]. Infants are supplemented with RTE foods, also known as weaning foods. Infant foods are semi-solid foods, which are formulated and available in various forms, such as powders, RTE, concentrated liquids; and are easily digestible by the infant. The infant food also includes cereals, fruits, and vegetables, milk, etc. Infant foods throughout the world have been accepted by all parents as a nutritious and healthy diet [2]. It is always advised that infant foods should not be kept at room temperature for more than 4 h as it causes contamination by the growth of microorganisms and spore-forming bacteria leading to spoilage and should be warmed before consumption and also should be consumed immediately after opening to prevent deterioration.

Infant food processing is very delicate as the consumers are infants, who are highly prone to any detrimental effects. Therefore, the main hurdle for infant foods is to prolong the keeping quality without degrading the nutritional value and decreasing the palatability along with inactivation of microbes.

Application of HPP in infant foods is a promising and emerging non-thermal technology, recently HHPP (high hydrostatic pressure processing) technology is in use [65]. HPP technique is used and experiments on powdered infant formulas have been performed and results have suggested that spores of *Bacillus cereus* spp. were reduced when high pressure was combined with heat treatment, which also minimized the degradation of nutritional quality attributes in infant formulas [30].

Cronobacter sakazakii infections are very common in powdered infant milk formulas, which give life-threatening forms of neonatal meningitis, sepsis, necrotizing enterocolitis (NEC) in new born and premature infants [22]. To meet powdered form infant milk nutritional quality, its health requirements along with a guarantee of food safety, HHPP application for *Cronobacter* species inactivation was focused on the wet mixture treatment before spray-drying by using HPP inactivation level between 2 and 6 log cycles.

The infant formula composition is a copy of human milk made as accurately as possible, but they're still exists some differences. Cow milk is the main component in infant formula composition. Therefore, cow milk allergy is a major issue in infant foods processing. The protein content and immunoglobulin E (IgE) response causes an adverse reaction, such as atopic disorders including atopic dermatitis, sepsis, and allergic rhinitis. Anti-allergy potential is very rare in childhood and infants below 3 years; and ß lactoglobulin is present in whey protein, which is most allergenic and is absent in human milk. Therefore, HPP is used, which is capable of denaturing the protein resulting in more sufficient hydrolysis. HPP will give transient conformations of the hydrophobic cores and, in turn, will provide new hydrolysis, which will result in reduced allergenicity [38].

Human breast milk is the ideal source of nutrition for infants. Human milk can be stored or banked and is processed using thermal processing methods, and thermal methods of processing results in denaturation of desirable components of milk, such as immunoglobulin A (IgA). Recent studies related to HPP have shown minimum destruction of macromolecules in food and also keeping intact its nutritional values. HPP is used to prolong the keeping quality without any deteriorating effect in the processed product as in milk banking. HPP is also beneficial in killing of microbes in animal milk, such as reducing *Enterobacteriaceae* counts to undetectable levels. Studies have reported that HPP increases the keeping quality of human milk along with maintaining the IgA and other nutritional value and providing immunological protection [61].

Despite having so many benefits of HPP in infant foods, still the developing countries are not using it. HPP is till now used for extending the storage life of human milk banking, inactivation of bacterial spores and infections, reducing hypoallergenic activity in cow milk protein without lowering the national value and palatability of infant foods.

Though HPP has numerous advantages over conventional food processing methods, yet different food manufacturers are not using this novel technology (Table 3.2).

TABLE 3.2 Current Issues in Food Processing Sectors Due to Conventional Food Processing Methods and Expected Benefits from HPP

Sector	Present Conventional Processing Technique(s)	Current Problem(s)	Expected Benefit(s) from HPP
Bakery industry	• Microwave heating.	• Puffiness • Mold growth	• Inactivates mold growth
Beverage industry	• Pasteurization	• Loss in sensory characteristics	• Improve sensory attributes
Dairy industry	• UHT pasteurization • LTLT pasteurization • HTST pasteurization	• Less shelf-life • Microbial growth • Change in flavor and aroma	• Reduce microbes growth • Reduce souring(loss of taste) • Increase the shelf-life.
Fruits and vegetables industry	• Blanching • Pasteurization	• Nutrition loss • Enzymatic browning	• Improved nutritional value • Inactivate enzymes
Infant food industry	• Pasteurization • Spray drying	• Increase the chances of microbial growth • Reduce the shelf-life	• Inactivates microbes • Increase the shelf-life
Meat industry	• Salting • Cooking	• Discoloration • Textural changes	• Prevent color loss • Retains texture
Seafood industry	• Boiling • Cooking	• Microorganisms growth	• Inactivates microorganism growth

3.5 FUTURE SCOPE OF HPP

The primary reason for processing is to stabilize the food products by protecting from any detrimental deviations in quality of food products. Food processing involves use of technologies, such as thermal and non-thermal.

Similarly, for the food preservation, the HPP is a substitute technology to thermal treatment promising safety attributes. To keep the nutritional value intact with an increase in shelf-life, HPP is used as a minimally processed technique [58]. High-pressure technology can keep the fresh quality of foods with-profits [48]. The main advantage of HPP is that it has almost instantaneous and isostatic pressure provided to food products regardless of size, shape, composition, and yield [16]. HPP processing can increase the keeping quality of the food product by inactivation of the enzymes, without degrading the sensory quality of the product. HPP processing is carried out in airtight packages, which are made in such a way that it can withstand the change in volume of the food [56].

Literature reviews have documented that consumer acceptance of HPP in various countries (America, Japan, Brazil, and Europe). The study reported that 67% of the participants accepted the HPP product positively by the influence of visual exposure and benefit statements [56]. The market trend of having natural minimally processed with no additives food products led to the evolution of technology, such as HPP.

Nowa-days, fresh-cut fruit juices, hams, meats condiments, die salads, dressings, soups have benefitted from the HPP. The world food production using HPP was 35 million-kg in 2012. HPP guacamole is the classic product in the market next to meat products. HPP maintains the color of the meat. HPP in future will highly in use due to the ability of minimal processing and as it can be applied post-packaging as well.

Nevertheless, HPP is a novel way, which can increase productivity, and reduce nutritional loss. It is a promising emerging technology, which can give added value to the product with microbial inactivation, which is the primary objective of every food company. Employing HPP in food processing synthesis of bioactive components leads to the nutritional benefits. HPP has now helped to replace the chemical preservatives used to increase the shelf-life. The HPP treatment reduced the use of benzoates (class one preservatives) and sorbates [25]. The installation of HPP in large volumes will also be helpful for the cost-reduction, as it only requires high pressure. HPP guarantees the microbial safety and increases in palatability of the products [32]. Future researches will make HPP the most commercially accepted non-thermal processing for producing minimally processed food products having higher nutritional value compared to conventionally processed food products without any chances of health hazards.

3.6   SPECIFIC AREAS OF HPP THAT NEED IMPROVEMENTS

Studies on consumer acceptance have been reported in which the consumers are loyal to the brand as they are not willing to switch to different food products. Consumers still do not have knowledge of the benefits of HPP treated products. A consumer-oriented study was carried out in Janeiro (Brazil) regarding consumer acceptance of HPP products. It was reported that the label does not have enough information on HPP, which gave a negative influence on the customer purchase of the product [16]. Marketing should be done to inform customers about the benefits of HPP products with its attributes and also emphasizing on the technology. Appropriate label information should be given as of its keeping quality, the nutritious value of foods, which help the consumers to know the advantages and benefits of HPP products.

There must be collaboration between the local food processors and the researchers or the food technologists for the benefit of the consumers. By this collaboration, the marketing department or the food processors will search for the kind of the products that are expected by the consumers. Local food processor acts as the middleman between the consumers and researchers. These processors can directly communicate to know the customer demands/ preferences and what they can improvise more in the HPP products so that there is profit along with health benefits. Food processors are yet to know the importance of HPP and its usage in the processing of food products. It would benefit both processors and consumers as HPP production is cost-effective because it is a physical processing, also it will increase the yield and health benefits, which will result in acceptance by the consumers resulting in high profit.

HPP equipment's are still not well developed to be used for processing of pork loins, hams, and meats. Research studies are still needed for this technology so that it can be used in bakery industries, beverage industries, etc. Modernized components such as automated parts are required so that it can easily be used by large companies. The change in chemical composition structure of the products still needs greater attention. There is much to be done in infant food industries for its preservation and processing.

3.7 SUMMARY

This chapter on the effect of HPP from various sources has revealed that it not only reduces or destroys microorganisms but also increases the storage life of the food product. It maintains the color, texture, quality, and chemical composition. It reduces the log cycle of pressure resistance microorganisms. This present review emphasizes on HPP feasibility more than thermal processing. Awareness and education on HPP of foods is lacking, especially in developing countries. Furthermore, studies to explore the potential use of HPP in all food sectors are needed.

KEYWORDS

- environment friendly
- food preservation
- high pressure
- microbial inactivation
- minimal processing
- non-thermal processing
- nutritional value
- thermal processing

REFERENCES

1. Ahmed, J., Ramaswamy, H. S., & Hiremath, N., (2005). The effect of high-pressure treatment on rheological characteristics and color of mango pulp. *International Journal of Food Science and Technology, 40*, 885–895.
2. Ahmed, J., & Hosahalli, S. R., (2006). Viscoelastic and thermal characteristics of vegetable puree-based baby foods. *Journal of Food Process Engineering, 29*, 219–221.
3. Balney, C., & Masson, P., (1993). Effects of high pressure on proteins. *Food Review International, 9*(4), 611–628.
4. Bridgman, P. W., (1909). An experimental determination of certain compressibility's. *Proceedings of the American Academy of Arts and Sciences, 44*(10), 255–279.
5. Bridgman, P. W., (1914). Change of phase under pressure, I: The phase diagram of eleven substances with especial reference to the melting curve. *Physical Review, 3*(3), 153–203.
6. Bridgman, P. W., (1923). The thermal conductivity of liquids under pressure. *Proceedings of the American Academy of Arts and Sciences, 59*(7), 141–169.

7. Brown, K., (1920). The manufacture of phenol in a continuous high-pressure autoclave. *Industrial and Engineering Chemistry, 12*(3), 279–280.

8. Buchheim, W., Schütt, M., & Frede, E., (1996). High pressure effects on emulsified fats. *Progress in Biotechnology, 13*, 331–336.

9. Campus, M., (2010). High pressure processing of meat, meat products, and seafood. *Food Engineering Reviews, 2*(4), 256–273.

10. Cao, X., Bi, X., Huang, W., Wu, J., Hu, X., & Liao, X., (2012). Changes of quality of high hydrostatic pressure processed cloudy and clear strawberry juices during storage. *Innovative Food Science and Emerging Technologies, 16*, 181–190.

11. Chakraborty, S., Kaushik, N., Rao, P. S., & Mishra, H. N., (2014). High-pressure inactivation of enzymes: A review on its recent applications on fruit purees and juices. *Comprehensive Reviews in Food Science and Food Safety, 13*(4), 578–596.

12. Chattong, U., Apichartsrangkoon, A., & Bell, A., (2007). Effects of hydrocolloid addition and high pressure processing on the rheological properties and microstructure of a commercial ostrich meat product *Yor* (Thai Sausage). *Meat Science, 76*(3), 548–554.

13. Cheah, P. B., & Ledward, D. A., (1997). Catalytic mechanism of lipid oxidation following high-pressure treatment in pork fat and meat. *Journal of Food Science, 62*(6), 1135–1139.

14. Cheftel, J. C., & Culioli, J., (1997). Effects of high pressure on meat: A review. *Meat Science, 46*(3), 211–236.

15. Daryaei, H., & Balasubramaniam, V. M., (2012). Microbial decontamination of food by high pressure processing: Chapter 13. In: Ali, D., & Michael, O. N., (eds.), *Microbial Decontamination in the Food Industry* (pp. 370–406). London, UK: Woodhead Publishing.

16. Deliza, R., Rosenthal, A., Abadio, F. B. D., Silva, C. H., & Castillo, C., (2005). Application of high-pressure technology in fruit juice processing: Benefits perceived by consumers. *Journal of Food Engineering, 67*(1/2), 243–245.

17. Duranton, F., Simonin, H., Chéret, R., Guillou, S., & Lamballerie, M. D., (2012). Effect of high pressure and salt on pork meat quality and microstructure. *Journal of Food Science, 77*(8), 188–194.

18. Eberhard, P., Strahm, W., & Eyer, H., (1999). High-pressure treatment of whipped cream. *Agrarforschung, 6*(9), 352–354.

19. Farkas, D. F., & Hoover, D. G., (2000). High pressure processing. *Journal of Food Science, 65*, 47–64.

20. Farkas, D. F., (2016). Short history of research and development efforts leading to the commercialization of high-pressure processing of food: Chapter 2. In: Balasubramaniam, V., Barbosa-Cánovas, G., & Lelieveld, H., (eds.), *High Pressure Processing of Food* (pp. 19–36). New York: Springer.

21. Fitria, A., Buckow, R., Singh, T., Hemar, Y., & Kasapis, S., (2015). Color change and proteolysis of skim milk during high pressure thermal processing. *Journal of Food Engineering, 147*, 102–110.

22. Friedemann, M., (2007). *Enterobacter sakazakii* in food and beverages (other than infant formula and milk powder). *International Journal of Food Microbiology, 116*(1), 1–10.

23. Fuentes, V., Ventanas, J., Morcuende, D., Estévez, M., & Ventanas, S., (2010). Lipid and protein oxidation and sensory properties of vacuum-packaged dry-cured ham subjected to high hydrostatic pressure. *Meat Science, 85*(3), 506–514.

24. Garcia-Palazon, A., Suthanthangjai, W., Kajda, P., & Zabetakis, I., (2004). The effects of high hydrostatic pressure on β-glucosidase, peroxidase, and polyphenol oxidase in red raspberry (*Rubus idaeus*) and strawberry (*Fragaria ananassa*). *Food Chemistry, 88*(1), 7–10.

25. Glass, K. A., McDonnell, L. M., Rassel, R. C., & Zierke, K. L., (2007). Controlling *Listeria monocytogenes* on sliced ham and turkey products using benzoate, propionate, and sorbate. *Journal of Food Protection, 70*(10), 2306–2312.

26. Goh, E. L., Hocking, A. D., Stewart, C. M., Buckle, K. A., & Fleet, G. H., (2007). Baroprotective effect of increased solute concentrations on yeast and moulds during high pressure processing. *Innovative Food Science and Emerging Technologies, 8*(4), 535–542.

27. Goyal, A., Sharma, V., Upadhyay, N., Sihag, M., & Kaushik, R., (2018). High pressure processing and its impact on milk proteins: A review. *Research and Reviews: Journal of Dairy Science and Technology, 2*(1), 12–20.

28. Haard, N. F., (2001). Enzymic modification of proteins in food systems: Chapter 7. In: Sikorski, Z. E., (ed.), *Chemical and Functional Properties of Food Proteins* (pp. 155–181). Lancaster, PA: Technomic Publishing Co.

29. Hayman, M. M., Anantheswaran, R. C., & Knabel, S. J., (2007). The effects of growth temperature and growth phase on the inactivation of *Listeria monocytogenes* in whole milk subject to high pressure processing. *International Journal of Food Microbiology, 115*(2), 220–226.

30. Cetin-Karaca, H., & Morgan, M. C., (2018). Inactivation of *Bacillus cereus* spores in infant formula by combination of high pressure and trans-cinnamaldehyde. *LWT-Food Science and Technology, 97,* 254–260.

31. Hite, B. H., (1899). *The Effect of Pressure in the Preservation of Milk: A Preliminary Report* (Vol. 58, pp. 15–35). Morgantown, West Virginia: West Virginia Agricultural Experiment Station.

32. Hogan, E., Kelly, A. L., & Sun, D. W., (2005). High pressure processing of foods: An overview: Chapter 1. In: Sun, D. W., (ed.), *Emerging Technology in Food Processing* (pp. 3–32). New York, USA: Academic Press.

33. Hoover, D. G., Metrick, C., Papineau, A., Farkas, D. F., & Knorr, D., (1989). Biological effects of high hydrostatic pressure on food microorganisms. *Food Technology, 43*(3), 99–107.

34. Huang, H. W., Wu, S. J., Lu, J. K., Shyu, Y. T., & Wang, C. Y., (2017). Current status and future trends of high-pressure processing in food industry. *Food Control, 72,* 1–8.

35. Hugas, M., Garriga, M., & Monfort, J. M., (2002). New mild technologies in meat processing: High pressure as a model technology. *Meat Science, 62*(3), 359–371.

36. Huppertz, T., Smiddy, M. A., Goff, H. D., & Kelly, A. L., (2011). Effects of high-pressure treatment of mix on ice cream manufacture. *International Dairy Journal, 21*(9), 718–726.

37. Hur, S. J., Park, G. B., & Joo, S. T., (2007). Formation of cholesterol oxidation products (COPs) in animal products. *Food Control, 18*(8), 939–947.

38. Jiang, Y. J., & Guo, M., (2014). Processing technology for infant formula: Chapter 8. In: Guo, M., (ed.), *Human Milk Biochemistry and Infant Formula Manufacturing Technology* (pp. 211–229). London, UK: Woodhead Publishing.

39. Kalagatur, N. K., Kamasani, J. R., & Mudili, V., (2018). Effect of high pressure processing on growth and mycotoxin production of *Fusarium graminearum* in maize. *Food Bioscience, 21*, 53–59.
40. Knorr, D., (1993). Effects of high-hydrostatic-pressure processes on food safety and quality. *Food Technology (Chicago), 47*(6), 156–161.
41. Kouniaki, S., Kajda, P., & Zabetakis, I., (2004). The effect of high hydrostatic pressure on anthocyanins and ascorbic acid in black currants (*Ribes nigrum*). *Flavor and Fragrance Journal, 19*(4), 281–286.
42. Krebbers, B., Matser, A. M., Koets, M., & Van Den-Berg, R. W., (2002). Quality and storage-stability of high-pressure preserved green beans. *Journal of Food Engineering, 54*(1), 27–33.
43. Krebbers, B., Matser, A., Koets, M., Bartels, P., & VanDen-Berg, R. W., (2002). High pressure-temperature processing as an alternative for preserving basil. *International Journal of High Pressure Research, 22*(3/4), 711–714.
44. Lambert, Y., Demazeau, G., Largeteau, A., & Bouvier, J. M., (1999). Changes in aromatic volatile composition of strawberry after high-pressure treatment. *Food Chemistry, 67*(1), 7–16.
45. Liepa, M., Zagorska, J., & Galoburda, R., (2016). High-pressure processing as novel technology in dairy industry: A review. *Research for Rural Development, 1*, 46–83.
46. Mandal, R., & Kant, R., (2017). High-pressure processing and its applications in the dairy industry. *Food Science and Technology: An International Journal (FSTJ), 1*(1), 33–45.
47. Meyer, R. S., Cooper, K. L., Knorr, D., & Lelieveld, H. L., (2000). High-pressure sterilization of foods. *Food Technology (Chicago), 54*(11), 67–72.
48. McClements, J. M. J., Patterson, M. F., & Linton, M., (2001). The effect of growth stage and growth temperature on high hydrostatic pressure inactivation of some psychrotrophic bacteria in milk. *Journal of Food Protection, 64*(4), 514–522.
49. Medina-Meza, I. G., Barnaba, C., & Barbosa-Cánovas, G. V., (2014). Effects of high pressure processing on lipid oxidation: A review. *Innovative Food Science and Emerging Technologies, 22*, 1–10.
50. Miranda, M., Maureira, H., Rodriguez, K., & Vega-Gálvez, A., (2009). Influence of temperature on the drying kinetics, physicochemical properties, and antioxidant capacity of aloe vera (*Aloe barbadensis*) gel. *Journal of Food Engineering, 91*(2), 297–304.
51. Mor-Mur, M., & Yuste, J., (2003). High pressure processing applied to cooked sausage manufacture: Physical properties and sensory analysis. *Meat Science, 65*(3), 1187–1191.
52. Navarro, M., Verret, C., Pardon, P., & El Moueffak, A., (2002). Changes in volatile aromatic compounds of strawberry puree treated by high-pressure during storage. *International Journal of High Pressure Research, 22*(3/4), 693–696.
53. Oey, I., Lille, M., Van, L. A., & Hendrickx, M., (2008). Effect of high-pressure processing on color, texture, and flavor of fruit-and vegetable-based food products: A review. *Trends in Food Science and Technology, 19*(6), 320–328.
54. Oh, H. E., Pinder, D. N., Hemar, Y., Anema, S. G., & Wong, M., (2001). Effect of high-pressure treatment on various starch-in-water suspensions. *Food Hydrocolloids, 22*(1), 150–155.
55. Okpala, C. O. R., Piggott, J. R., & Schaschke, C. J., (2009). Effects of high-pressure processing (HPP) on the microbiological, Physico-chemical and sensory properties of fresh cheeses: A review. *African Journal of Biotechnology, 8*(25), 7391–7398.

56. Olsen, N. V., Grunert, K. G., & Sonne, A. M., (2010). Consumer acceptance of high-pressure processing and pulsed-electric field: A review. *Trends in Food Science and Technology, 21*(9), 464–472.

57. O'Reilly, C. E., Murphy, P. M., & Kelly, A. L., (2002). The effect of high-pressure treatment on the functional and rheological properties of mozzarella cheese. *Innovative Food Science and Emerging Technologies, 3*(1), 3–9.

58. Palou, E., Lopez-Malo, A., & Welti-Chanes, J., (2002). Innovative fruit preservation using high pressure: Chapter 43. In: Welti-Chanes, J., Barbosa-Canovas, G. V., & Aguilera, J. M., (eds.), *Engineering and Food for the 21st Century* (pp. 715–726). Boca Raton-FL: CRC Press.

59. Palou, E., Lopez-Malo, A., Barbosa-Canovas, G. V., & Swanson, B. G., (2007). High-pressure treatment in food preservation: Chapter 34. In: Rahman, M. S., (ed.), *Handbook of Food Preservation* (pp. 833–872). Boca Raton – FL: CRC Press.

60. Pedras, M. M., Tribst, A. A. L., & Cristianini, M., (2014). Effects of high-pressure homogenization on physicochemical characteristics of partially skimmed milk. *International Journal of Food Science and Technology, 49*(3), 861–866.

61. Permanyer, M., Castellote, C., & Ramírez-Santana, C., (2010). Maintenance of breast milk immunoglobulin-A after high-pressure processing. *Journal of Dairy Science, 93*(3), 877–883.

62. Pou, K. R. J., (2015). Recent advances in the application of non-thermal technologies as effective food processing techniques. *International Journal of Agriculture Science, 7*(12), 801–808.

63. Pratt, D. A., Tallman, K. A., & Porter, N. A., (2011). Free radical oxidation of polyunsaturated lipids: New mechanistic insights and the development of peroxyl radical clocks. *Accounts of Chemical Research, 44*(6), 458–467.

64. Rastogi, N. K., Raghavarao, K. S. M. S., & Balasubramaniam, V. M., (2007). Opportunities and challenges in high pressure processing of foods. *Critical Reviews in Food Science and Nutrition, 47*(1), 69–112.

65. Reineke, K., Ellinger, N., Berger, D., & Baier, D., (2013). Structural analysis of high pressure treated *Bacillus subtilis* spores. *Innovative Food Science and Emerging Technologies, 17*, 43–53.

66. Rivalain, N., Roquain, J., & Demazeau, G., (2010). Development of high hydrostatic pressure in biosciences: Pressure effect on biological structures and potential applications in biotechnologies. *Biotechnology Advances, 28*(6), 659–672.

67. Schaich, K. M., (2005). Lipid oxidation: Theoretical aspects: Chapter 7. In: Shahidi, F., (ed.), *Bailey's Industrial Oil and Fat Products* (Vol. 1, pp. 269–355). New York, USA: John Wiley & Sons, Inc.

68. Sfakianakis, P., & Tzia, C., (2014). Conventional and innovative processing of milk for yogurt manufacture, development of texture and flavor: A review. *Foods, 3*(1), 176–193.

69. Smelt, J. P., (1998). Recent advances in the microbiology of high pressure processing. *Trends in Food Science and Technology, 9*(4), 152–158.

70. Stolt, M., Oinonen, S., & Autio, K., (2000). Effect of high pressure on the physical properties of barley starch. *Innovative Food Science and Emerging Technologies, 1*(3), 167–175.

71. Suarez, J. C., & Morrissey, M. T., (2006). Effect of high pressure processing (HPP) on shelf-life of albacore tuna (*Thunnus alalunga*) minced muscle. *Innovative Food Science and Emerging Technologies, 7*(1/2), 19–27.

72. Tauscher, B., (1995). Pasteurization of food by hydrostatic high pressure: Chemical aspects. *ZeitschriftfürLebensmittel-Untersuchung und Forschung [Journal of Food Study and Research], 200*(1), 3–13.

73. Ting, E. Y., & Marshall, R. G., (2002). Production issues related to UHP food: Chapter 7. In: Welti-Chanes, J., Barbosa-Canovas, G. V., & Aguilera, J. M., (eds.), *Engineering and Food for the 21st Century* (pp. 727–738). Boca Raton-USA: CRC Press.

74. Tokuşoğlu, Ö., Alpas, H., & Bozoğlu, F., (2010). High hydrostatic pressure effects on mold flora, citrinin mycotoxin, hydroxytyrosol, oleuropein phenolics and antioxidant activity of black table olives. *Innovative Food Science and Emerging Technologies, 11*(2), 250–258.

75. Trujillo, A. J., Capellas, M., Saldo, J., Gervilla, R., & Guamis, B., (2002). Applications of high-hydrostatic pressure on milk and dairy products: A review. *Innovative Food Science and Emerging Technologies, 3*(4), 295–307.

76. Udabage, P., Augustin, M., & Versteeg, C., (2010). Properties of low-fat stirred yoghurt made from high-pressure-processed skim milk. *Innovative Food Science and Emerging Technologies, 11*(1), 32–38.

77. Vallons, K. J., & Arendt, E. K., (2009). Effects of high pressure and temperature on the structural and rheological properties of sorghum starch. *Innovative Food Science and Emerging Technologies, 10*(4), 449–456.

78. Van, B. M., Fogliano, V., & Pellegrini, N., (2010). Review on the beneficial aspects of food processing. *Molecular Nutrition and Food Research, 54*(9), 1215–1247.

79. Vazquez, P. A. L., Qian, M. C., & Torres, J. A., (2007). Kinetic analysis of volatile formation in milk subjected to pressure-assisted thermal treatments. *Journal of Food Science, 72*(7), E389–E398.

80. Yuste, J., Capellas, M., Pla, R., Fung, D. Y., & Mor-Mur, M., (2001). High pressure processing for food safety and preservation: A review. *Journal of Rapid Methods and Automation in Microbiology, 9*(1), 1–10.

81. Zhang, H., & Mittal, G. S., (2008). Effects of high-pressure processing (HPP) on bacterial spores: An overview. *Food Reviews International, 24*(3), 330–351.

CHAPTER 4

POTENTIAL OF PULSED ELECTRIC FIELDS IN FOOD PRESERVATION

PRANALI NIKAM, SUVARTAN RANVIR, JOHN DAVID,
THEJUS JACOB, and RAMAN SETH

ABSTRACT

Today, the pulsed electric field (PEF) treatment is a popular nonthermal process used for the preservation of foods and food products. PEF processing of liquid foods keeps the freshness of foods and retains the vital components, such as vitamins. The literature review for PEF treatment reveals its potential use in the preservation of food products. The PEF technique is considered superior than the traditional thermal techniques for food processing.

4.1 INTRODUCTION

Ohmic heating, microwave heating, and PEF are some novel methods, where electricity has been used in the treatment of foods. Among these, ohmic heating is the oldest method in which the heat generated by passing current throughout food is used for killing microorganisms. In 1960s, first report on pulsed electric fields (PEFs) processing was published by Heinz Doevenspeck [26, 113]. They reported the cell disruption effect and recommended its application for improving mass transfer processes, such as separation, and drying of foods, and inactivation of microbes without a considerable amount of heating.

In the 1980s, Werner Sitzmann and Heinz Doevenspeck installed the first industrial-scale system in the fish industry. Unfortunately, power-switching technology was not developed far enough. Milk pasteurization using electric current was first carried out with electro-pure method, in which thermal sterilization of milk was achieved using generation of heat by an alternating electrical current (220–4200 V) through milk. The electro-pure method was successfully used to inactivate the two most harmful microorganisms (such as *Mycobacterium tuberculosis* and *Escherichia coli*). After 1980, the scientific community tried to commercialize the PEF processing system. The scientist from different streams started working together and series of patents were filed between 1987 and 2000.

To avoid spoilage of food products by microorganisms, to increase the shelf-life and food quality, many novel processing methods with prospective industrial applicability have been examined [89, 91]. Conventional thermal treatments in the food industry have provided necessary safety profiles and for increasing the shelf-life. However, thermal treatments may cause nutrients losses, alternate in physicochemical characteristics, and development of undesirable flavor and color, which are unacceptable by consumers. Many variations in thermal treatment techniques were proposed to overcome this, but they are not sufficiently efficient to provide the desired results. Many technologies have been examined to keep the food safe without affecting their nutritional quality or altering their characteristics [51]. Therefore, nonthermal technologies (such as pulse light, high voltage electric discharge, PEF, high-pressure processing (HPP), ultrasound processing) have been extensively studied in food processing and preservation [90]. The term nonthermal processing is frequently allied with the methods, which work at ambient temperature [136].

The PEF treatment is a unique, nonthermal preservation technique, which results in finished products with good nutritional value, high retention of organoleptic properties, and extended shelf-life. The PEF processing is done at sub-ambient, ambient, or little above the ambient temperature for shorter than one second are, and is attained by several short-time pulses usually <5 µs. The unwanted energy losses due to thermal processing of foods are minimized with addition to the development of undesirable color and aroma of food [50]. PEF technology provides superior food quality characteristics compared with the traditional heat processing methods [87].

Some reports have documented that PEF processing conserves the nutritional quality of food. Although, effect of PEF treatment on physicochemical and nutritional properties of foods should be well understood and evaluated prior to processing [84]. Although PEF treatment can kill the bacteria and yeasts in different types of foods, yet the bacterial spores are resistant to PEF treatment. The principle applications of PEF treatment particularly for low pH food products are to destroy spoilage causing microorganisms and foodborne pathogens [6, 49, 52]. The goal of this chapter is to target the potential of PEF in food processing industry.

4.2 PRINCIPLE OF PULSED ELECTRIC FIELD TREATMENT

In PEF processing, the liquid foods are subjected to short pulses with high voltage while maintaining the temperature below 30–40°C without any additional heating components. PEF technology employs PEFs ranging from 10 to 80 kV/cm, which disrupt the cell membrane through electroporation. The development of transmembrane potentials beyond a threshold value across the cell frequently results in damage cell and death [137].

In PEF processing, electrodes deliver a high voltage electric field. When food is kept in between a set of two electrodes and pulsing power is delivered to the food inside the treatment chamber. Food consists of several ions, which enable it to transfer the electricity. Applied high voltage electric field brings unfavorable changes in the microbial cells, which later cause microbial inactivation. The PEF treated product is then packed aseptically and stored at <7°C. The phenomenon due to the presence of charged molecules in food allows the transfer of electrical current throughout the liquid food. The electric field can be considered as an exponentially decaying, oscillatory pulses, square wave at sub-ambient, ambient, or slightly above the ambient temperature [49, 130].

4.3 APPLICATIONS OF PULSE ELECTRIC FIELD IN FOOD PRESERVATION

Several research and review articles have indicated that PEF techniques are suitable for improvement of drying rates, alteration of enzyme activity, enhancement of metabolite extraction, and genetic engineering (GE), etc. The principal focus of the usages of PEF techniques in food industry remains as a feasible substitute pasteurization and sterilization of foods through heat. Numerous reports have proved the suitability of PEF techniques to produce qualitative, fresh, and nutritive foods products mostly milk and milk-based products [78]; and inactivation of microbes in water, milk, juice, yogurt, eggs, etc., [11].

4.3.1   PULSE ELECTRIC FIELD PROCESSING OF MILK AND DAIRY BEVERAGES

Among beverages, milk is considered an important drink for humans, right from one's infancy to geriatrics. Several milk beverages available in markets are: medicated milk, flavored milk, yogurt, kefir, and fermented drinks, etc. The milk contains a high amount of nutrients and its low acidity makes it a suitable breeding ground for microorganisms, including those causing food poisoning, because milk must be processed immediately after receiving.

Traditionally several thermal heat treatments (such as pasteurization, sterilization, and ultra-high temperature (UHT) treatment, etc.), have been adapted to process milk and milk-based beverages. The major aim of thermally treated milk is a desirable degree of killing of microorganisms, inactivation of enzymes and making it safe for human consumption. However, thermal heat treatment causes undesirable changes in physicalchemical properties, denaturation of proteins, decrease in nutritional quality, decrease in sensory properties, and losses of vitamins, etc. The PEF technique has several advantages compared to the thermal treatment. The PEF treated food products are able to retain the flavor, color, and nutritional properties.

4.3.1.1   MILK

Water, carbohydrate, fat, protein, and vitamins are the key components of milk; and these have a suitable atmosphere for the growth of microbes, which are responsible for the spoilage of milk. Many research studies have reported the inactivation of microbes under PEF treatment. Fernandez-Molina et al. [35] and US-FDA [115] reported that the shelf-life of skimmed milk was 2 weeks at refrigerated temperature when subjected to PEF treatment at 40 kV/cm, 30 pulses/2 µs using exponential decaying pulses. Raw skim milk was heat-treated at 80°C/6s followed by PEF treatment to 30 kV/cm, 30 pulses/2 µs; and the shelf-life of the milk was up to 2 weeks. Qin et al. [85] monitored the impact of 2 step-PEF treatment on milk and they revealed that shelf-life of milk was increased by 2 weeks when stored at 5°C. In addition, they found no apparent change in physicochemical characteristics, higher retention of organoleptic properties compared to pasteurized treated milk. This study recommended that PEF processing under the standardized conditions was suitable for killing pathogenic microbes of milk at least similar to the thermal treated (HTST pasteurization).

Dunn et al. [28] assessed the shelf-life of a homogenized milk inoculated with *Salmonella Dublin* and subjected to the PEF treatment at 36.7 kV/cm and 40 pulses for 25 minutes. *Salmonella Dublin* was found to be absent in milk after the storage at 7°C for eight days. In another study, Dunn [27] observed the minimum losses of flavor, no changes in physicochemical in milk quality parameters for cheese making after PEF treatment. Bermudez et al. [14] and Sharma et al. [106] assessed the shelf-life of skim milk and whole milk subjected to the PEF treatment of 46.15 kV/cm, at 20 to 60°C and 30 pulses; and in PEF treated milk stored at 4°C showed higher stability, whereas faster spoilage was noted in milk stored at 21°C. It was also observed that the growth of mesophilic count was under control in both samples under PEF treatment.

Physicochemical properties and shelf stability of PEF treated milk have also been examined by investigators [13, 17, 45]. The skim milk and whole milk subjected to electric field strength from 31 to 534 kV/cm. Minimal changes in physicochemical properties of milk were observed due to the changes in the fat and protein composition after treatment. Calderon-Miranda [20] and US-FDA [115] monitored the impact of PEF processing on inactivation of *Listeria innocua* in skimmed milk. The PEF treatments at 30 to 50 kV/cm showed a decline in growth of *Listeria innocua* by 2.5 log microbial.

4.3.1.2 MILK-BASED FUNCTIONAL DRINKS

Fruits are rich source of various antioxidants, ascorbic acid, carotenoids, and phenolic compounds [7, 104]. The milk-based fruit juices provide several health benefits and are popular in markets of Europe, Japan, and USA [83]. Usually, the milk-based fruit beverages are formulated with citric acid as an acidifier, pectin as stabilizer, sucrose, and water [17, 63]. The commercialization of these types of beverages needs preservation [40]. PEF treatment is the novel nonthermal treatment for the preservation of milk-based beverages [92].

The examples of PEF treated milk-based beverages include: milk drinks, fruit juice-milk blends, and fermented milk drinks [17]. It was reported that *E. coli* inactivation in an orange juice (50%) + milk (20%) mixed beverage of 3.83 log cycles [93]. Results indicate that the PEF processing can achieve the same level of pectin methylesterase (PME) inactivation as the thermal treatment with orange juice + milk beverage [37, 97].

4.3.1.3 YOGHURT-BASED BEVERAGES

Traditionally yogurt-based beverages are considered healthy food and can be used in several formulations [25]. The shelf-life of yogurt based beverages is near 3 weeks; and after that it often spoiled because of attack by yeasts and bacteria especially *lactobacilli* species as these microbes can grow under low pH [66]. The PEF treatment could be a suitable treatment for the preservation of non-frozen dairy desserts by reducing the loss of quality that occurs due to attack of microbes, thus resulting in an increase in shelf-life [98].

The yogurt inoculated with yeast and PEF at 45°C resulted in the increased in shelf-life up to ten days at 4°C; while the PEF treatment at 55°C led to increase up to one-month storage at 4C [28]. The mild heating with PEF treatment reported a considerable decrease in total aerobic bacteria count, yeast, and mold with no impact on the color and organoleptic properties of the finished product [43].

Flavored milk and yogurt drinks by treating with a combination PEF + heating did not found any change in organoleptic characteristics of the products [17]. In another study, flavored yogurt (strawberry, grape, and blueberry) treated with a combination of heat treatment at 60°C for 30s with PEF 30 kV/cm had a shelf-life of 90 days at 4°C because of reduction in total viable counts and molds counts by 2 to 4 log cycles [17, 127].

Cueva and Aryana [24] assessed the suitability of PEF treatments (5, 15, and 25 kV/cm) to control growth of *Lactobacillus acidophilus LA-K*. The authors noted that the growth rate of targeted probiotic was considerably influenced by the nature of pulse, the strength of the electric field, and pulse duration, which slowed down the log phase growth rate of the bacterium. They suggested that PEF technology is a new avenue for regulating the growth of culture bacteria and maintaining flavor and the texture of yogurt and ripened cheese [17, 75].

4.3.2 PROCESSING OF EGGS

It is necessary to pasteurize liquid egg products for ensuring food safety and storage stability [76, 128]. There are numerous studies carried out for assessing the impact of PEF on inactivation of numerous target microorganisms in an egg, such as *Listeria innocua, Salmonella enteritidis, E. coli,* and *Pseudomonas fluorescens* [46, 99].

Qin et al. [12], Ma et al. [57] and Liu et al. [55] carried out the study on liquid eggs treated with PEF treatment; and they observed that PEF treatment led to significantly lower viscosity, whereas color (especially based on β-carotene percentage) of liquid eggs in comparison with fresh eggs was significantly increased. With a sensory panel assessment using a triangle test, Qin et al. [85] reported that there was non-significant changes between scrambled eggs produced from PEF treated eggs and fresh eggs [115]; and the latter was favored over a commercial brand.

Along with the color assessment of egg products, Ma et al. [57] analyzed the density of PEF treated and fresh liquid egg, together with strength of baked cake with PEF treated eggs. They reported that the systematic method does not cause any impact on the whiteness or density between the PEF treated eggs and fresh liquid eggs. Statistical analysis of sensory assessment concluded that there was no variation between the cake made either from PEF treated eggs or fresh liquid eggs. Researchers concluded that eggs with PEF treatment at 25 kV/cm up to 800 μs did not cause any impact on the oxidation of egg white protein (EWP) [58, 120–122]. Zhao and Yang [134] reported that there was no influence on the structure and functional properties of lysozyme with PEF treatment at 35 kV/cm for 300 μs.

4.3.3 MEAT AND FISHES

Effect of PEF processing on meat products is multifaceted, as changes are represented together by the PEF treating parameters and structural and composition variability of meat. This structural as well as compositional variation is based on age, species, and gender of the animal, on-farm animal production processes, type of meat cut, handling of animal's prior- and after-slaughtering, meat aging regime, and thawing process [5].

The electro-permeabilization of cell membranes causing a high progression in rates of mass transfer can be used to increase the drying rates of cellular tissues [50]. An increase in rates of mass transfer caused faster water transport to the surface of meat product and consequently decreased the drying time after a pre-treatment, which resulted in good usage of production capacities and saving of energy through convective air-drying [79]. The little energy input is needed for a PEF processing of animal or plant tissue (2 to 20 kJ/kg) that indicates the potential to decrease the total energy input for drying of product. PEF treatment also reduces the size of muscle cell and persuades visible gaps between the muscle cells [33, 46].

4.3.4   PEA SOUP

The PEF treatment inhibits the growth of *B. subtilis* and *E. coli* suspended in pea soup based on several pulses, pulsing rate, electric field intensity, and flow rate [42, 50, 116, 117]. Vega-Mercado et al. [116] treated pea soup with two steps of 16 pulses at 35 kV/cm for avoiding a rise in temperature above 55°C. The authors revealed that the shelf-life of PEF processed pea soup at 5°C exceeded 4 weeks, whereas the sample stored at 32°C was unsuitable for storage. In addition, they did not find any changes in the physicochemical parameters, sensory characteristics of the pea soup after storage of four weeks at refrigeration temperature.

4.3.5   APPLE JUICE

Simpson et al. [107] reported that the ultra-filtrated apple juice under PEF treatments did not any impact on acidity, pH, vitamin C content, glucose content, fructose content, and sucrose content. The PEF treated apple juice at 50 kV/cm at 10 pulses, 2 µs pulse width, and extreme processing temperature of 45°C had a shelf-life of 4 weeks compared to the shelf-life of 3 weeks of freshly squeezed apple juice [50]. In addition, no impact of PEF treatment was observed on sensory attributes, physicochemical parameters, and changes in vitamin C or sucrose content.

4.3.6   GRAPE JUICE

Grimi [41] assessed the influence of PEF processing on grape juice at 400 V/cm pulses. The yield of PEF pretreated juice was increased about 67–75% than the control sample [6, 50].

4.4   EFFECT OF PEF ON NUTRITIONAL COMPONENTS IN FOODS

4.4.1   EFFECT ON PROTEIN

Proteins in food are major important ingredients that not only provide nutrients but also give desirable textural attributes to foods. Protein also improves the techno-functional characteristics of foods, such as foaming, emulsifying, gelation, and solubility [28, 67]. All these functional properties are based on molecular structure, concentration, and behavior of proteins in foods. The inherent protein in foods is usually held together with delicate stability of various non-covalent bonds (such as hydrophobic, Vander-wall interaction, ionic, hydrogen, and disulfide bond) [119]. The functionality of proteins is dependent on several environmental factors during PEF treatment, such as pH, temperature, ionic strength, the composition of food, which can disturb the delicate equilibrium for maintaining the inherent structure of proteins [108].

PEF processing for liquid foods causes the conformational changes in the structure of proteins owing to the alteration of the balanced forces that stabilize the inherent structure. The effect of PEF processing could be responsible for the ionization of certain chemical groups, which may break the electrostatic interaction inside a large polypeptide chain or between two monomeric units of protein. However, the impact of PEF on the structure of proteins is quite descriptive, and possible mechanisms behind the structural modification are not being reported. In recent years, the effect of PEF processing on the structure of protein has gained considerable attention [10, 39, 53, 81].

The effect of PEF on food protein depends on the nature of PEF conditions in use and processing parameters, such as intensity of electric field, period of pulses treatment, and width of pulse. Most of the investigators have reported the effect of PEF treatment on proteins mainly in egg white and β-Lg in cow milk whey, probably due to the gelling behavior in many food formulations. The effects of PEF on the structure of proteins are summarized in Table 4.1.

TABLE 4.1   Effect of PEF on Different Food Proteins

Source	Media	PEF Parameter	Effect	References
Cow immunoglobulin	–	41.1 kV/cm, 84 μs	No change in secondary structure and its thermal stability	[53]
Egg white	2% Ovalbumin solution	31.5 kV/cm 180 μs	Increase in the reactivity of sulfhydryl group	[34]
Liquid whole egg	–	48 kV/cm, 120 μs	No effect on globulin	[59]
	–	26 kV/cm, 2–4 μs	No protein coagulation	[60]

TABLE 4.1   *(Continued)*

Source	Media	PEF Parameter	Effect	References
Soy protein	–	41.1 kV/cm, 91 μs	Significant increase in functional properties and denaturation of protein	[54, 56]
Whey protein	β-Lg (2–12% solution)	30 kV/cm	Non-significant aggregation or unfolding of β-Lg	[10]

Ovalbumin is the most heat-stable, phosphoglyco protein containing four thiol groups inside the structure and they are exposed during the unfolding of proteins. Fernandez-Diaz et al. [34] monitored the impact of PEF processing on 2% ovalbumin solution at 31.5 kV/cm of PEF, which caused partial unfolding of ovalbumin and exposed all four SH⁻ groups on the surface, which may due to the enhance ionization of SH⁻ group into highly sensitive S⁻ form. The significant impact on SH⁻ group reactivity is dependent on the energy applied and a number of pulses [109]. Nevertheless, this attraction appeared to transit, when ovalbumin PEF treated solution was kept at 30 minutes at 4°C that resulted in the reactivity of thiol groups. Protein coagulation in the liquid entire egg was not observed when exposed at an electric field 26 kV/cm at 100 square wave pulses of 2–4 μs [123].

Among the whey proteins, β-Lg is the primary gelling agent that forms thermal-induced gels on heating [81]. No alterations were observed in the fourth derivative of UV spectra or native and PEF processed (250 μs at 30 kV/cm) β-Lg solution. Barsotti et al. [7] and Blight [19] studied the impact of PEF on β-Lg protein solution of various concentrations ranging from 2–12% w/v. These changes were also supported with electrophoresis, which was identical for native and PEF processed β-Lg solutions [135]. PEF treatments with 200 exponential decay pulses of 31.5 kV/cm at 1 Hz, 2.35/pulses resulted in inactivation of *E. coli NCTC 9001* with 2–5 log cycles but did not induce the unfolding of β-Lg molecules [109].

Perez and Pilosof [81] suggested partial alteration up to 40% in the native structure of β-Lg using ten exponentially decaying pulses at 2 μs of 12.4 kV/cm. This treatment might have induced the protein denaturation and enhanced gelling properties of β-Lg [21]. The effect of PEF on protein involves several mechanisms: (a) polarization of protein molecules; (b) dissociation of the quaternary structure of protein molecule by disturbing covalent bond; (c) modifications in protein structure, which leads to exposure of hydrophobic and thiol group which are previously buried inside of the protein core; and (d) if the time of PEF treatment was quite high, then it may lead to aggregation of the proteins. However, all these proposed mechanisms are not proven yet and need further research for clarification.

Odriozola et al. [70] reported significant difference between the heat-treated and PEF treated milk for α-La, β-Lg, and serum albumin. Impact of PEF processing on bovine immunoglobulin for 41.1 kV/cm at 54 μs did not show any noticeable change in secondary structure or the thermal stability [53, 109]. Liu et al. [39] and Hadden et al. [44] investigated the changes in structure and protein profile of soybean after giving PEF treatment at 0–50 kV/cm intensity for 40 μs by using differential scanning calorimetric (DSC) and Fourier transformer infrared spectroscopy (FTIR) techniques. PEF treatment brought the alteration in bond vibration in side chain of amino acid, antiparallel sheets, β-turns and β-sheets structure, which may due to the unfolding of protein molecules.

Li and Chen [54] also observed a significant enhancement of techno-functional characteristics in emulsibility, solubility, hydrophobicity, foaming, and degree of denaturation of protein that was isolated by the increased electric field up to 4.1 kV/cm for 91 μs. Odriozola-Serrano et al. [69] monitored PEF effect on the amino acid profile of tomato and strawberry juices when compared with thermally processed. PEF treated samples retained a higher amount of free amino acid (2.5%) as compared to the heat-treated sample. PEF treatment increased the concentration of alanine, glutamic acid, valine, serine by 4.8%, 27%, 6.8%, 6.3%, and 5.5%, respectively. During the storage of PEF processed tomato juice samples, there was a considerable increase in total free amino acid compared to that of heat-treated samples [71, 74]. The favorable potential effect on PEF on preservation and processing must cause stability of the juice samples and increase the nutritional value by enhancing the free amino acid concentration.

4.4.2 EFFECT ON FATTY ACIDS

Monounsaturated fatty acids (MUFA) and polyunsaturated fatty acids (PUFA) are the most important nutritional components in food due to their several positive health benefits [103]. Consumption of these compounds is now gaining attention by both public and food processing industries. However, neither heat treatments nor PEF processing affect the concentration of free fatty acids in whole milk [84]. On the other hand, a significant increase in the free fatty acid of freshly PEF processed (35.5 kV/cm with 7 μs monopolar square wave pulses) whole milk, when stored at refrigerated temperature for 12 days [22, 38, 53]. All of these changes may be due to the spoilage of milk by microorganisms that may lead to lipolysis because of the release of esters and may contribute to the formation of free fatty acids [114].

Zulueta et al. [139], Salvia-Trujillo et al. [95], Sharma et al. [105] observed insignificant changes in SFA, MUFA, and PUFA on PEF treated orange milk beverages, which were processed at 30–40 kV/cm for 40–180 μs, bipolar square waves pulses 2.5 μs when compared to the non-treated sample. However, small decrease in fat content was observed due to the PEF treatment, rupture the fat globule aggregates into smaller globules; however, the extent of droplet dispersion was too small [13] that was caused by a reduction in oil in water (O/W) interface stability after PEF processing.

Morales-de-la Peña et al. [62, 64] reported slight improvement of free fatty acids between PEF treated (35 kV/cm, 4 μ bipolar pulses at 200 Hz for 800 and 400 μs) and thermally processed (90C for 60s) fruit juice + soy milk beverage when stored at refrigerated temperature. This may happen due to biochemical changes in volatile components during the course of the time.

Garde-Cerdán et al. [36] observed a parallel percentage of total fatty acid in grape juice at intense PEF treatment of 35 kV/cm with 4 μs bipolar pulses at 100 hz for 1 minute as compared to untreated samples. The decrease of lauric acid content was observed in PEF treated grape juices. The content of saturated fatty acid and PUFA in peanut oils were found to decrease in PEF treated oil as compared to non-treated oil when stored for 100 days at 40°C due to oxidation reaction [129].

4.4.3   EFFECT ON VITAMINS

Effect of PEF treatments on vitamins has been observed to cause relatively minor changes as compared to thermal treatments. Effects of PEF on vitamins are reported in Table 4.2. Several scientists have investigated ascorbic acid retention after PEF treatment compared to thermal treatment in different fruit juices, such as orange, apple, grape, tomato, strawberry, watermelon, etc., [29]. More losses in vitamin C were observed in juice, if the pulse intensity of PEF is increased and high retention of vitamin C was observed; when pulse frequency and pulse width are lowered then the PEF treatment maintained the vitamin C in the monopolar mode as compared to bipolar mode [15, 77].

TABLE 4.2   Effect of PEF Treatments on Vitamins

Vitamin	Media	PEF Parameter	Effects	References
Ascorbic acid (Vitamin C)	Protein substituted orange-based juices	28 kV/cm, 100–300 µs	4–13% losses when processing period increased	[101]
	Orange juice	87 kV/cm, 40 reversal pulses at 50°C	No reduction	[47]
	Orange juice	15–35 kV/cm, 100 to 1000 µs	1.8 to 12.5% losses	[29]
	Milk	18–27 kV/cm, 400 µs	30% losses	[12]
	Orange juice	35 kV, 60 to 87 µs	96% retention	[86]
	Orange juice	40 kV/cm, 97 µs, 2.6 µs width at 58°C/5 s	Retention of vitamin C	[60]
	Apple juice	23 to 34 kV/cm, 166 µs, bipolar square wave from pulses, 4 µs width	Retention of vitamin C	[32]
	Grape juice	Reverse pulse at 50°C	Very low reduction	[122]
	Gazpacho soup	15 to 35 kV/cm, 100 to 1000 µs	2.9 to 15.7% losses	[29]
	Tomato juice	35 kV/cm, 1000 µs	58.2 to 99% retention	[68]
	Strawberry juice	35 kV/cm, 1700 µs	98% retention	[73]
	Carrot juice	35 kV/cm, 2000 µs	95.1% retention	[88]
	Brocolli juice	35 kV/cm, 2000 µs	74.6% retention	[102]
	Watermelon juice	30 to 35 KV/cm, 50 to 2050 µs	Reduction in vitamin C	[77]
	Red bell pepper	0.5 to 2.5 KV/cm, 20 pulses, 400 µs pulse width	Decrease of vitamin C (87 to 93%)	[2]
	Red bell pepper	2 KV/cm, 1 to 50 pulses	Decrease of vitamin C (35 to 44%)	[3]

TABLE 4.2   *(Continued)*

Vitamin	Media	PEF Parameter	Effects	References
Vitamin B1	Fruit beverage	35 KV/cm, 1800 µs	No reduction	[94]
	Milk	18.3–27.1 kV/cm, 400 µs	Very low reduction	[12]
Vitamin B2	Milk	18–27 kV/cm, 400 µs	Very low reduction	[12]
Vitamin B3	Fruit beverage	35 KV/cm, 1800 µs	No reduction	[94]
Vitamin D	Milk	18 to 27 kV/cm, 400 µs	No reduction	[12]

PEF treated protein-orange juice-based beverage heated at 28 kV/cm showed 4–13% losses as compared to non-PEF treated juice [29, 101]. However, orange juice treated at 87 kV/cm at 40 instant charge reversal pulses with added bacteriocin did not report no reductions or very low reduction in vitamin C [47, 122, 125]; also orange juice treated at 35 kV/cm, 60–87 µs and 40 kV/cm, 97 µs showed greater retention of vitamin C [61, 86]. While the decrease of 7.52% of vitamin A in orange juice was found in PEF processing at 30 kV/cm, 100 µs as compared to losses of 15.62% of vitamin A, pasteurized at 90°C for 20s [23]. PEF processed carrot juice and broccoli juice reported greater retention of 95% and 74% of vitamin C, respectively, as compared with thermally treated samples [88, 102].

However, ascorbic acid retention was significantly lower in thermally treated strawberry juice (94%) than that in PEF processed juice (98%) [73]. The difference in stability of ascorbic acid between PEF treated strawberry juice might be due to low pH in comparison with tomato and carrot juice; because a high acidic environment helps to stabilize vitamin C [69, 111]. Vitamin C is a heat-labile bioactive component in the presence of air (oxygen) [1]. Therefore, the high temperature of the processing of fruit juices affects the rates of degradation over an aerobic pathway [96]. Most workers suggested that retention of vitamin C is better in PEF processed juices as compared to thermally treated juices [110].

Ascorbic acid depletion kinetics in PEF treated fruit juices as a function of storage time followed first-order kinetic model (R^2 = 0.968–0.838) with degradation rates K_1 from 1.7×10^{-2} to 3.5×10^{-2} days^{-1} in carrot juice; while K_1 from 2.4×10^{-2} to 3.0×10^{-2} days^{-1} was for ascorbic acid depletion in tomato juice [88, 73].

The vitamin C degradation was better fitted by zero-order kinetics model than that of first-order for PEF processed orange juice and orange juice milk beverages as a function of storage time [48, 118, 139, 140]. Salvia-Trujillo et al. [94] and Tiwari et al. [112] reported that Weibull model describes the vitamin C degradation kinetics accurately during refrigerated storage with R^2 = 0.955 at Af values ranges from 1.01 to 1.11.

Consequently, most of the researchers have reported a reduction of vitamin C content in watermelon juice at PEF treatment of 30–35 kV/cm for 50–2050 μs at 50–250 Hz [77]; in apple juice at PEF treatment of 0–35 kV/cm for 0.2–2 μs [16, 77] with highest losses of ascorbic acid up to 36.6%. Similarly, Elez-Martinez et al. [29, 31] also reported a higher loss of ascorbic acid up to 2.9–15.7% in gazpacho and orange juices in a bipolar PEF treatment of 15–35 kV/cm for 100–1000 μs.

Zhang et al. [133] monitored PEF effect on the structure of ascorbic acid and also on antioxidant properties and they stated that effect of PEF treatment had a favorable effect on the antioxidant properties of ascorbic acid because of development of free hydroxyl (OH⁻) radicals, which may modify the structure of ascorbic acid. PEF treatment might have supported the transformation of ascorbic acid from "enol to keto" form, which was determined by fluorescence intensity of ascorbic acid prior- and post-the PEF treatment, which resulted in the conversion of ascorbic acid-induced by hydroxyl radicals [124].

Bendicho et al. [12] monitored the impact of PEF process on the retention of vitamins D and E (fat-soluble) and vitamins B1, B2, and C (water-soluble) in milk and simulated milk ultra-filtrate (SMUF) at 18.3 to 27.1 kV/cm up to 400 μs and these were compared with mild or intense heating process at 63°C for 30 min and 75°C for 15s [30, 106]. They observed that fat-soluble vitamins and water-soluble vitamins (B1 and B2) were stable except the ascorbic acid content was found to be slightly decreased [12]. However, under more intense treatment conditions of 27 kV/cm for 400 us, ascorbic acid content losses were 20%. This treatment was also suitable for the inactivation of *Listeria monocytogenes* at 2 log cycles and inactivation of *Salmonella Dublin* at 4 log cycles in milk [80]. Therefore, milk processed at PEF treatment with 22.6 kV/cm, 400 μs is more effective than the mild and intense pasteurization processes for preserving the native contents of vitamins [109].

4.4.4 EFFECT ON POLYPHENOLS

The stability of polyphenol compounds under PEF processing in food matrices is not so easy to investigate because of the intrinsic complexity of most food matrices, and the treatment intensities given to foods may undergo deleterious reactions. Polyphenols have complex chemical structures and also diverse in molecular size as a result they undergo distinct complex biosynthetic metabolisms, such as methoxylation, glycosylation, and hydroxylation; but changes greatly depend on phenolic content of the treated food products. Hence, the molecular structure of polyphenols is an important factor in the impact of PEF processing [126].

Effect of PEF treatment on polyphenols content is summarized in Table 4.3. Odriozola-Serrano et al. [72] explored the influence of PEF treatment on the tomato juice samples by storing for 56 days at 4°C and suggested to exhibit greater percentage of phenolic acids (chlorogenic acid) and flavanols (quercetin) than that of conventionally heat-treated samples at 90°C for 60s; whereas negligible amount of phenolic acids (such as ferulic, p-coumaric, and caffeic acids) were observed during the storage time probably because of action of residual enzymes [72].

Caffeic acid concentration was marginally improved over the period, while PEF treated and thermally treated tomato juices undergo a significant decrease in p-coumaric acid throughout the storage period. The escalation in caffeic acid content in tomato juices after storage of 4 weeks might be due to the remaining hydroxylase activities, which transform p-coumaric acid into caffeic acid [69]. While PEF and thermally treated samples contain p-coumaric acids and ellagic in higher amounts over the storage period in case of strawberry juices. PEF-treated strawberry juices reported similar amounts of kaempferol and quercetin as compared to those heat-treated for 7 weeks at a refrigerated temperature [73].

TABLE 4.3   Effect of PEF on Polyphenols

Product	PEF Parameter	Effects	References
Fruit juice milk	35 kV/cm, 1800 µs with Bipolar 4 µs pulses at 200 Hz	Improved retention of total phenolic	[65]
Fruit juice + soy milk	35 kV/cm, 1800 µs with Bipolar 4 µs pulses at 200 Hz	Increase in total phenolic content especially hesperidin	[62]
Orange juice	25.26 kV/cm, 1206.2 µs with Bipolar square-wave pulses 35 kV/cm, 750 µs	Increase in flavonoid levels No changes	[4, 100]

TABLE 4.3 *(Continued)*

Product	PEF Parameter	Effects	References
Strawberry juice	35 kV/cm, 1700 μs, Bipolar 4 μs pulses, 100 Hz	1. Retention of phenolics, such as 67% p-hydroxybenzoic acid, 88% p-coumaric acid, 117% ellagic acid	[73]
		2. Retention of flavonols: 44% quercetin, 89% Kaemfferol	
Tomato juice	35 kV/cm, 1700 μs, Bipolar 4 μs pulses, 100 Hz	1. Retention of phenolic acids: 86% chlorogenic acid, 67% ferulic acid, 53% p-coumaric acid, 132% caffeic acid	[72]
		2. Retention of flavanols: 80% quercetin; and 82% kaempferol	

Effect of PEF processing on natural color compounds ((such as anthocyanins), which provides colors to several vegetables and fruits (such as onions, beets, berries, and apples)) have been investigated by some workers. In few studies, PEF processing at high electric field strength led to more retention of anthocyanin content in strawberry juice [8, 9]. In contrast, Zhang et al. [132] revealed that degradation of cyanidin-3-glucoside in a methanolic solution when treatment time and intensity of electric field are increased [69]. Also, under low pressure and low moisture environment, proanthocyanins are converted into anthocyanins with PEF treatment at 35 kV/cm at 37°C [82].

Odriozola-Serrano et al. [73] studied the influence of PEF treatment (35 kV/cm for 1700 μs, at <40°C) and severe heat treatment (90°C for 30/60s) on the content of anthocyanins in strawberry juices during storage at 4°C. They observed that untreated strawberry juices contained a slightly higher amount of anthocyanins as compared to PEF and thermally treated juice. Symmetry occurs between four anthocyanin species, containing quinonoidal base, flavylium cation, pseudobase, or carbitol; and chalcone in fresh juice samples were reported [18].

Zhang et al. [131, 132] reported that after treating a cyanidin-3-glucoside in aqueous methanolic solution at 1.2–3 kV/cm, 300 pulses at 47°C, the formation of chalcone compound took place after the treatment, which is the first step of anthocyanins degradation together by opening of pyrilium ring. The PEF and thermal treatments caused depletion of total anthocyanins throughout the storage of strawberry juices [73]. These changes could be concomitant with the improvement of activity of certain enzymes, viz., β-glucosidase.

Sánchez-Moreno et al. [101] studied orange juice treated with PEF treatment (35 kV/cm for 750 μs with 4-μs bipolar pulses at 800 Hz) [138]. The authors observed that there was no influence of PEF treatment on total flavanones compounds, hesperidin, aglycones, and naringenin. While, Agcam et al. [4] found enhancement of flavonoid levels in PEF-processed orange juice than the thermally treated juice. Morales-de la Peña et al. [62] treated fruit juice-soymilk with PEF treatment (35 kV/cm for 1400 μs using 4-μs bipolar pulses at 200). They found a non-significant impact of PEF treatment on total isoflavone content.

4.5   SUMMARY

Compared to the traditional thermal treatments, the PEF processing is a most appealing nonthermal method for inactivating enzymes, vegetative microbes, including pathogens causing spoilage in liquid foods. The process is widely studies in beverages, meat, milk, eggs, etc. The researchers have revealed that PEF processed foods can retain more aroma, organoleptic characteristics, and nutritional components as compared to thermal treated samples. Being a nonthermal treatment, PEF treatment is gaining interest in safeguarding heat-labile foods.

KEYWORDS

- **differential scanning calorimetric**
- **Fourier transformer infrared spectroscopy**
- **nutritional components**
- **polyphenols**
- **pulsed electric field**
- **shelf-life**
- **yogurt**

REFERENCES

1. Aadil, R. M., Roobab, U., Sahar, A., Rahman, U., & Khalil, A. A., (2019). Functionality of bioactive nutrients in beverages: Chapter 7. In: Grumezescu, A. M., & Holban, A. M., (eds.), *Nutrients in Beverages* (Vol. 12, pp. 237–276). Cambridge-MA, USA: Academic Press.

2. Ade-Omowaye, B. I., Rastogi, N. K., Angersbach, A., & Knorr, D., (2002). Osmotic dehydration of bell peppers: Influence of high intensity electric field pulses and elevated temperature treatment. *Journal of Food Engineering, 54*, 35–43.

3. Ade-Omowaye, B. I., Taiwo, K. A., & Eshtiaghi, N. M., (2003). Comparative evaluation of the effects of pulsed electric field and freezing on cell membrane permeabilization and mass transfer during dehydration of red bell peppers. *Innovative Food Science and Emerging Technologies, 4*, 177–188.

4. Agcam, E., Akyıldız, A., & Evrendilek, G. A., (2014). Comparison of phenolic compounds of orange juice processed by pulsed electric fields (PEF) and conventional thermal pasteurization. *Food Chemistry, 143*, 354–361.

5. Alahakoon, A. U., Faridnia, F., Bremer, P. J., Silcock, P., & Oey, I., (2016). Pulsed electric fields effects on meat tissue quality and functionality: Chapter 1. In: Miklavcic, D., (ed.), *Handbook of Electroporation* (pp. 1–21). New York, USA: Springer.

6. Andreou, V., Dimopoulos, G., Dermesonlouoglou, E., & Taoukis, P., (2020). Application of pulsed electric fields to improve product yield and waste valorization in industrial tomato processing. *Journal of Food Engineering, 270*, 109778.

7. Anwar, H., Hussain, G., & Imtiaz, M., (2018). Antioxidants from natural sources: Chapter 1. In: Shalaby, E., & Azzam, G. M., (eds.), *Antioxidants in Foods and its Applications* (pp. 3–28). London, UK: IntechOpen.

8. Barba, F. J., Galanakis, C. M., Esteve, M. J., Frigola, A., & Vorobiev, E., (2015). Potential use of pulsed electric technologies and ultrasounds to improve the recovery of high-added value compounds from blackberries. *Journal of Food Engineering, 167*, 38–44.

9. Barba, F. J., Parniakov, O., & Pereira, S. A., (2015). Current applications and new opportunities for the use of pulsed electric fields in food science and industry. *Food Research International, 77*, 773–798.

10. Barsotti, L., Dumay, E., Mu, T. H., Diaz, M. D. F., & Cheftel, J. C., (2001). Effects of high voltage electric pulses on protein-based food constituents and structures. *Trends in Food Science and Technology, 12*(3/4), 136–144.

11. Bendicho, S., Barbosa-Cánovas, G. V., & Martín, O., (2002). Milk processing by high intensity pulsed electric fields. *Trends in Food Science and Technology, 13*(6/7), 195–204.

12. Bendicho, S., Espachs, A., Arantegui, J., & Martín, O., (2002). Effect of high intensity pulsed electric fields and heat treatments on vitamins of milk. *Journal of Dairy Research, 69*(1), 113–123.

13. Bermúdez-Aguirre, D., Fernández, S., & Esquivel, H., (2011). Milk processed by pulsed electric fields: Evaluation of microbial quality, physicochemical characteristics, and selected nutrients at different storage conditions. *Journal of Food Science, 76*(5), S289–S299.

14. Bermúdez-Aguirre, D., Yáñez, J. A., & Dunne, C. P., (2010). Study of strawberry flavored milk under pulsed electric field processing. *Food Research International, 43*(8), 2201–2207.
15. Bhattacharjee, C., Saxena, V. K., & Dutta, S., (2019). Novel thermal and nonthermal processing of watermelon juice. *Trends in Food Science and Technology, 93*, 234–243.
16. Bi, X., Liu, F., Rao, L., Li, J., Liu, B., Liao, X., & Wu, J., (2013). Effects of electric field strength and pulse rise time on physicochemical and sensory properties of apple juice by pulsed electric field. *Innovative Food Science and Emerging Technologies, 17*, 85–92.
17. Birwal, P., Deshmukh, G. P., & Ravindra, M. R., (2019). Nonthermal processing of dairy beverages: Chapter 12. In: Grumezescu, A., & Holban, A. M., (eds.), *Milk-Based Beverages* (pp. 397–426). London, UK: Woodhead Publishing.
18. Brouillard, R., (1982). Chemical structure of anthocyanins. In: Markakis, P., (ed.), *Anthocyanins as Food Colors* (pp. 1–40). London, U K: Food Science and Technology Division, Academic Press.
19. Blight, A. R., (2011). Treatment of walking impairment in multiple sclerosis with dalfampridine. *Therapeutic Advances in Neurological Disorders, 4*(2), 99–109.
20. Calderón-Miranda, M. L., Barbosa-Cánovas, G. V., & Swanson, B. G., (1999). Inactivation of *Listeria innocua* in skim milk by pulsed electric fields and nisin. *International Journal of Food Microbiology, 51*(1), 19–30.
21. Campbell, L., Raikos, V., & Euston, S. R., (2003). Modification of functional properties of egg-white proteins. *Food (Nahrung), 47*(6), 369–376.
22. Chen, B., Huang, Y., Zheng, D., Ni, R., & Bernards, M. A., (2019). Dietary fatty acids alter lipid profiles and induce myocardial dysfunction without causing metabolic disorders in mice. *Nutrients, 10*(1), 106.
23. Cortés, C., Torregrosa, F., Esteve, M. J., & Frígola, A., (2006). Carotenoid profile modification during refrigerated storage in untreated and pasteurized orange juice and orange juice treated with high-intensity pulsed electric fields. *Journal of Agricultural and Food Chemistry, 54*(17), 6247–6254.
24. Cueva, O., & Aryana, K. J., (2012). Influence of certain pulsed electric field conditions on the growth of *Lactobacillus acidophilus* LA-K. *Journal of Microbial and Biochemical Technology, 4*(7), 137–140.
25. Datta, N., & Tomasula, P. M., (2015). *Emerging Dairy Processing Technologies: Opportunities for the Dairy Industry* (1st edn., p. 360). New York, USA: John Wiley and Sons.
26. Doevenspeck, H., (1960). *Method and Device for Obtaining the Individual Phases from Disperse Systems* (p. 13). German Patent: DE1237541.
27. Dunn, J. E., & Pearlman, J. S., (1987). Washington, DC: U.S. Patent and Trademark Office. *Patent 4,695,472* (p. 15).
28. Dunn, J. E., (1996). Pulsed light and pulsed electric field for foods and eggs. *Poultry Science, 75*(9), 1133–1136.
29. Elez-Martinez, P., & Martin-Belloso, O., (2007). Effects of high intensity pulsed electric field processing conditions on vitamin C and antioxidant capacity of orange juice and gazpacho, cold vegetable soup. *Food Chemistry, 102*(1), 201–209.
30. Elez-Martínez, P., Sobrino-López, Á., & Soliva-Fortuny, R., (2012). Pulsed electric field processing of fluid foods: Chapter 4. In: Cullen, P. J., Tiwari, B. K., & Valdramidis, V. P., (eds.), *Novel Thermal and Non-Thermal Technologies for Fluid Foods* (pp. 63–108). Cambridge-MA, USA: Academic Press.

31. Elez-Martínez, P., Suárez-Recio, M., & Martín-Belloso, O., (2007). Modeling the reduction of pectin methylesterase activity in orange juice by high intensity pulsed electric fields. *Journal of Food Engineering, 78*(1), 184–193.

32. Evrendilek, G. A., Jin, Z. T., Ruhlman, K. T., & Qiu, X., (2000). Microbial safety and shelf-life of apple juice and cider processed by bench and pilot-scale PEF systems. *Innovative Food Science and Emerging Technologies, 1*(1), 77–86.

33. Faridnia, F., Bekhit, A. E. D. A., Niven, B., & Oey, I., (2014). Impact of pulsed electric fields and post-mortem vacuum ageing on beef *Longissimus Thoracic* Muscles. *International Journal of Food Science and Technology, 49*(11), 2339–2347.

34. Fernandez-Diaz, M. D., Barsotti, L., Dumay, E., & Cheftel, J. C., (2000). Effects of pulsed electric fields on ovalbumin solutions and dialyzed egg white. *Journal of Agricultural and Food Chemistry, 48*(6), 2332–2339.

35. Fernandez-Molina, J. J., & Barkstrom, E., (1999). Shelf-life extension of raw skim milk by combining heat and pulsed electric fields. *Food Research International, 8* https://doi.org/10.1111/j.1745-4549.2005.00029.x.

36. Garde-Cerdán, T., & Arias-Gil, M., (2007). Effects of thermal and nonthermal processing treatments on fatty acids and free amino acids of grape juice. *Food Control, 18*(5), 473–479.

37. Giner, J., Gimeno, V., Espachs, A., & Elez, P., (2000). Inhibition of tomato (*Lycopersicon Esculentum Mill*) pectin methylesterase by pulsed electric fields. *Innovative Food Science and Emerging Technologies, 1*(1), 57–67.

38. Giner, J., Grouberman, P., Gimeno, V., & Martín, O., (2005). Reduction of pectinesterase activity in a commercial enzyme preparation by pulsed electric fields: Comparison of inactivation kinetic models. *Journal of the Science of Food and Agriculture, 85*(10), 1613–1621.

39. Giner, J., Ortega, M., Mesegué, M., & Gimeno, V., (2002). Inactivation of peach polyphenol oxidase by exposure to pulsed electric fields. *Journal of Food Science, 67*(4), 1467–1472.

40. Granato, D., Branco, G. F., Cruz, A. G., Faria, J. D., & Shah, N. P., (2010). Probiotic dairy products as functional foods. *Comprehensive Reviews in Food Science and Food Safety, 9*(5), 455–470.

41. Grimi, N., Lebovka, N. I., Vorobiev, E., & Vaxelaire, J., (2009). Effect of a pulsed electric field treatment on expression behavior and juice quality of chardonnay grape. *Food Biophysics, 4*(3), 191–198.

42. Gudmundsson, M., & Hafsteinsson, H., (2001). Effect of electric field pulses on microstructure of muscle foods and roes. *Trends in Food Science and Technology, 12*(3/4), 122–128.

43. Guo, M., Jin, T. Z., Geveke, D. J., Fan, X., Sites, J. E., & Wang, L., (2014). Evaluation of microbial stability, bioactive compounds, physicochemical properties, and consumer acceptance of pomegranate juice processed in a commercial scale pulsed electric field system. *Food and Bioprocess Technology, 7*(7), 2112–2120.

44. Hadden, J. M., Bloemendal, M., Haris, P. I., Srai, S. K., & Chapman, D., (1994). Fourier transform infrared spectroscopy and differential scanning calorimetry of transferrins: Human serum transferrin, rabbit serum transferrin and human lactoferrin. *Biochimica et Biophysica Acta (BBA)-Protein Structure and Molecular Enzymology, 1205*(1), 59–67.

45. Halpin, R. M., Cregenzán-Alberti, O., Whyte, P., Lyng, J. G., & Noci, F., (2013). Combined treatment with mild heat, manothermo-sonication and pulsed electric fields reduces microbial growth in milk. *Food Control, 34*(2), 364–371.
46. Hermawan, N., Evrendilek, G. A., Dantzer, W. R., Zhang, Q. H., & Richter, E. R., (2004). Pulsed electric field treatment of liquid whole egg inoculated with *Salmonella enteritidis. Journal of Food Safety, 24,* 71–85.
47. Hodgins, A. M., Mittal, G. S., & Griffiths, M. W., (2002). Pasteurization of fresh orange juice using low-energy pulsed electrical field. *Journal of Food Science, 67*(6), 2294–2299.
48. Jiang, L., Zheng, H., & Lu, H., (2014). Use of linear and Weibull functions to model ascorbic acid degradation in Chinese winter jujube during postharvest storage in light and dark conditions. *Journal of Food Processing and Preservation, 38*(3), 856–863.
49. Kotnik, T., & Miklavčič, D., (2006). Theoretical evaluation of voltage inducement on internal membranes of biological cells exposed to electric fields. *Biophysical Journal, 90*(2), 480–491.
50. Kumar, S., Agarwal, N., & Raghav, P. K., (2016). Pulsed electric field processing of foods: A review. *International Journal of Engineering Research and Modern Education, 1*(1), 111–1118.
51. Lado, B. H., & Yousef, A. E., (2002). Alternative food-preservation technologies: Efficacy and mechanisms. *Microbes and Infection, 4*(4), 433–440.
52. Lebovka, N., & Vorobiev, E., (2009). *Electro-Technologies for Extraction from Food Plants and Biomaterials* (1st edn., p. 281). New York: Springer.
53. Li, S. Q., Bomser, J. A., & Zhang, Q. H., (2005). Effects of pulsed electric fields and heat treatment on stability and secondary structure of bovine immunoglobulin G. *Journal of Agricultural and Food Chemistry, 53*(3), 663–670.
54. Li, Y., & Chen, Z., (2006). Effect of high intensity pulsed electric field on the functional properties of protein isolated from soybean. *Transactions of the Chinese Society of Agricultural Engineering, 22*(8), 194–198.
55. Liu, T., Lv, B., Zhao, W., Wang, Y., Piao, C., & Dai, W., (2020). Effects of ultra-high temperature pasteurization on the liquid components and functional properties of stored liquid whole eggs. *BioMed Research International, 10.* Article ID: 3465465.
56. Liu, Y. Y., Zeng, X. A., Deng, Z., Yu, S. J., & Yamasaki, S., (2011). Effect of pulsed electric field on the secondary structure and thermal properties of soy protein isolate. *European Food Research and Technology, 233*(5), 841–850.
57. Ma, L., Chang, F. J., & Barbosa-Cánovas, G. V., (1997). Inactivation of *E. coli* in liquid whole eggs using pulsed electric fields technologies. In: *New Frontiers in Food Engineering: Proceedings of the Fifth Conference of Food Engineering* (pp. 216–221). American Institute of Chemical Engineers.
58. Ma, L., Chang, F. J., Gongora-Nieto, M. M., & Barbosa-Canovas, G. V., (2001). Comparison study of pulsed electric fields, high hydrostatic pressure, and thermal processing on the electrophoretic patterns of liquid whole eggs. In: Barbosa-Canovas, G. V., & Zhang, Q. H., (eds.), *Pulsed Electric Fields in Food Processing* (pp. 225–239). Lancaster PA, USA: Technomic Publishing.
59. Martín-Belloso, O., & Elez-Martínez, P., (2005). Enzymatic inactivation by pulsed electric fields. In: Sun, D. W., (eds.), *Emerging Technologies for Food Processing* (pp. 155–181). London-UK: Elsevier.

60. Michalac, S., Alvarez, V., Ji, T., & Zhang, Q. H., (2003). Inactivation of selected microorganisms and properties of pulsed electric field processed milk. *J. Food Process. Preserv., 27*, 137–151.

61. Min, S., Jin, Z. T., Min, S. K., Yeom, H., & Zhang, Q. H., (2003). Commercial-scale pulsed electric field processing of orange juice. *Journal of Food Science, 68*(4), 1265–1271.

62. Morales-De, L. P. M., & Salvia-Trujillo, L., (2012). High intensity pulsed electric fields or thermal treatments effects on the amino acid profile of a fruit juice - soymilk beverage during refrigeration storage. *Innovative Food Science and Emerging Technologies, 16*, 47–53.

63. Morales-De, L. P. M., Salvia-Trujillo, L., & Rojas-Graü, M. A., (2010). Isoflavone profile of a high intensity pulsed electric field or thermally treated fruit juice. *Innovative Food Science and Emerging Technologies, 11*(4), 604–610.

64. Morales-De, L. P. M., Salvia-Trujillo, L., Rojas-Graü, M. A., & Martín-Belloso, O., (2011). Impact of high intensity pulsed electric fields or heat treatments on the fatty acid and mineral profiles of a fruit juice-soymilk beverage during storage. *Food Control, 22*(12), 1975–1983.

65. Morales-de, L. P. M., Salvia-Trujillo, L., Rojas-Graü, M. A., & Martín-Belloso, O., (2017). Effects of high intensity pulsed electric fields or thermal pasteurization and refrigerated storage on antioxidant compounds of fruit juice-milk beverages. Part I: Phenolic acids and flavonoids. *Journal of Food Processing and Preservation, 41*(3), 7. Article ID: 12912.

66. National Research Council (NRC), (1992). *Applications of Biotechnology to Fermented Foods: Report of an Ad. Hoc. Panel of the Board on Science and Technology for International Development* (p. 9). Washington, D.C.: National Academies Press. PMID: 25121339.

67. Nisov, A., Ercili-Cura, D., & Nordlund, E., (2020). Limited hydrolysis of rice endosperm protein for improved techno-functional properties. *Food Chemistry, 302*, 10. Article ID: 125274.

68. Odriozola-Serrano, I., & Aguiló-Aguayo, I., (2007). Lycopene, vitamin C, and antioxidant capacity of tomato juice as affected by high-intensity pulsed electric fields critical parameters. *Journal of Agricultural and Food Chemistry, 55*(22), 9036–9042.

69. Odriozola-Serrano, I., Aguiló-Aguayo, I., & Soliva-Fortuny, R., (2013). Pulsed electric fields processing effects on quality and health-related constituents of plant-based foods. *Trends in Food Science and Technology, 29*(2), 98–107.

70. Odriozola-Serrano, I., Bendicho-Porta, S., & Martín-Belloso, O., (2006). Comparative study on shelf-life of whole milk processed by high-intensity pulsed electric field or heat treatment. *Journal of Dairy Science, 89*(3), 905–911.

71. Odriozola-Serrano, I., Garde-Cerdán, T., Soliva-Fortuny, R., & Martín-Belloso, O., (2013). Differences in free amino acid profile of non-thermally treated tomato and strawberry juices. *Journal of Food Composition and Analysis, 32*(1), 51–58.

72. Odriozola-Serrano, I., Soliva-Fortuny, R., Hernández-Jover, T., & Martín-Belloso, O., (2009). Carotenoid and phenolic profile of tomato juices processed by high intensity pulsed electric fields compared with conventional thermal treatments. *Food Chemistry, 112*(1), 258–266.

73. Odriozola-Serrano, I., Soliva-Fortuny, R., & Martín-Belloso, O., (2008). Phenolic acids, flavonoids, vitamin C and antioxidant capacity of strawberry juices processed by

high-intensity pulsed electric fields or heat treatments. *European Food Research and Technology, 228*(2), 239–245.

74. Ohnedaa, H., & Haranoa, A., (2002). Improvement of NOx removal efficiency using atomization of fine droplets into corona discharge. *Journal of Electrostatics, 55*(3/4), 321–332.

75. Ohshima, T., Okuyama, K., & Sato, M., (2002). Effect of culture temperature on high-voltage pulse sterilization of *Escherichia coli*. *Journal of Electrostatics, 55*(3/4), 227–235.

76. Olatunde, O. O., & Benjakul, S., (2018). Nonthermal processes for shelf-life extension of seafoods: A review. *Comprehensive Reviews in Food Science and Food Safety, 17*(4), 892–904.

77. Oms-Oliu, G., Odriozola-Serrano, I., Soliva-Fortuny, R., & Martín-Belloso, O., (2009). Effects of high-intensity pulsed electric field processing conditions on lycopene, vitamin c and antioxidant capacity of watermelon juice. *Food Chemistry, 115*(4), 1312–1319.

78. Ortega-Rivas, E., (2011). Critical issues pertaining to the application of pulsed electric fields in microbial control and quality of processed fruit juices. *Food and Bioprocess Technology, 4*(4), 631–645.

79. Ozuna, C., Cárcel, J. A., García-Pérez, J. V., & Mulet, A., (2011). Improvement of water transport mechanisms during potato drying by applying ultrasound. *Journal of the Science of Food and Agriculture, 91*(14), 2511–2517.

80. Patil, S., Valdramidis, V. P., Cullen, P. J., Frias, J. M., & Bourke, P., (2010). Ozone inactivation of acid stressed *Listeria Monocytogenes* and *Listeria innocua* in orange juice using a bubble column. *Food Control, 21*(12), 1723–1730.

81. Perez, O. E., & Pilosof, A. M., (2004). Pulsed electric fields effects on the molecular structure and gelation of β-lactoglobulin concentrate and egg white. *Food Research International, 37*(1), 102–110.

82. Porter, L. J., Hrstich, L. N., & Chan, B. G., (1985). The conversion of procyanidins and prodelphinidins to cyanidin and delphinidin. *Phytochemistry, 25*(1), 223–230.

83. Pszczola, D. E., (2005). Ingredients-making fortification. *Food Technology, 59*, 44–61.

84. Qin, B. L., Chang, F. J., Barbosa-Cánovas, G. V., & Swanson, B. G., (1995). Nonthermal inactivation of *Saccharomyces cerevisiae* in apple juice using pulsed electric fields. *LWT-Food Science and Technology, 28*(6), 564–568.

85. Qin, B. L., Pothakamury, U. R., Vega, H., Martin, O., Barbosa-Cánovas, G. V., & Swanson, B. G., (1995). Food pasteurization using high intensity pulsed electric fields. *Journal of Food Technology, 49*(12), 55–60.

86. Qiu, X., Sharma, S., Tuhela, L., Jia, M., & Zhang, Q. H., (1998). Integrated PEF pilot plant for continuous nonthermal pasteurization of fresh orange juice. *Transactions of the ASAE, 41*(4), 1069.

87. Quass, D. W., (1997). *Pulsed Electric Field Processing in the Food Industry: Status Report on PEF* (p. 76). Palo Alto, CA: Electric Power Research Institute; Report No. CR-109742.

88. Quitão-Teixeira, L. J., Odriozola-Serrano, I., & Soliva-Fortuny, R., (2009). Comparative study on antioxidant properties of carrot juice stabilized by high intensity pulsed electric fields or heat treatments. *Journal of the Science of Food and Agriculture, 89*(15), 2636–2642.

89. Rawat, S., (2015). Food spoilage: Microorganisms and their prevention. *Asian Journal of Plant Science and Research, 5*, 47–56.

90. Rawson, A., Patras, A., Tiwari, B. K., Noci, F., Koutchma, T., & Brunton, N., (2011). Effect of thermal and nonthermal processing technologies on the bioactive content of exotic fruits and their products: Review of recent advances. *Food Research International, 44*(7), 1875–1887.

91. Ricci, A., Parpinello, G., & Versari, A., (2018). Recent advances and applications of pulsed electric fields (PEF) to improve polyphenol extraction and color release during red winemaking. *Beverages, 4*(1), 18–26.

92. Riganakos, K. A., Karabagias, I. K., Gertzou, I., & Stahl, M., (2017). Comparison of UV-C and thermal treatments for the preservation of carrot juice. *Innovative Food Science and Emerging Technologies, 42*, 165–172.

93. Rivas, A., Sampedro, F., Rodrigo, D., Martínez, A., & Rodrigo, M., (2006). Nature of the inactivation of *Escherichia coli* suspended in an orange juice and milk beverage. *European Food Research and Technology, 223*(4), 541–545.

94. Salvia-Trujillo, L., Morales-de, L. P. M., Rojas-Graü, A., & Martín-Belloso, O., (2011). Changes in water-soluble vitamins and antioxidant capacity of fruit juice-milk beverages as affected by high-intensity pulsed electric fields (HIPEF) or heat during chilled storage. *Journal of Agricultural and Food Chemistry, 59*(18), 10034–10043.

95. Salvia-Trujillo, L., Morales-de, L. P. M., & Rojas-Graü, A., (2017). Mineral and fatty acid profile of high intensity pulsed electric fields or thermally treated fruit juice-milk beverages stored under refrigeration. *Food Control, 80*, 236–243.

96. Sampedro, F., Fan, X., & Rodrigo, D., (2010). High hydrostatic pressure processing of fruit juices and smoothies: Research and commercial application. In: Doona, C. J., Kustin, K., & Feeherry, F. E., (eds.), *Case Studies in Novel Food Processing Technologies: Innovations in Processing, Packaging, and Predictive Modeling* (pp. 34–72). Cambridge, England: Woodhead Publishing.

97. Sampedro, F., Geveke, D. J., Fan, X., & Zhang, H. Q., (2009). Effect of PEF, HHP and thermal treatment on PME inactivation and volatile compounds concentration of an orange juice-milk based beverage. *Innovative Food Science and Emerging Technologies, 10*(4), 463–469.

98. Sampedro, F., & Rodrigo, D., (2015). Pulsed electric fields (PEF) processing of milk and dairy products: Chapter 5. In: Datta, N., & Tomasula, M., (eds.), *Emerging Dairy Processing Technologies: Opportunities for the Dairy Industry* (pp. 115–148). New York: USA: John Wiley and Sons.

99. Sampedro, F., Rodrigo, D., Martínez, A., Barbosa-Cánovas, G. V., & Rodrigo, M., (2006). Application of pulsed electric fields in egg and egg derivatives. *Food Science and Technology International, 12*(5), 397–405.

100. Sanchez-Moreno, C., Plaza, L., Elez-Martınez, P., & De Ancos, B., (2005). Impact of high pressure and pulsed electric fields on bioactive compounds and antioxidant activity of orange juice in comparison with traditional thermal processing. *Journal of Agricultural and Food Chemistry, 53*(11), 4403–4409.

101. Sánchez-Vega, R., Elez-Martínez, P., & Martín-Belloso, O., (2015). Influence of high-intensity pulsed electric field processing parameters on antioxidant compounds of broccoli juice. *Innovative Food Science and Emerging Technologies, 29*, 70–77.

102. Sánchez-Vega, R., Rodríguez-Roque, M. J., & Elez-Martínez, P., (2019). Impact of critical high-intensity pulsed electric field processing parameters on oxidative enzymes and color of broccoli juice. *Journal of Food Processing and Preservation, 44*(3), 12. https://doi.org/10.1111/jfpp.14362.

103. Sereflisan, H., & Altun, B. E., (2018). Amino acid and fatty acid composition of freshwater mussels, *Anodonta pseudodopsis* and *Unio tigridis*. *Pakistan Journal of Zoology, 50*(6), 2153–2158.

104. Shahdadi, F., Mirzaei, H. O., & Garmakhany, A. D., (2015). Study of phenolic compound and antioxidant activity of date fruit as a function of ripening stages and drying process. *Journal of Food Science and Technology, 52*(3), 1814–1819.

105. Sharma, P., Oey, I., Bremer, P., & Everett, D. W., (2018). Microbiological and enzymatic activity of bovine whole milk treated by pulsed electric fields. *International Journal of Dairy Technology, 71*(1), 10–19.

106. Sharma, S. K., Zhang, Q. H., & Chism, G. W., (1998). Development of a protein-fortified fruit beverage and its quality when processed with pulsed electric field treatment. *Journal of Food Quality, 21*(6), 459–473.

107. Simpson, M. V., Barbosa-Cánovas, G. V., & Swanson, B. G., (1995). The combined inhibitory effect of lysozyme and high voltage pulsed electric fields on the growth of *Bacillus subtilis* spores. In: *IFT Annual Meeting: Book of Abstracts* (p. 267). Institute of Food Technology.

108. Singh, H., & Ye, A., (2020). Interactions and functionality of milk proteins in food emulsions. In: Thompson, A., Boland, M., & Singh, H., (eds.), *Milk Proteins: From Expression to Food* (pp. 467–497). UK: Academic Press.

109. Soliva-Fortuny, R., Balasa, A., Knorr, D., & Martín-Belloso, O., (2009). Effects of pulsed electric fields on bioactive compounds in foods: A review. *Trends in Food Science and Technology, 20*(11, 12), 544–556.

110. Suna, S., Tamer, C. E., & Sayın, L., (2014). Impact of innovative technologies on fruit and vegetable quality. *Bulgarian Chemical Communications, 46,* 131–136.

111. Tannenbaum, S. R., Archer, M. C., & Young, V. R., (1985). Vitamins and minerals. In: Fennema, O. R., (ed.), *Food Chemistry* (pp. 488–493). New York: Marcel Dekker Inc.

112. Tiwari, B. K., O'Donnell, C. P., Muthukumarappan, K., & Cullen, P. J., (2009). Ascorbic acid degradation kinetics of sonicated orange juice during storage and comparison with thermally pasteurized juice. *LWT-Food Science and Technology, 42*(3), 700–704.

113. Toepfl, S., & Heinz, V., (2007). Application of pulsed electric fields to improve mass transfer in dry-cured meat products. *Fleischwirtschaft International* (*International Meat Industry), 22*(1), 23–30.

114. Toepfl, S., Heinz, V., & Knorr, D., (2006). Applications of pulsed electric fields technology for the food industry. In: Raso, J., & Heinz, V., (eds.), *Pulsed Electric Fields Technology for the Food Industry* (pp. 197–221). Boston, MA: Springer.

115. US Food and Drug Administration (FDA), (2000). *Kinetics of Microbial Inactivation for Alternative Food Processing Technologies: Pulsed Electric Fields: Report for the Institute of Food Technologists* (p. 34). Washington, D.C.: Food and Drug Administration of the US Department of Health and Human Services.

116. Vega-Mercado, H., Martin-Belloso, O., Chang, F. J., & Barbosa-Cánovas, G., (1996). Inactivation of *Escherichia coli* and *Bacillus subtilis* suspended in pea soup using pulsed electric fields. *Journal of Food Processing and Preservation, 20*(6), 501–510.

117. Vega-Mercado, H., Pothakamury, U. R., Chang, F. J., Barbosa-Cánovas, G. V., & Swanson, B. G., (1996). Inactivation of *Escherichia coli* by combining pH, ionic strength, and pulsed electric fields hurdles. *Food Research International, 29*(2), 117–121.

118. Vervoort, L., Van, D. P. I., & Grauwet, T., (2011). Comparing equivalent thermal, high pressure and pulsed electric field processes for mild pasteurization of orange juice. Part

II: Impact on specific chemical and biochemical quality parameters. *Innovative Food Science and Emerging Technologies, 12*(4), 466–477.

119. Vieille, C., & Zeikus, G. J., (2001). Hyperthermophilic enzymes: Sources, uses, and molecular mechanisms for thermostability. *Microbiology and Molecular Biology Reviews, 65*(1), 1–43.

120. Walayat, N., Xiong, Z., Xiong, H., & Moreno, H. M., (2020). The effectiveness of egg white protein and β-cyclodextrin during frozen storage: Functional, rheological, and structural changes in the myofibrillar proteins of *Culter alburnus*. *Food Hydrocolloids, 105*, 105842.

121. Wang, L. H., Pyatkovskyy, T., Yousef, A., Zeng, X. A., & Sastry, S. K., (2020). Mechanism of *Bacillus subtilis* spore inactivation induced by moderate electric fields. *Innovative Food Science and Emerging Technologies, 9.* Article ID: 102349.

122. Wu, L., Zhao, W., Yang, R., & Chen, X., (2014). Effects of pulsed electric fields processing on stability of egg white proteins. *Journal of Food Engineering, 139*, 13–18.

123. Wu, L., Zhao, W., Yang, R., & Yan, W., (2015). Pulsed electric field (PEF)-induced aggregation between lysozyme, ovalbumin, and Ovo transferrin in multi-protein system. *Food Chemistry, 175,* 115–120.

124. Wu, X., Diao, Y., Sun, C., Yang, J., Wang, Y., & Sun, S., (2003). Fluorimetric determination of ascorbic acid with O-phenylenediamine. *Talanta, 59*(1), 95–99.

125. Wu, Y., Mittal, G. S., & Griffiths, M. W., (2005). Effect of pulsed electric field on the inactivation of microorganisms in grape juices with and without antimicrobials. *Biosystems Engineering, 90*(1), 1–7.

126. Yamamoto, L. Y., De Assis, A. M., & Roberto, S. R., (2015). Application of abscisic acid (S-ABA) to Cv. Isabel grapes (*Vitis Vinifera*) for color improvement: Effects on color, phenolic composition, and antioxidant capacity of their grape juice. *Food Research International, 77*, 572–583.

127. Yeom, H. W., Evrendilek, G. A., Jin, Z. T., & Zhang, Q. H., (2004). Processing of yogurt-based products with pulsed electric fields: Microbial, sensory, and physical evaluations. *Journal of Food Processing and Preservation, 28*(3), 161–178.

128. Yogesh, K., (2016). Pulsed electric field processing of egg products: A review. *Journal of Food Science and Technology, 53*(2), 934–945.

129. Zeng, X. A., Han, Z., & Zi, Z. H., (2010). Effects of pulsed electric field treatments on quality of peanut oil. *Food Control, 21*(5), 611–614.

130. Zhang, Q., Barbosa-Cánovas, G. V., & Swanson, B. G., (1995). Engineering aspects of pulsed electric field pasteurization. *Journal of Food Engineering, 25*(2), 261–281.

131. Zhang, S., Yang, R., Zhao, W., Liang, Q., & Zhang, Z., (2011). The first ESR observation of radical species generated under pulsed electric fields processing. *LWT-Food Science and Technology, 44*(4), 1233–1235.

132. Zhang, Y., Liao, X., Ni, Y., Wu, J., Hu, X., Wang, Z., & Chen, F., (2007). Kinetic analysis of the degradation and its color change of cyanidin-3-glucoside exposed to pulsed electric field. *European Food Research and Technology, 224*(5), 597–603.

133. Zhang, Z. H., Zeng, X. A., Brennan, C. S., Brennan, M., Han, Z., & Xiong, X. Y., (2015). Effects of pulsed electric fields (PEF) on vitamin C and its antioxidant properties. *International Journal of Molecular Sciences, 16*(10), 24159–24173.

134. Zhao, W., & Yang, R., (2008). The effect of pulsed electric fields on the inactivation and structure of lysozyme. *Food Chemistry, 110*(2), 334–343.

135. Zhao, W., Yang, R., & Zhang, H. Q., (2012). Recent advances in the action of pulsed electric fields on enzymes and food component proteins. *Trends in Food Science and Technology, 27*(2), 83–96.
136. Zhao, Y. M., De Alba, M., Sun, D. W., & Tiwari, B., (2019). Principles and recent applications of novel nonthermal processing technologies for the fish industry: A review. *Critical Reviews in Food Science and Nutrition, 59*(5), 728–742.
137. Zimmermann, U., (1986). Electrical breakdown, electro-permeabilization, and electrofusion. In: *Reviews of Physiology, Biochemistry, and Pharmacology* (Vol. 105, pp. 234–256). Berlin, Heidelberg: Springer.
138. Zulueta, A., Barba, F. J., Esteve, M. J., & Frígola, A., (2010). Effects on the carotenoid pattern and vitamin a of a pulsed electric field-treated orange juice-milk beverage and behavior during storage. *European Food Research and Technology, 231*(4), 525–534.
139. Zulueta, A., Barba, F. J., Esteve, M. J., & Frígola, A., (2013). Changes in quality and nutritional parameters during refrigerated storage of an orange juice-milk beverage treated by equivalent thermal and nonthermal processes for mild pasteurization. *Food and Bioprocess Technology, 6*(8), 2018–2030.
140. Zulueta, A., Esteve, M. J., Frasquet, I., & Frígola, A., (2007). Fatty acid profile changes during orange juice-milk beverage processing by high-pulsed electric fields. *European Journal of Lipid Science and Technology, 109*(1), 25–31.

CHAPTER 5

PULSED LIGHT TECHNOLOGY IN FOOD PROCESSING AND PRESERVATION

MAHENDRA GUNJAL, HUMEERA TAZEEN, PAVAN M. GUNDU, PREETI BIRWAL, and ABILA KRISHNA

ABSTRACT

Among the new preservation non-thermal methods of food preservation, pulsed light technology (PLT) has been prominent in the microbial decontamination. This technology decontaminates food and packaging materials with photochemical and photothermal effects. The PLT involves discharging electrical pulses of higher voltage (up to 70 kV/cm) using a flash lamp containing xenon gas for few seconds through the sample between two electrodes. This chapter discusses the principles, components, different equipments, process parameters, operation, mechanism of inactivating microorganisms and action on enzymes and its applications in food processing and packaging.

5.1 INTRODUCTION

Over the past 15 years, pulsed light technology (PLT) has attracted attention as a minimum process for improving microbial safety and extending the shelf-life of processed foods [12, 21, 44, 49]. PLT's other names are: Pulsed ultraviolet (UV) light, High-intensity broad-spectrum pulsed light and pulsed white light [30, 42, 43, 45]. The PLT has been known since 1980, and in 1996 the U. S Food and Drug Administration (FDA) implemented this technology for various food processing unit operations [19]. It requires discharging of high voltage (up to 70 kV/cm) electrical pulses through two electrodes in the food products or on packaging materials for few seconds [1]. Now in the

U.S.A., PLT is used in different foods and packaging materials as an efficient way of killing the microbes present in various foods [22, 36, 52].

This technology utilizes Nobel gasses into a flash lamp that transforms light into the short-lived and higher power electrical pulses, and it requires a very short period to sterilize the products [5]. For the process, radiations are in the spectral range of UV: 200–400 nm, visible (VL): 400–700 nm, and infrared light (IR) 700–1100 nm [4, 32, 37].

PLT involves power utilization to achieve fast and effective disinfection or sterilizing of food products. Nowadays, this technology (UV Light) is extensively used for the sterilization of food-grade packaging materials [8]. The antimicrobial mechanism of pulsed light is to injure to the cell wall and membranes of microbes; however, DNA damage (UV-C region, 200–280 nm) is the main effect of microbial death by using pulsed light [39]. With the help of PLT, one can minimize the use of chemicals for sterilization purpose, and it can be easily incorporated in industrial sectors in different processing lines without much modification [13].

This chapter focuses on the applications of PLT in the processing and preservation of food products.

5.2   THEORY OF PULSED LIGHT

5.2.1   PRINCIPLE OF PULSED LIGHT TECHNOLOGY (PLT)

The waves of electromagnetic radiations have different wavelength, frequency, and also its energy, which is emitted and propagated. The radiation of light in use is UV, visible, and infrared, varying from 200 to 1100 nm in wavelength. Light can be produced from different sources with different mechanisms, due to the spontaneous transition of certain atoms from such an excited state to a condition of low energy [38]. This energy is released as single, zero-mass packets (photons), the energy of which is calculated as below:

$$E = hv = \frac{hc}{\lambda} \tag{1}$$

where; h is the Planck constant, c is the speed of light in a vacuum and λ is frequency.

It can be observed from Eqn. (1) that the energy released is at the lower wavelength (λ) and the higher frequency (v). UV (particularly UVC) light has more energy than visible or infrared rays, that is why, UV rich treatments are more successful than other treatments, such as light-based technology and also PLT.

If light radiation of energy E_0 falls on the outside of a material body (food or packaging materials), then a portion of its energy (rE_0, whereby γ is materials reflection coefficient) is reflected by materials of top layers and captured by layers by which it penetrates and part of it is transferred to an innermost layers [38].

According to the Lambert-Beer law, the energy $E(x)$ of light transmitted from to a distance 'X' below inside the surface of a material body decreases with 'X' as shown below [37]:

$$E(x) = (1-r)E_0 e^{-[\alpha]X} \tag{2}$$

The opacity and transparency of the material for the λ is calculated using Eqn. (2) during which the extinction coefficient is indicated by α. Lot of liquids and every one of the gasses are transparent ($\alpha \to 0$) and that they do not absorb any amount of energy. Most of the solids are opaque ($\alpha \to \infty$), they do not transmit radiations. While penetrating bulk materials (include food) the asperity of light rapidly decreases. The absorbed energy E_d by a layer at inside distance x is given by:

$$E_d = E(x)\left(e^{-[\alpha]d}\right) \tag{3}$$

The temperature increases due to heat dissipated by the absorbed light energy and it is given by:

$$\Delta T = \frac{E_d}{\rho C_p A d} \tag{4}$$

where; ρ, A, and Cρ stand for density, the surface area and specific heat of the material, respectively.

The conductive heat transfer of heat inside the material is caused due to ΔT (the difference in temperature) between both the materials outer and inner layers. The rate of both heat transfer and material temperature increases depending on the time duration of the incident radiation, intensity, and thermal properties of materials.

The energy is calculated in kJ/m² that is delivered to a unit material surface defined by the energy density or fluence F that measures the radiation effects on the material body. The light can be supplied continuously or in pulse form.

Besides the number of pulses period, the power provided to pulses is higher than the same amount of energy provided by continuous light (CL) radiation. The higher pulse power, which is important characteristics of the energy supplied in the form of pulses, shows a much higher penetration capacity through the materials [15].

5.2.2   PRINCIPLE OF OPERATION OF PULSED LIGHT TECHNOLOGY (PLT)

The effects of the shorter time duration of light pulses with minimum time available to thermal conductivity within the material result in very quick heat on the surface layer compared to the constant temperature obtained by the CL radiation of the same total energy without changing the bulk temperature [14, 15, 38]. The pulsed light system primarily involves generating electrical pulses of higher power and converting them into light pulses of a higher power. The general working process of pulsed light system is shown in Figure 5.1. Continuous low-power energy is:

- Collected from a source of primary energy;
- Collected and stored partially;
- Quickly released and processed into large-power pulsed electrical energy, which is then transformed into high-power pulsed light energy, and finally;
- Supplied to the desired target.

The information on this process is explained below:

- Any type of power of electrical supplier is often used to transform low-voltage AC power into the high-voltage DC power.
- Energy mostly storage into the capacitor bank, which is a collection of parallel-connected high-voltage capacitors, which collect energy from the electrical power supplier during the charging phase and its release during the discharge phase, thereby supplying the high amounts of current. Additionally, Marx generators may be used in it, which vary from capacitor banks only during the discharged phase when all capacitors are linked temporarily in series; thus, Marx generators also work as voltage amplifiers providing large amounts of higher-voltage current [35].

POWER LINE
Lower-power lower-voltage continuous
electrical current at low AC

**ELECTRIC ENERGY SUPPLY
(CONVERTER)**
low power high voltage high DC
continuous electric current

**ELECTRICAL ENERGY STORAGE
(CAPACITORS)**
Low power high voltage high DC
continuous electric current

**ELECTRIC PULSE FORMING
(SWITCHES)**
High power high voltage high DC pulse
electric current

**PLUSED LIGHT SOURCE
(INERT-GAS FLASH LAMPS)**
High power pulsed light

TARGET MATERIAL

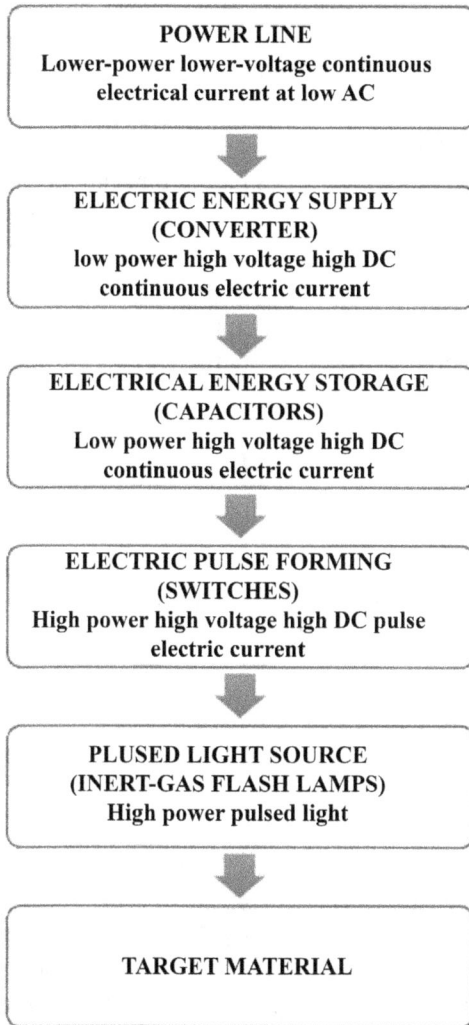

FIGURE 5.1   Flow of a general pulsed light system.

- The transformation of a continuous low power into the higher pulsed electric power is achieved by means of specialized switches able to handle very high power and performing opening/closing cycles of very short period passing instantly from a perfect condition of insulation to a perfect conducting of conduction. The operation of the switches is controlled by the controller, which determines the shape of the pulse and the conditions of electrical operation to provide a yield of optimum pulsed light wavelength for a specific application [35].

- Use of gas-filled flash lamps or other pulsed light sources. The higher-power pulsed electric energy supplied by switches is typically converted to high-power light pulses. The current connected with the higher-power electrical pulses passes through the gas in lamp, passing energy to certain gas atoms that are transported in an 'excited-state'; later, they usually return spontaneously in situations of lower energy-giving in the form of intense light pulses [38].
- Depending on the different applications, the energy produced is finally delivered to the target by the various systems.

5.2.3   INACTIVATION MECHANISM

The fatal effects of pulsed light vary with different wavelengths. Therefore, full-spectrum or high-intensity wavelengths are applied to treat the foods. The efficacy of pulsed light application for microbial inactivation relays with the distinctive effects of broad-spectrum and high peak power the flash. Due to higher energy levels at the short wavelengths of the UV region (200–320 nm); they are more efficient when compared to higher wavelengths [18].

Light pulses lead to photochemical or photothermal reactions in foods. Effects of pulsed light on microbial inactivation are basically conciliated through absorption by highly conjoined carbon-to-carbon double bond structures in nucleic acids and proteins. Chemical modifications and cleavage of the DNA on pulsed light treatment gives the antimicrobial effects for this novel application [3]. As DNA is a target molecule for the UV wavelengths at 200–320 nm, therefore it is considered as one of the primary reason for microbial termination and inactivation through structural modification of the DNA. Generally, traditional UV application affects DNA by the phenomenon that is reversible under certain conditions. The cell repairing system is divided as either "light enzymatic repair" or "dark enzymatic repair" [40]. Experiments conducted in this domain has showcased that, enzymatic repair of DNA does not occur after pulsed light treatment.

The effects of pulsed light on cell membranes and other cellular components probably take place simultaneously with the destruction of nucleic acids. The extent of the damage exhibited by pulsed light may also be too intensive for the repair phenomenon to get affected. Pulsed light at higher intensity significantly increases the mechanisms of cellular components destruction caused by individual wavelengths of light at higher energy levels. The overall damage caused by the broad-spectrum light is thought to induce extensive irreversible damage to DNA, proteins, and other macromolecules [3].

5.2.4   PROCESSING METHOD AND EQUIPMENTS

The major components of any pulsed light system consist of a lamp and the power unit. The power unit is used to generate high voltage; and this, in turn, is used to produce high current pulses in the lamp. At first, the AC power is transmuted into high voltage DC power, which in turn is used to charge the capacitor. When the capacitor gets charged to a certain voltage, a high-voltage switch discharges the charge from the capacitor to the lamp. The complete process of the system is adequately contained and sealed to protect the personnel from high voltages. In addition, cooling water is used to reduce the heating of the treated product.

Treatment unit consists of one or more inert gas lamps. Once high current pulses are introduced to the lamps, an intense pulse of light is emitted by the gas contained within the lamp. The number of lamps, frequency of flashing, and flashing configuration in the system largely depend on the mode of application. Surveilling of the light system is extremely crucial to assure that the treatment chamber is properly maintained. The lamps current and the lamps output (fluence) are the two types of diagnostic monitors used in the system. Lamp fluency is assessed to assure whether the lamp emits adequate UV radiation to inactivate the microbes.

In addition, silicon photodiode is used in the system, which is capable of detecting whether the lamp has the required output of UV light. A decreasing output would indicate the need for replacement of the lamp. This monitoring unit is needed in the system to stop the process if products do not receive any treatment above some predetermined threshold [28, 34]. A secondary monitoring unit is also adapted in the system to measure the lamp current for every flash. Here, the current denotes the intensity and spectrum of the radiation. The application device includes an illuminated treatment chamber with $0.1–3$ J/cm^2 per flash, with overall cumulated fluence of $0.1–12$ J/cm^2. Generally, flashes are applied at a rate of $0.5–10$ Hz for several hundred microseconds.

Although several private firms have already developed pulsed light devices for applications other than food processing and preservation, yet PLT is not incorporated at an industrial level in the food sector [15].

5.3   PULSED LIGHT TECHNOLOGY (PLT): FACTORS AFFECTING THE INACTIVATION OF MICROORGANISMS

5.3.1   TYPES OF MICROORGANISMS

The optical properties of cells, such as the absorption of light and degree of scattering are the major factors, which influence the efficiency of pulsed light application. The incident beam of light undergoes refraction due to variation in optical density between the product and the surrounding air. Meanwhile, certain microorganisms also exhibit resistance to pulsed light treatments [49].

5.3.2   RELATIONSHIP AMONG LIGHT, SUBSTRATE AND MICROBIAL CELLS

This factor is very crucial from the point of view of the efficiency of pulsed light application. The extent of reflection, refraction, scattering, and absorption of the light depends on the constituents of the medium and on the wavelength of the incident light. Meanwhile, the reflection and refraction of light has been noted as indispensable for surface treatments. The phenomenon of refraction is particularly relevant for transparent and colored food materials, whereas reflection is the prevailing phenomenon for opaque food materials.

Diffused or specular reflection can occur depending on the roughness or smoothness of the surface of the products. The incident light gets reflected on the smooth surface and exhibits at the same angle of the incident beam, along with the same spectral distribution of energy, which is referred to as specular reflection. On the other hand, incident light is partially absorbed, when light travels through the outer layers of the rough surface material; and this phenomenon is referred to as diffuse reflection.

However, the reflection of light significantly decreases the efficacy of the pulsed light application. Spectral distribution of incident and the diffused light spreads out in all directions due to variation in the absorption at different wavelengths. Some portion of the incident light interacts with the internal structure of the translucent materials and leads to the multiple internal reflections, redirections that lead to scattering effect. In addition, scattering, and absorption are the most relevant types of light-substrate interactions in the biological tissues.

5.3.3   THE DISTANCE FROM THE SOURCE LIGHT

As the substrate depth increases and the distance from the light source increases, then the scattering and absorption of light decrease. This is due to a decrease in the intensity of light when it passes through the substrate. The quantitative distribution of light dose within a substrate is defined by the depth of optical penetration that represents the distance over which light decreases to 37% of its initial value. Depth of optical penetration depends on wavelength variation, and it was noted that shorter-wavelength has better penetration in foods compared to longer wavelengths [11].

5.3.4   DESIGN OF PULSED LIGHT EQUIPMENT

The design of pulse electric light equipment depends from manufacturer to manufacturer. An important part of the equipment is the flash lamp as it converts 45–50% of the electrical energy into pulsed radiant energy. It is loaded with either xenon or krypton, which are inert gasses. Mostly xenon is used as it has higher energy conversion efficiency and proved to suppress microbial activities. The envelope, in flash lamps (two main structural components) that seals and electrodes, which is made with clear fused quartz called suprsil, with a thickness of 1 mm.

The metal electrode is deferred to each end of the envelope, which in turn is connected to high voltage charged capacitors to provide electric current. Cathode is another important component of the flash lamp as it determines the lifetime of the flash lamp. The function of the electrode is to provide an excessive and adequate quantity of electrons; and due to sputtering, it may lead to hot spots engendered during peak power supply that may increase the chances of corrosion on cathode material. Anode on the other end should have adequate mass or surface area to withstand power loading, which induces electrons bombardment from the electrical arc; and all of the assemblies need to be properly sealed. Rod seal, ribbon seal, or solder seal is frequently used.

Other problems may occur during the operation, such as overheating that can be avoided using cooling devices (cooling fans). Other types of pulsed light sources in use are solid-state Marx generator for pulsing an UV lamp used for microbial inactivation, static discharge lamps with spectral outputs similar to the flash lamps, and sparker technology to generate light and sonic sound pulse [20].

5.4   EFFECTS OF PULSED LIGHT TECHNOLOGY (PLT) ON FOODS AND PACKAGING MATERIALS

The effects of PLT on microorganisms and their related parameters on microorganisms are listed in Table 5.1. FDA has provided preliminary requisites for optimizing the pulsed light treatments, which indicates that "*any type of food materials treated with pulsed light shall receive the mild treatments plausibly required to accomplish the intended effect.*"

In designing of Pulsed light treatments for the foods and packaging materials items, both sources (duration and number of the pulses, the interval between pulses, as light wavelength, energy density) and target (as the product transparency, color, size, smoothness, and cleanliness of surface) are very important parameters for process optimization to increase effectiveness towards microbial inactivation and to minimize the products properties diversion [38]. The diversion in the properties of foods due to Ultraviolet C (UVC) can mostly be determined by excessive temperature increase causing thermal damage to the food materials that could result in some of the undesired photochemical damage to foods or foods packaging materials.

TABLE 5.1   Effects of Pulsed Light on Solid and Liquid Food Materials, Some Food Items and Microbiological Media

Material	Experimental	Results/Remarks	References
Cake	BSPL $F = 1.5$ J/cm^2 n = 1–16	lcr of *Aspergillus niger* sp. (inoc.) ranging from 3 to 6 with n ranging from 2 to 6; Shelf-life increased from 26 days for untreated samples to 6 months for treated samples	[31]
Chicken wings	BSPL	2 lcr of *Salmonella* (inoc.)	[16]
Clover honey	UVPL $F = 5.6$ J/cm^2 n = 15–540	Samples 2 mm deep: reduction of sp. of *Clostridium sporogenes* (inoc.) ranging from 39.5 to 73.9% with n ranging from 135 to 405. Samples 8 mm deep: reduction of sp. of *Clostridium sporogenes* (inoc.) ranging from 0 to 89.4% with n ranging from 15 to 540.	[23]
Commercial or raw eggs	BSPL $F = 0.5$ J/cm^2 n = 8	Up to 8 lcr of *Salmonella enteriditis* (inoc.); Inactivation effect observed on eggshells and a little extended into the egg pores.	[14]

TABLE 5.1   *(Continued)*

Material	Experimental	Results/Remarks	References
Curds of dry cottage cheese	BSPL $F = 16$ J/cm² n = 1–2	1.5 lcr of *Pseudomonas sp.*[c]	[15]
Dry PE surfaces	BSPL F up to 1.5 J/cm² n = 1	Lcr[a] of *Bacillus pumilus* sp. (inoc.) ranging from 1 to 5.5 with F ranging from 0.26 to 1.5; lcr of *Bacillus subtilis* sp. (inoc.) ranging from 1 to 5.5 with F ranging from 0.28 to 1.5; lcr of *Bacillus stearothermophilus* sp. (inoc.) ranging from 1 to 5 with F ranging from 0.29 to 1.5; lcr of *Aspergillus niger* sp. (inoc.) ranging from 1 to 7 with F ranging from 0.12 to 1	[2]
Eggshells	BSPL $F = 1.5$ J/cm² n = 1–6	lcr of *Bacillus subtilis* sp. (inoc.) ranging from 3 to 6 with n ranging from 2 to 6	[31]
Flowing water	BSPL $F = 0.25$ J/cm² n = 1	4.6 lcr of *Cryptosporidium parvum* (inoc.) 7.5 lcr of *Klebsiella terrigena* (inoc.) 4.9 lcr of Simian rotavirus SA11 (inoc.) 6.2 lcr of Poliovirus type 1 (inoc.) 4.3 lcr of Bacteriophage MS-2 (inoc.) 5.4 lcr of Bacteriophage PRD-1 (inoc.) 7.7 lcr of Pseudomonas fluorescens (inoc.) 6.0 lcr of *Bacillus stearothermophilus* (inoc.)	[2]
Frankfurters	BPSL F up to 30 J/cm²	2 lcr of *Lysteria innocua* (inoc.)	[16]
Freshly baked cakes packaged in clear plastic containers	BSPL $F = 16$ J/cm² n = 3	Under controlled condition, molds are absent in treated samples after the storage at room temperature for 11 days and untreated samples are more moldy.	[15]
Hard crusted white bread rolls	BSPL $F = 16$ J/cm² n = 1–2	1.5 lcr of mold sp. with n = 1 and 2.7 lcr of mold sp. with n = 2	[15]

TABLE 5.1   *(Continued)*

Material	Experimental	Results/Remarks	References
HDPE Prepackaged catfish filets	BSPL $F = 0.25–0.50$ J/ cm^2 n = 2–4	Psychotropic (PPC) and coliform (TCC) bacteria were not reduced initially by any treatment. After one week of storage, PPC were 1 (in treated samples with $F = 0.25$) or 2 (in treated samples with F = 0.50) log cfu/g lower than untreated samples; TCC were reduced from about 50 to less than 10 cfu/g	[46]
Liquids in plastic recipients	BSPL $F = 4.5$ J/cm^2 n = 2	About 6 lcr of a variety of inoc[b]. sp.[c] and veg.[d] microorganisms	[31]
Meat	BSPL	Reduction of *Listeria* and *Salmonella* population	[41]
Microbiological media	BSPL $F = 0.05–12$ J/cm^2 n = 1–35	10 lcr[a] of *Escherichia coli* (inoc.[b]) with $F = 1.5$ and n = 2 or with $F = 4$ and n = 1 10 lcr of *Bacillus subtilis* veg.[d] (inoc.) with $F = 1$ and n = 4 or with $F = 4$ and n = 2 10 lcr of *Bacillus subtilis* sp.[c] (inoc.) with $F = 1.5$ and n = 2 or with $F = 4$ and n = 1 8 lcr of *Staphylococcus aureus* (inoc.) with $F = 0.2$ and n = 4, 10 lcr with $F = 0.75$ and n = 2 10 lcr of *Saccharomyces cerevisiae* (inoc.) with $F = 0.4$ and n = 4 10 lcr of *Aspergillus niger* sp. (inoc.) with $F = 4$ and n = 4 or with $F = 12$ and n = 1 VPL was far less effective in all cases, requiring much more F and n for comparable lcr	[15]
Microbial media	BSPL with either low or high content of UV n = 100–200	lcr of *Escherichia coli, Listeria monocytogenes, Salmonella enteriditis, Pseudomonas aeruginosa, Bacillus cereus* and *Staphylococcus aureus* ranging from 3 to 6 with n ranging from 50 to 300; with high-UV light no significant lcr with any n of low-UV light	[43]

TABLE 5.1 *(Continued)*

Material	Experimental	Results/Remarks	References
Microbiological media	PL with λ = 200–530 nm F = 3 J/cm² n = 1–512	lcr of *Escherichia coli* and *Listeria monocytogenes* ranging from 1 to 5 with n ranging from 64 to 512; lcr of *Escherichia coli* O157:H7 ranging from 1 to 5 with n ranging from 16 to 512	[29]
Microbiological media	UVPL n = 1–4	lcr of Bacillus subtilis sp. (inoc.) ranging from about 2 to about 5 with n ranging from 1 to 4	[47]
Packed slices of bread	BSPL F = 1.5 J/cm²	Shelf-life increased from 16 days for untreated samples to 5 months for treated samples	[31]
Packed white bread slices	BSPL	Freshness of treated samples is more than two weeks as compared to untreated samples and without growth surface on mold	[41]
PET pieces and glass plates	BSPL or VPL F up to 5 J/cm² n up to 5	BSPL: lcr of *Aspergillus niger* sp. ranging from 2 to 6 with F ranging from 1 to 5 and n = 1 VPL: lcr of *Aspergillus niger* sp. ranging from 0.2 to 1.3 with F ranging from 1 to 5 and n = 1 BSPL: lcr of Bacillus subtilis sp. ranging from 1.7 to 4.6 with F ranging from 0.25 to 5 and n = 1 VPL: lcr of *Bacillus subtilis* sp. 1 with F ranging from 1 to 5 and n = 1	[51]
Physiological solutions in PE pouches	BSPL n = 8	>6 lcr of *Clostridium sporogenes* sp. (inoc.) and *Bacillus pumilus* sp. (inoc.)	[10]
Plastic lids and cup surfaces	BSPL F = 0.25–3 J/cm² n = 1	No growth of *Staphylococcus aureus* (inoc.[b]) with F>1 No growth of *Bacillus cereus* sp.[c] (inoc.) with F >1.75 No growth of *Aspergillus niger* (inoc.) with F >2	[15]
Potable and ingredient water	BPSL	6–7 lcra/ml of *Klebsiella terrigena* with F = 0.5 and n = 2 6–7 lcr/ml of *Cryptosporidium parvum* with F = 1 and n = 1	[16]

TABLE 5.1   *(Continued)*

Material	Experimental	Results/Remarks	References
Retail meat	BSPL	1–3 lcr of total aerobic, lactic, enteric bacteria and Pseudomonas	[16]
Shrimp	BSPL $F = 1$–2 J/cm^2 n = 4–8	1–3 lcr[a] of *Listeria* (inoc.[b]), resulting in a shelf-life extension of 1 week versus untreated samples	[15]
Transparent model liquids	BSPL n = 1–8	6–8 lcr of *E. coli, Staphylococcus aureus* and *E. faecalis* with n = 1 lcr of ascospores of *Aspergillus niger* ranging from 3 to 4 with n ranging from 1 to 2 lcr of *Bacillus subtilis* sp. ranging from 5 to 6 with n ranging from 1 to 2	[48]
Washing water	BSPL	>7 lcr of a microbial population (inoc.) with $F_{tot} = 6$	[31]
Water in PE containers	BSPL $F = 1$ J/cm^2 n = 10–20	>6 lcr of *Aspergillus niger* (inoc.) *Bacillus subtilis* sp. (inoc.), *Bacillus pumilus* sp. (inoc.)	[10]
Water in plastic bottles	BSPL	About 5 lcr of *Aspergillus niger* (inoc.) and *Bacillus stearothermophilus* (inoc.) with $F_{tot} = 3$	[31]
Wax-coated strawberries	BSPL $F = 0.5$ J/cm^2 n = 4	No mold growth was observed when stored for two weeks at room temperature	[14]
White plastic packaging materials	BSPL $F = 0.34$–1.3 J/cm^2 n = 1–40	No survivors of inoc. population of *Aspergillus niger* with $F = 1.3$ and n = 1, 0.48 log survivals with $F = 1$ and n = 1, 0.5 log survivals with $F = 0.75$ and n = 7, 1.67 log survivals with $F = 0.53$ and n = 20, 1.73 log survivals with $F = 0.3$ and n = 40	[10]

Legend: a = log cycle reduction; b = inoculated; c = spores; d = vegetative.

The changes in sensory quality induced by intense light pulses in meat products depend on the animal's breed, meat type, and intense light pulse dose applied. The changes in the odor of the meat, poultry significantly changes after the pulsed light treatment [26]. PLT has been significantly effective for the reducing the activity of various types of enzymes (such as oxidoreductases, hydrolases, lipases, isomerases, proteinases, etc.), that are presented in food products (fruit, vegetables, meats, fish, and shellfish) having the surface depth of 0.1 mm. In addition, the chances of a reduction in vitamin B_2 (riboflavin) and also other micronutrients remain same but greatly reduce the spoilage enzymes and no effects on the sensory properties in fruits and vegetables [15].

5.5   APPLICATIONS OF PULSED LIGHT TECHNOLOGY (PLT)

The use of PLT as short time and high-power light pulses has been a combination of photochemical and photothermal effects to inactivate microbes in foods. The PLT achieves efficient bacterial inactivation with much lower processing time compared to other treatments like conventional methods. Thus, it helps to preserve the nutritional as well as sensory parameters of the foods. This technology requires high investment, so it gets difficult to use the technology for all types of food products. However, the main limitation of this technology is its low penetrating power and the requirement of food products to be treated for transparency and smoothness. Therefore, PLT finds applications in the areas of primary surface sterilization or decontamination of solid unpacked food materials, packaged solid food requiring terminal sterilization or decontamination using pulsed light-compatible packages, unpacked liquid foods flowing through the treatment chambers and in liquid foods packaged into pulsed light-compatible packages [25].

The interesting application of PLT is for sterilization of packaging films and materials, especially, which is used in aseptic technology, as a practical alternative to using hydrogen peroxide. The core applications of PLT have been tested successfully for the breadsticks, pizza, chocolate cupcakes, tortillas, and bagels resulting in its longer shelf-life by retarding the mold growth. Another application of this technology is in meat and meat poultry products where, it helps to increase the shelf-life.

5.6   ADVANTAGES OF PULSED LIGHT TECHNOLOGY (PLT)

PLT helps to retain the quality and nutrient of the food materials as its light intensity is 20,000 more compared to sunlight; and importantly being a non-thermal technique, it finds its applications in many foods [6]. PLT is eco-friendly compared with mercury vapor lamps as in UV treatments [35]. The other thermal and non-thermal treatments need the addition of some chemical preservatives and also ionizing radiation, but PLT does not involve any toxic chemicals and photolytic by-products (The used wavelengths can too cause ionization of small molecules) [38]. The pulsed white light is not accurately non-thermal but it is of shorter duration of time showing no undesirable effects on the nutritional profile [33].

5.7   DISADVANTAGES OF PULSED LIGHT TECHNOLOGY (PLT)

A possible problem with the method of preserving PLT is fold or fissure in food products, which can protect microorganisms by being exposed to pulsed light [6]. Some strains of microorganisms (such as *Listeria monocytogenes*) are resistant to the pulsed light treatment [7]. Pulsed light technologies are more useful in the case of liquid food products and for decontamination of foods. Pulsed light treatment is the possibility of shadowing occurrence, when microorganisms absorb the rays, as in case of *Aspergillus niger* and are present one upon the another and it makes organisms in the lower layers very hard to destroy in contrast to those in the top of the layer [24]. However, the use of relatively higher peak powers could overcome the effect of shadowing.

5.8   SUMMARY

PLT is a non-thermal processing technology, where heat and chemicals are not used during processing for food products to inactivate microorganisms and enzymes. Similar to other non-thermal techniques, PLT has a negligible effect on food products, which may affect organoleptic properties and nutritional properties. The utilization of PLT to inactivate the microbe's even bacterial cells has gained high acceptance in the medical and food processing industries compared to the thermal technology. Pulsed light obtained from xenon flash lamp is mercury-free and does not require a warm-up period (i.e., they are instant-on). The areas that need more research on the PLT are:

equipment manufacturing for industries, development of specific processes and standards, and clarify the consumer misunderstanding.

KEYWORDS

- decontamination
- infrared light
- non-thermal technology
- preservation
- pulsed light
- sterilization

REFERENCES

1. Abida, J., Rayees, B., & Masoodi, F. A., (2014). Pulsed light technology: A novel method for food preservation. *International Food Research Journal, 21*(3), 115–120.
2. Anonymous PureBright® Sterilization Systems, (2000). PurePulse Technologies; Now Maxwell Technologies, San Diego, USA, p. 19.
3. Barbosa-Canovas, G. V., Schaffner, D. W., Pierson, M. D., & Zhang, Q. H., (2000). Pulsed light technology. *Journal of Food Science, 65*(8), 82–85.
4. Bhavya, M. L., & Umesh, H. H., (2017). Pulsed light processing of foods for microbial safety. *Food Quality and Safety, 1*(3), 187–202.
5. Bohrerova, Z., Shemer, H., Lantis, R., Impellitteri, C. A., & Linden, K. G., (2008). Comparative disinfection efficiency of pulsed and continuous-wave UV irradiation technologies. *Water Research, 42*(12), 2975–2982.
6. Brown, C. A., (2008). *Understanding Food: Principles and Preparation* (3rd edn., p. 546). Boston - MA: Thomson Learning (Cengage now).
7. Caminiti, I. M., Noci, F., Muñoz, A., & Whyte, P., (2011). Impact of selected combinations of non-thermal processing technologies on the quality of an apple and cranberry juice blend. *Food Chemistry, 124*(4), 387–1392.
8. Cerny, G., (1977). Sterilization of packaging materials in aseptic packaging. *Verpackungs Rundschau [Packing Rundschau], 28*(10), 77–82.
9. Cheigh, C. I., Hwang, H. J., & Chung, M. S., (2013). Intense pulsed light (IPL) and UV-C treatments for inactivating *Listeria* monocytogenes on solid medium and seafoods. *Food Research International, 54*(1), 745–752.
10. Clark, R. W., Lierman, J. C., Lander, D., & Dunn, J. E., (2003). *U.S. Patent No. 6,566,659* (p. 21). Washington, DC: U.S. Patent and Trademark Office.
11. Dagerskog, M., & Osterstrom, L., (1979). Infra-red radiation for food processing. Part I: A study of the fundamental properties of infra-red radiation. *Lebensmittel-Wissenschaftu and Technologie (Food Science and Technology), 12,* 237–242.

12. Darghahi, M. S. Y. H., (2006). Inactivation of pathogenic bacteria using pulsed UV-light and its application in water disinfection and quality control. *Acta Medica Iranica, 44*(5), 305–309.
13. Dos, S. A. J. G., (2019). Pulsed light treatment in food. *Chemical Reports, 1*(1), 108–111.
14. Dunn, J., (1996). Pulsed light and pulsed electric field for foods and eggs. *Poultry Science, 75*(9), 1133–1136.
15. Dunn, J. E., Clark, R. W., Asmus, J. F., Pearlman, J. S., Boyer, K., Painchaud, F., & Hofmann, G. A., (1989). *U.S. Patent No. 4,871,559* (p. 18). Washington, DC; U.S. Patent and Trademark Office.
16. Clark, W., & Ott, T., (1995). Pulsed-light treatment of food and packaging. *Food Technology, 49*(9), 95–98.
17. Elmnasser, N., Guillou, S., Leroi, F., Orange, N., Bakhrouf, A., & Federighi, M., (2007). Pulsed-light system as a novel food decontamination technology: A review. *Canadian Journal of Microbiology, 53*(7), 813–821.
18. Farkas, J., (1997). Physical methods of food preservation. In: Doyle, M., & Beuchat, L., (eds.), *Food Microbiology: Fundamentals and Frontiers* (3rd edn., pp. 685–712). Washington, DC: ASM Press.
19. Food and Drug Administration (FDA), (1996). *Pulsed Light for the Treatment of Food* (p. 15). Report 21CFR179.4.
20. Giese, N., & Darby, J., (2000). Sensitivity of microorganisms to different wavelengths of UV light: Implications on modeling of medium pressure UV systems. *Water Research, 34*(16), 4007–4013.
21. Gomez-Lopez, V. M., Ragaert, P., Debevere, J., & Devlieghere, F., (2007). Pulsed light for food decontamination: A review. *Trends in Food Science and Technology, 18*(9), 464–473.
22. Heinrich, V., Zunabovic, M., Bergmair, J., Kneifel, W., & Jaeger, H., (2015). Post-packaging application of pulsed light for microbial decontamination of solid foods: A review. *Innovative Food Science and Emerging Technologies, 30*, 145–156.
23. Hillegas, S. L., & Demirci, A., (2003). *Inactivation of Clostridium Sporogenes: Clover Honey by Pulsed UV-Light Treatment* (p. 6). Presented at 2003 ASABE Annual Meeting: American Society of Agricultural and Biological Engineers.
24. Hiramoto, T., (1984). *U.S. Patent No. 4,464,336* (p. 17). Washington, DC: U.S. Patent and Trademark Office.
25. https://www.myprocessexpo.com/blog/expert-in-residence/past-present-future-light-basedtechnologies-foods/ (accessed on 27 January 2021).
26. Ignat, A., Manzocco, L., Maifreni, M., Bartolomeoli, I., & Nicoli, M. C., (2014). Surface Decontamination of fresh-cut apple by pulsed light: Effects on structure, color, and sensory properties. *Postharvest Biology and Technology, 91*, 122–127.
27. Jun, S., Irudayaraj, J., Demirci, A., & Geiser, D., (2003). Pulsed UV light treatment of cornmeal for inactivation of *Aspergillus niger* spores. *International Journal of Food Science and Technology, 38*(8), 883–888.
28. Luksiene, Z., Buchovec, I., Kairyte, K., Paskeviciute, E., & Viskelis, P., (2012). High-power pulsed light for microbial decontamination of some fruits and vegetables with different surfaces. *Journal of Food Agriculture Environment, 10*(3/4), 162–167.
29. MacGregor, S. J., Rowan, N. J., McIlvaney, L., Anderson, J. G., Fouracre, R. A., & Farish, O., (1998). Light inactivation of food-related pathogenic bacteria using a pulsed power source. *Letters in Applied Microbiology, 27*(2), 67–70.

30. Marquenie, D., Michiels, C. W., Van, I. J. F., Schrevens, E., & Nicolaï, B. N., (2003). Pulsed white light in combination with UV-C and heat to reduce storage rot of strawberry. *Postharvest Biology and Technology, 28*(3), 455–461.

31. Mimouni, A., (2000). Applications de la lumière pulsée en agroalimentaire: Process alimentaire (Applications of pulsed light in agricultural foods: Food processing). *Industries Alimentaires Et agricoles (Agricultural Food Industries), 117*(10), 37–39.

32. Moreaua, M., Nicorescua, I., Turpina, A. S., Agoulonb, A., Chevaliera, S., & Orangea, N., (2011). Decontamination of spices by using a pulsed light treatment. *Food Process Engineering in a Changing World, 2011*, 22–26.

33. Ohlsson, T., & Bengtsson, N., (2002). *Minimal Processing Technologies in the Food Industry* (pp. 34–57). Cambridge, England: Woodhead Publishing.

34. Ortega-Rivas, E., (2012). Pulsed light technology. In: *Non-Thermal Food Engineering Operations* (pp. 263–273). Boston, MA: Springer.

35. Pai, S. T., & Zhang, Q., (1995). *Introduction to High Power Pulse Technology* (Vol. 10, pp. 237–277). World Scientific Publishing Company.

36. Palgan, I., Caminiti, I. M., Muñoz, A., Noci, F., Whyte, P., Morgan, D. J., & Lyng, J. G., (2011). Effectiveness of high intensity light pulses (HILP) treatments for the control of *Escherichia coli* and *Listeria innocua* in apple juice, orange juice and milk. *Food Microbiology, 28*(1), 14–20.

37. Palmieri, L., Cacace, D., & Dall'Aglio, G., (1999). Non-thermal methods of food preservation based on electromagnetic energy. *Food Technology and Biotechnology, 37*(2), 145–149.

38. Palmieri, L., & Cacace, D., (2005). High intensity pulsed light technology. In: *Emerging Technologies for Food Processing* (pp. 279–306). Boston - USA: Academic Press.

39. Pedrós-Garrido, S., Condón-Abanto, S., Clemente, I., Beltrán, J. A., Lyng, J. G., Bolton, D., & Whyte, P., (2018). Efficacy of ultraviolet light (UV-C) and pulsed light (PL) for the microbiological decontamination of raw salmon (*Salmo salar*) and food contact surface materials. *Innovative Food Science and Emerging Technologies., 50,* 124–131.

40. Pure Pulse Technologies, (1999). *Pure Bright COOLPURE Advanced Sterilization, Decontamination and Preservation Technology for the Food and Food Packaging Industry.* San Diego, CA: PurePulse Technologies Inc.; Brochure; http://www.packaging2000.com/purepulse/Purepulse.html (accessed on 27 January 2021).

41. Rice, J., (1994). Sterilizing with light and electrical impulses. *Food Processing, 7,* 66.

42. Roberts, P., & Hope, A., (2003). Virus inactivation by high intensity broad spectrum pulsed light. *Journal of Virological Methods, 110*(1), 61–65.

43. Rowan, N. J., MacGregor, S. J., Anderson, J. G., Fouracre, R. A., McIlvaney, L., & Farish, O., (1999). Pulsed-light inactivation of food-related microorganisms. *Applied and Environmental Microbiology, 65*(3), 1312–1315.

44. Rowan, N. J., (2019). Pulsed light as an emerging technology to cause disruption for food and adjacent industries-quo Vadis. *Trends in Food Science and Technology, 88,* 316–332.

45. Sharma, R. R., & Demirci, A., (2003). Inactivation of *Escherichia coli* O157: H7 on inoculated alfalfa seeds with pulsed ultraviolet light and response surface modeling. *Journal of Food Science, 68*(4), 1448–1453.

46. Shuwaish, A., Figueroa, J. E., & Silva, J. L., (2000). Pulsed light treated pre-packaged catfish fillets. In: *IFT Annual Meeting* (pp. 10–14).

47. Sonenshein, A. L., (2003). Killing of Bacillus spores by high-intensity ultraviolet light. In: *Sterilization and Decontamination Using High-Energy Light* (pp. 15–19). Xenon Corporation, Woburn, Mass.

48. Tonon, F., & Agoulon, A., (2003). Lumiere pulse, principe et application au cas des solutions liquides (Pulsed light: principles of applications in liquid solutions). In: *Industries Agro-Alimentaires, La Conservation De Demain (Agricultural Food Industries)* (4th edn., pp. 239–258). Talence, France: Academic Press.

49. Turtoi, M., & Nicolau, A., (2007). Intense light pulse treatment as alternative method for mould spores destruction on paper-polyethylene packaging material. *Journal of Food Engineering, 83*(1), 47–53.

50. Vladimirov, Y. A., Roshchupkin, D. I., & Fesenko, E. E., (1970). Photochemical reactions in amino acid residues and inactivation of enzymes during UV irradiation: A review. *Photochemistry and Photobiology, 11*(4), 227–246.

51. Wekhof, A., Trompeter, F. J., & Franken, O., (2001). Pulsed UV disintegration (PUVD): A new sterilization mechanism for packaging and broad hospital applications. In: *The First International Conference on Ultraviolet Technologies* (pp. 1–15). New York: USA.

52. Zhang, Z. H., Wang, L. H., Zeng, X. A., Han, Z., & Brennan, C. S., (2019). Non-thermal technologies and its current and future application in the food industry: A review. *International Journal of Food Science and Technology, 54*(1), 1–13.

CHAPTER 6

POTENTIAL OF GREEN NANOTECHNOLOGY IN FOOD PROCESSING AND PRESERVATION

SHIKHA PANDHI, ARVIND KUMAR, SADHNA MISHRA, and DINESH CHANDRA RAI

ABSTRACT

Increasing environmental concerns have led to a paradigm shift from various chemical-based technologies to a greener and sustainable approach. Recent green nanotechnological interventions for application in food processing, packaging, and preservation make use of biological sources, safer solvents, recyclable materials, and energy-saving processes for the creation of nanoparticles (NPs) to surpass the detrimental ecological impacts and high costs associated with conventional methods. Various advanced nanoencapsulation technologies have also been developed that utilize various biocompatible delivery systems as a carrier for various bioactive and nutrient components for their controlled release and enhanced stability for food processing and preservation. This chapter provides comprehensive information on various prospects and approaches exhibited by green nanotechnology and possible opportunities for its application in the food domain in enhancing the existing performance.

6.1 INTRODUCTION

Nanotechnology possesses extensive applications in various fields of science and technology (including food process engineering) for the manipulation of materials at the nanoscale level [8]. The organization of materials at the nanoscale level ranging from 1 to 100 nm gained a tremendous impetus over the years as it offers inventive frontiers associated with their entirely improved characteristics attributed to the change in shape, dimension, and arrangement [37, 64].

Increasing environmental trepidations have directed the attention of various chemical-based technologies to an eco-friendly and more sustainable approach. Entailing this vision into the emergent field of nanotechnology has stimulated the need to examine green resources and processes for the creation of green nanoparticles (NPs). Therefore, the concept of "green nano-technology" comes as salvage with two major objectives: (a) the creation of nanomaterials that minimizes the detrimental effect on the ecosystem or human health; and (b) an effective solution to environmental problems [38].

The recent green nanotechnology approach assimilates dogmas of green chemistry and green engineering that comprise the utilization of biological sources, non-toxic medium, eco-friendly, and recyclable materials, and energy-efficient methodologies for the creation of NPs to surpass the detri-mental ecological impacts and higher costs concomitant with conventional methods [10, 58, 83].

Insight of the above-mentioned perks and perquisites offered by green nanotechnology, various fields have been scouted for probable applications. Amongst these, the Food Industry is a less explored domain and has begun to discover the possible applications of green nanotechnology to embark on the vision of enhancing the efficiency of existing food processing and preservation methodologies using a sustainable approach. Nanotechnology for intended food applications can be projected as two distinct approaches: either "top-down" or "bottom-up" [6, 55] (Figure 6.1):

- **Top-Down Approach:** In this approach, particle size manipulation at the molecular and atomic level is favored by breaking down the bulk material to its nanosize counterpart; and
- **Bottom-Up Approach:** This promotes size manipulation through self-assembly of atoms and molecules as building blocks to give a larger size dimension.

Nanotechnology is a promptly upcoming field in the food domain that may serve various purposes to provide safe and good quality products through effective food packaging and efficient processing technologies [54]. Nanotechnological application in the field of food can be broadly classified into two forms of application [54, 64]:

- • Nano-inside, as in case of food additives or a bioactive component incorporated using the nanotechnological approach; and
- • Nano-inside, which provides effective outer protective coverings to food in terms of active and smart packaging systems.

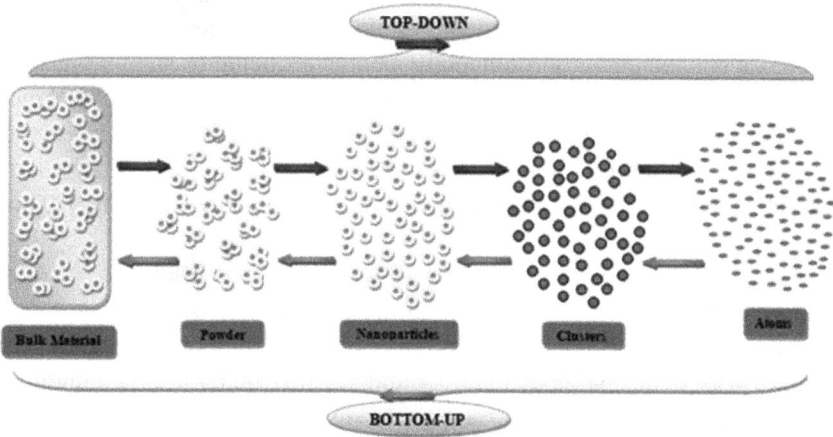

FIGURE 6.1   Bottom-up and top-down approach for particle size manipulation [6].

Nano-dimension food additives in the form of a preservative, nutrient supplement, processing-aid, a bioactive component could influence product characteristics, such as color, flavor, texture, shelf-life, nutrient composition, or else can also be utilized to detect foodborne pathogens as food quality pointers. Enhancing the bioavailability of nutraceuticals, nutrients, bioactive components, as well as enhances the organoleptic parameters such as taste, flavor, texture, and stability of food products through advanced nanoencapsulation technologies has conquered prodigious attention these days [11, 12, 50].

In the food packaging perspective, nanotechnology could be functionalized through antimicrobial packaging or intelligent packaging to indicate spoilt components and deteriorative changes in the product. Antimicrobial characteristics of green synthesized NPs can be utilized for incorporation into the packaging matrix of food to extend its shelf-life and make it safe for intended consumer use [8]. Apart from this, nanotechnology also facilitates the detection of various chemical and microbial-derived contaminants in foods through making use of highly specific biosensors.

Biogenic synthesis of NPs through plants and microbes route follows a bottom-up approach, wherein bioreduction of metal salts takes place to give nanosized metal particles. These green NPs possess potent antimicrobial activity against a broad range of microbes that could be effectively used for food application as a preservative and serve as a workable substitute to the conventional chemical preservatives [60].

NPs of silver, gold, palladium, iron, and zinc oxide have been effortlessly synthesized via green route by means of biological sources. The utilization of non-toxic solvent and environmentally benign reducing or stabilizing agents is major considerations in the green creation of metallic NPs [69]. Among these, the creation of NPs via plant route had conquered several advantages due to their availability, non-toxic nature, better stability, and elimination of maintenance of cell cultures, which will be mainly addressed in this chapter [39, 69].

Bioactive components and nutrients (phenols, essential oils, insoluble vitamins, and minerals) serve as essential components of nutraceutical and functional foods. The key challenge associated with their use in food formulation lies with their low bioavailability and stability under normal conditions. Nanoencapsulation can assist as a promising alternative to preserve these sensitive compounds from external factors and enhance their bioavailability and stability for various food applications. Nanoencapsulation provides a barrier to these components using several advanced techniques (such as electrospinning, electrospraying, nano-spray drying, etc.), utilizing various different biodegradable delivery systems as nanocarriers for these bioactive and nutrient components [7].

In spite of the tremendous opportunities offered by nanotechnology for food applications, there are certain parameters associated with the advent of this stimulating technology that needs to be inquired critically. Safety as well as the regulatory aspects need to be addressed or evaluated critically while making, processing, packaging, and consumption of these nano-system processed food products to facilitate the safe marketing.

This chapter documents comprehensive information on novel green nanotechnology (such as biocompatible nanoencapsulation methodologies and green route synthesis of metallic NPs) and plethora of opportunities it offers in various domains related to the food sector (such as food processing, packaging, and preservation) for enhancing the existing performance. Further, related concerns, challenges, and future prospects are also addressed.

6.2   METALLIC NANOPARTICLES (NPS) VIA GREEN ROUTE

Recent advancement in the area of nanotechnology has encouraged the utilization of cost-effective and eco-friendly synthesis methodologies, in which green synthesis of NPs has received tremendous attention. Green synthesis is an eco-friendly and benign mode of fabrication of NPs through the biological route using natural sources. These NPs exhibiting a nanosize dimension ranging from 1–100 nm serve as a link between massive materials and their nanostructured counterparts at the atomic and molecular levels [34].

Manipulation of particle size to nano-dimension imparts remarkable and interesting properties to these NPs owed by their small sizes, more surface area, and high reactivity as compared to their massive counterparts [10]. The ability of biological entities as an efficient reducing agent has received great attention over the years, but the mechanism behind this is still needed to be explored. Green route permits the use of natural reducing, capping, as well as stabilizing agents that eliminate the need for harmful, costly chemicals and energy-intensive processes for the creation of NPs.

Conventional chemical methods of nanoparticle synthesis are costlier as well as make use of harmful solvents, chemicals that may lead to the generation of hazardous byproducts and hence it has aroused the need for the development of an environmentally-benign alternative, where the advent of green route has served as a solution and offers tremendous research opportunities to facilitate its further expansion [42]. The self-aggregation ability of biomolecules is efficiently exploited in the bottom-up approach of particle synthesis.

Green synthesis follows a bottom-up approach that utilizes various plant parts (such as leaves, bark, flower, stem, peel, flower, seed) and microorganisms (such as bacteria, fungi, algae) for clean, eco-friendly, non-toxic creation of NPs [29, 31]. Among these, the creation of nanosize particles using plant-extract as a reducing agent offers several benefits over microbe-mediated route as they are easily available, non-toxic in nature possess superior stability and abolish maintenance of cell cultures (Figure 6.2).

A variety of living organisms specifically plants generate metabolites that can alter or manipulate the creation of NPs via green route and are in general categorized as primary (proteins, vitamins, polysaccharides) or secondary metabolites (phenols, flavonoids, tannins, alkaloids) that secure potential reducing and capping capabilities [75]. NPs created via the green route have exposed to exhibit bactericidal action hostile to both Gram-positive as well as Gram-negative bacteria, encouraging their application in various domains, among which the food domain needs to be explored a bit more [60]. Extensive literature exists that supports the antimicrobial activity exhibited by silver and gold NPs fabricated via the green route [3, 62, 79].

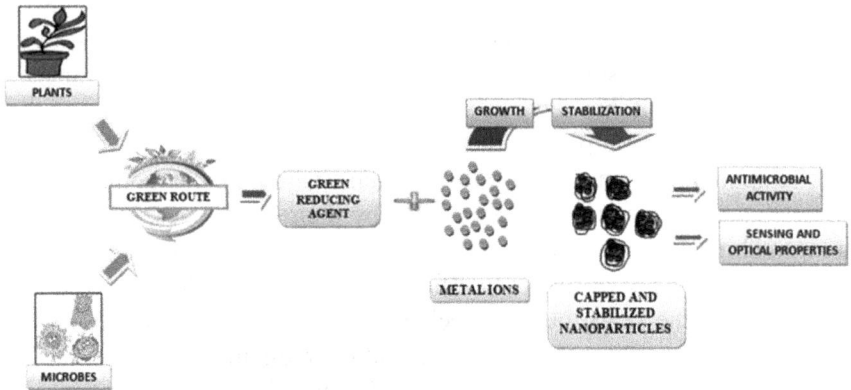

FIGURE 6.2   Green route synthesis of metallic nanoparticles.

Analysis using UV-visible spectroscopy is engaged to substantiate the creation of a variety of NPs, as these NPs show explicit absorbance bands in characteristic spectra when light incident on their surface is due to surface plasmon resonance. Various factors (such as pH, reaction temperature, and the concentration of plant extract/metal solution) govern and stimulate the creation and characterization of NPs as they affect the shape and size of the resultant NPs along with the reaction rate [16].

Coupling of green route creation with assisting technologies, such as microwave irradiation and ultrasonication have received considerable attention due to superior yields [14, 24, 26]. Numerous research studies have been conducted for the creation of NPs via biological entities, but their application in the food domain is still limited and needs to be explored. Some of the potential applications have been discussed in this chapter.

6.3   APPLICATIONS IN FOOD PROCESSING AND PRESERVATION

6.3.1   FOOD PACKAGING

Food packaging holds an essential position in governing safety and preserving the quality of food. The active packaging system has come-up as a novel packaging approach with enhanced function to continuously meet the demand of consumers for safe and convenient food with longer shelf-life. The active packaging system possesses the capability to alter the composition as well as the atmosphere around when kept in contact with food [65].

Antimicrobial packaging is a kind of active system of packaging based on the interaction between a package and the food surface or package atmosphere to alleviate the expansion of microbes residing on the food surface. Incorporation of antimicrobial components into the package facilitates bactericidal or bacteriostatic action through steady transmission within the food matrix, which abolishes the necessity of additional preservative [18].

The advent of green nanotechnology has offered great opportunities in terms of green NPs as effective antimicrobial agents. Nanosize antimicrobials even at low concentration showcase high antimicrobial activity due to their large area to volume ratio and exceptional elemental and physical characteristics. Among all the noble metal NPs, silver nanoparticle stands as a superior antimicrobial agent due to its better chemical stability and good conductivity, which facilitates its incorporation into various packaging films [2].

A study was conducted to combine the silver nanoparticles (AgNPs) creation and film-forming solution preparation for the development of an active film exhibiting antimicrobial property and it showed that films with AgNPs concentration >71.5 ppm repressed the multiplication of *E. coli* ATCC and *Salmonella* spp. Whereas, films with 143 ppm AgNPs concentration were chosen as they effectively maintain their integrity against microbial action and prolonged safe consumption period of fresh cheese samples up to 21 days [52].

Apart from this, another domain of active packaging system deals with the development of a nanocomposite system that set outs deterioration catalyzing components from an inherent food system (such as moisture), gasses (such as ethylene, oxygen, carbon dioxide, etc.), that may negatively affect the quality of food, hence, prolong the shelf-life. The use of AgNPs for absorption and decomposition of ethylene has been shown to retain the freshness of fruits and vegetables for an extended period of time [30].

Nanosized titanium dioxide nanoparticle has been effectively used for the packaging and processing of foods for their oxygen eliminating property [40]. A poly(3-hydroxybutyrate) (PHB) based film incorporated with palladium NPs possessing oxygen scavenging property was developed using the electrospinning technique accompanied by annealing at 160°C, which could be used as an innovative packaging alternative for ensuring food quality [13].

The application of NPs is not limited to active food packaging systems, but it has also been extended to other versions, such as intelligent packaging. Intelligent packaging is a sort of tagging, which continuously scrutinize the condition of a food inside a package or the surrounding conditions of the package to give necessary information regarding the current physico-chemical properties of food and its surrounding condition within the food package [61].

Nanosize particles of zinc and titanium have been used to inspect the presence of volatile constituents, e.g., a gaseous form of amines, ethanol (associated with fish and meat deterioration) using nanosize fibers of perylene-derived fluorophores. In addition, nanocomposite fabricated using tungsten oxide and titanium dioxide NPs enable the rapid detection of ripening catalyzing ethylene gas in fruits [59]. Incorporation of nanosize iron oxide particles as one of the components of the composite film has shown to possess humidity sensing ability [76].

6.3.2   NANOPARTICLE BASED-BIOSENSORS

The functionalized nanomaterials can serve as an effective bio-sensing agent due to high stability, selectivity, a sensitivity, which could be utilized for the development of various catalytic, immobilized in addition to the most frequent optical biosensors. Moreover, these nano-bio sensors can also be employed for the creation of innovative food detection approaches and promote food safety. Nanotechnological interventions in the food domain have generated an array of opportunities for researchers to appraise the probable remunerations of this evolving technology as it is getting remarkable consideration equally from the communal and the private sector.

In this perspective, various 'nanosized' and nanomaterial-based biosensors have been fabricated as advance and effective detection systems to promote food quality and food safety with quick tests [56]. Two colorimetric assays were successfully fabricated using Rutin/curcumin reduced AgNPs for rapid and reliable detection of melamine in milk with a detection limit of 0.01 ppm (79 nM) for rutin reduced AgNPs and 0.24 ppm (1900 nM) for Curcumin reduced AgNPs [63]. Colorimetric assays have been developed using silver and gold NPs for the rapid evaluation of sugars and polyphenols in apples that showed comparable results as conventional methods ABTS and ion chromatography [72].

6.4 NANOENCAPSULATION

An ample amount of attention has been drawn by nutraceuticals and functional foods with improved nutritional and functional properties as a result of the incorporation of a bioactive or nutrient component [46]. These bioactive components, due to their low stability, solubility, and bioavailability limit their direct incorporation into foods [48]. Hence it necessitates the requirement of an efficient delivery system for their targeted conveyance, where nanoencapsulation has open up as a possible solution. Nanoencapsulation is a key arena of nanotechnology (Figure 6.3), which involves the entrapment of bioactive agent and nutrient to be encapsulated within a carrier matrix exhibiting nanosize dimension that enhances the stability and functionality with improved sensory properties during processing and storage [21, 33, 44].

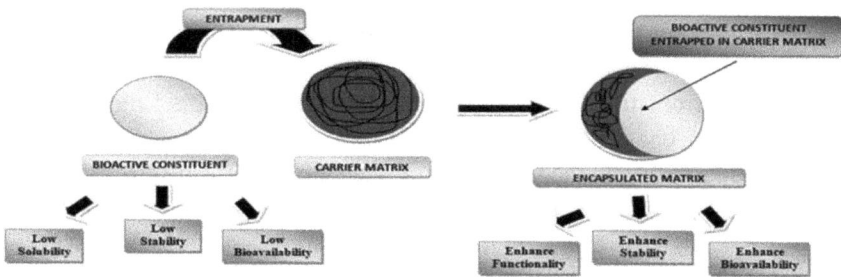

FIGURE 6.3 Schematic representation of nanoencapsulation of a bioactive constituent in a carrier matrix.

Nanoencapsulation includes the confinement of a substance in nanosized conveyors via inclusion, integration, chemical interaction, or diffusion [7]. Several advance nanoencapsulation methodologies have been recently reported for the confinement of bioactive compounds and nutraceuticals within a carrier matrix using diverse delivery systems and carrier matrixes that can be broadly classified into various categories, such as lipid-derived, protein-derived, and carbohydrate-derived carrier systems. In light of this range of nanosystems, diverse equipments can visibly be used to nanoencapsulation food constituents, such as electro spinners, electrospray, nano-spray dryers, and microfluidic devices [32].

6.4.1   NANOENCAPSULATION DELIVERY SYSTEMS

The delivery system is employed as an entrapment matrix, wherein a bioactive substance is captured in a carrier for its controlled release. These nano-carriers act as a protective covering for the bioactive substance that prevents it from hostile environment, e.g., oxidation, and degradation [19, 28, 89]. Nano delivery systems exhibit large surface area and acquire the possibility to improve solubility and bioavailability and upgrade restricted and targeted release of the captured ingredient, more efficiently than the micro-dimension matrix [47, 51]. Basically, the targeted and restricted release of encapsulated ingredients is facilitated through two possible mechanisms [41]:

- **Sustained-Release:** It is engaged for lagged liberation of the captured material to maintain the stable concentration of a biologically active ingredient at the place where liberation is required, e.g., flavors, chewing gums.
- **Delayed-Release:** This in which the discharge of a bioactive ingredient is postponed from a restricted "delay time" to a certain point when/where its discharge is favored, e.g., color release in beverages. Apart from these, there are several factors that affect the choice of suitable encapsulating ingredients and methods, such as stability, compatibility, biodegradability, and economic-feasibility.

6.4.1.1   PROTEIN-DERIVED DELIVERY SYSTEMS

The use of the protein-based delivery system for encapsulation has gained tremendous attention over the years due to their explicit nutritional and functional characteristics, such as the ability to assemble into different structures [49], inhibitory action against oxidation [70] and capacity to adsorb on the surface of colloidal particles and protects them from accumulation [47].

Besides this, proteins offer an array of functional groups on their surfaces that facilitate their interaction with a variety of diverse substances, thus enabling the fabrication of a nano-delivery system to capture both water-soluble as well as the water-insoluble active constituents. The proteins that are commonly used for nanoencapsulation are composed of those derived from an animal source (e.g., casein, gelatin, and albumin) or botanical source (e.g., gliadin, zein, or soy proteins), etc.

6.4.1.2 LIPID-DERIVED DELIVERY SYSTEMS

Lipid-derived delivery system for encapsulation is a promising technology due to the supreme advantages it offers, such as the capability to trap material having diverse solubilities, good barrier properties to degradation, and encourage targeted delivery via active or passive route. They can be utilized for entrapment of both water-loving as well as lipid-loving ingredients and sometimes may provide a synergistic effect through the use of emulsion-based delivery systems [82].

Liposomal and Niosomal nanovesicles are self-assembled vesicular entities that are renowned nano-conveyors engaged for the encapsulation of an active ingredient. They are prepared using a variety of different components and methodologies responsible for their distinct morphology and structure, charge distribution and technological properties. Phospholipids and non-ionic surfactants form key constituents of liposomal and niosomal nanovesicles system, respectively [1, 73]. This structure, when incorporated as a functional component in the food matrix, does not affect the esthetics and sensory properties [53].

Liposomal and niosomal nanocarriers loading approaches offer the most suitable technical solution [33, 66]. Niosomes propose a number of benefits over liposomes, such as higher chemical constancy and low cost. The effortlessness in transfer and superior drug-carrying adequacy of niosomes makes them a flexible and appropriate deliverance system for lycopene as indicated in a study [74].

Nanoemulsions are nanosize droplets consisting of multiple phase colloidal distribution fabricated by dispersing two immiscible liquid by physical share-induced rupturing (Figure 6.4).

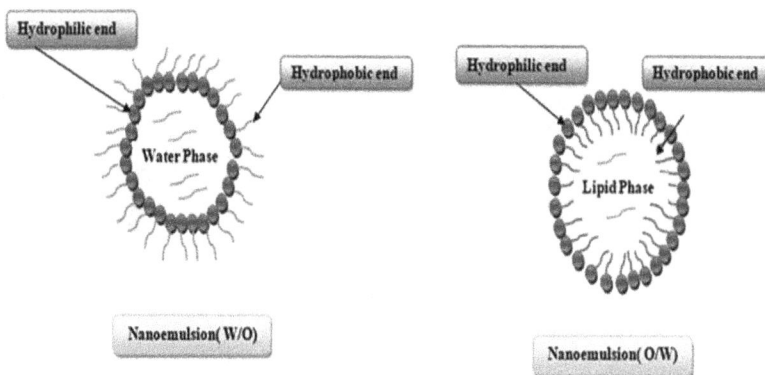

FIGURE 6.4 Diagrammatic representation of nanoemulsion.

Nanosize emulsions can be fabricated by either mechanical or non-mechanical techniques [22, 82]. The mechanical approach utilizes two different techniques: homogenization and microfluidization as a fabrication technology (Figure 6.5). The non-mechanical approach makes use of solvent diffusion techniques. Nanoemulsions are decent nominees for the conveyance of water-insoluble food constituents (such as fish oil), due to their capability to enhance solubilization of bioactive ingredient and adequacy for facilitating their assimilation in the gut, governed by surfactant persuaded permeability changes and encourage extensive allocation of the encapsulated bioactive inside the gastrointestinal (GI) tract.

Solid lipid nanoparticles (SLNs) have received great attention over the years for food applications [5, 25, 78]. SLNs consist of a solid lipid matrix that acts as a shell for entailing particles inside them. As compared to the use of liposome and nanoemulsion based lipid delivery system SLNs have some distinctive benefits [22], such as elevated encapsulation competence, eliminates the need for organic solvent, can be easily scale-up, slow degradation rate and facilitate the flexible release of bioactive component (Figure 6.6). An additional advantage may include a longer release rate of bioactive as compared to other carriers, such as nanoemulsion which release ingredient too fast based on the partition coefficient [81]. Two basic methods that are widely used for the fabrication of SLN are: hot homogenization and cold homogenization [71].

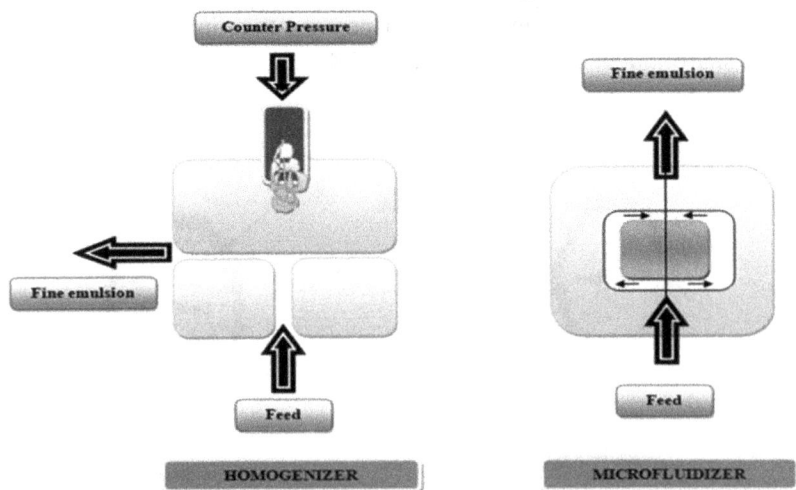

FIGURE 6.5   Schematic diagram of a homogenizer and microfluidizer.

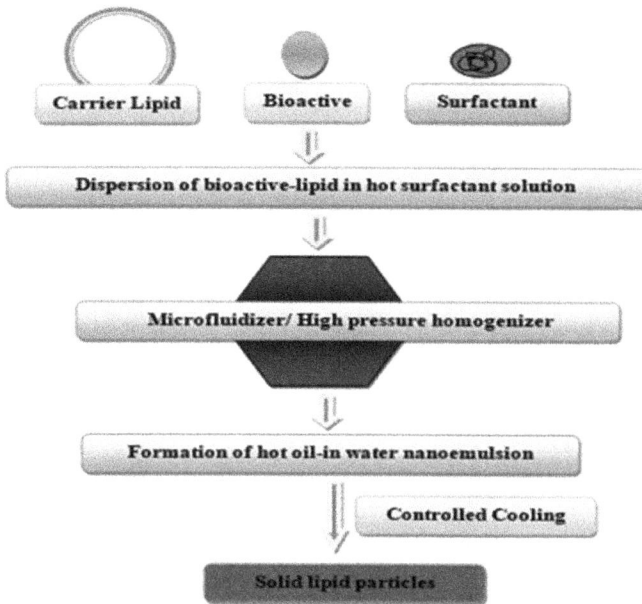

FIGURE 6.6   Schematic diagram for the formation of solid lipid particles.

Confinement of bioactive components within SLNs generally follows three basic approaches, such as homogenous matrix, bioactive-augmented shell, and bioactive-augmented core model. The selection of the approach relies on the composition of formulation material as well as fabrication conditions.

6.4.1.3   CARBOHYDRATE-DERIVED DELIVERY SYSTEMS

Polysaccharide-derived delivery systems provide an appropriate candidate with numerous industrial applications due to their non-toxic, biodegradable, and easily modifiable behavior to attain the desired properties. As compared to the lipid-derived system, carbohydrate-based nanocarriers can efficiently intermingle with a broad array of functional entities and serve as a versatile candidate for encapsulation of diverse water-soluble and water-insoluble bioactive ingredients [20].

Conversely, they also provide an efficient system that remains stable during high-temperature treatments during processing as compared to other protein and lipid-based delivery systems, which may either get melted, oxidized or denatured. Usually, carbohydrate-based delivery systems are suitably characterized based on their origins, such as those obtained from plant origin (starch, cellulose, pectin, and gums) or animal origin (e.g., chitosan) and from microbial origin (e.g., carrageenan, alginate, dextran), etc., [17].

6.4.2   NANOENCAPSULATION TECHNOLOGIES

6.4.2.1   COACERVATION

Coacervation brings about the separation of phases/fractions within a polymer solution on the application of adequately high attractive force and gives two fractions, such as a polymer-enriched fraction and a polymer-exhausted fraction [77]. Coacervation occurs due to electrostatic attraction among different oppositely charged molecules of biopolymer [20, 36, 86]. When it occurs between a bioactive component and oppositely charged polymer, it is called simple coacervation; whereas when two oppositely charged biopolymers and a bioactive component experience electrostatic pull, the process complex coacervation occurs. The process of encapsulating a water-insoluble bioactive constituent via a complex coacervation route generally follows series of steps as indicated in (Figure 6.7).

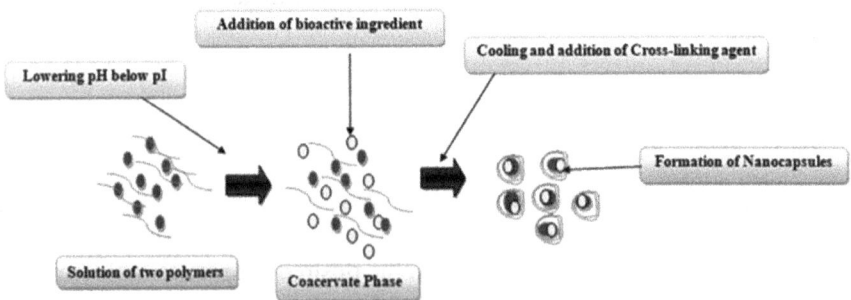

FIGURE 6.7   Diagrammatic representation of complex coacervation process.

6.4.2.2   COLD GELATION

Cold gelation is a multistep process that gives protein-hydrogels at ambient temperatures [4]. The first step involves protein heat treatment to a temperature where denaturation occurs under controlled pH and ionic strength that prompt uncoiling of protein. The second step involves mixing a bioactive agent, which is then subjected to association with heat-denatured proteins to give a hydrogel in the third step by adjusting pH and shifting ionic strength to eradicate electrostatic repulsion amongst the protein molecules (Figure 6.8). It is most appropriate for heat-labile protein compounds in the case of protein-based encapsulation system such as soy and whey proteins [20].

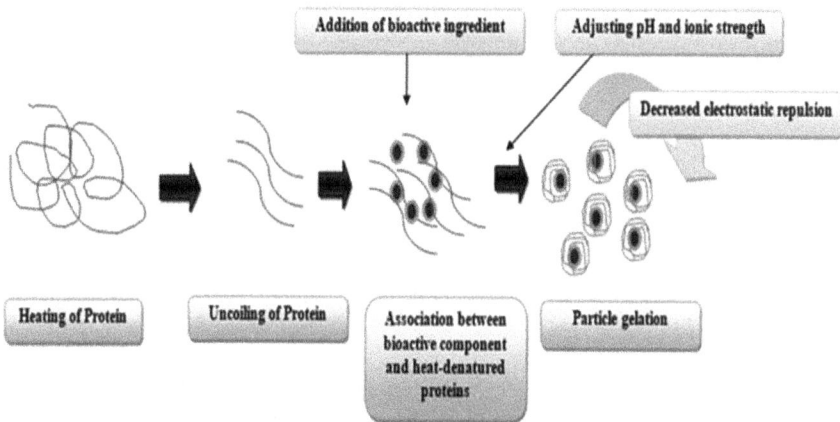

FIGURE 6.8   Diagrammatic representation of a cold gelation process.

6.4.2.3   SPRAY-DRYING

Spray drying is the most widely used drying and encapsulation technique that gives quick, relatively economical, and reliable output [87]. It works by dissolving or dispensing active ingredient in a biopolymer solution and then subjected to an atomizer that facilitates the rapid removal of solvent giving dried particles with the embedded active ingredient in a porous matrix (Figure 6.9). Loss of heat-sensitive bioactive limits their application in certain cases. The size of encapsulated particles and effectiveness of the process relies on numerous parameters like type of material used, the viscosity of the solution, spray-dryer features, and type of atomizer, airflow rate, and inlet/outlet temperature of the spray drier. The common spray drying process gives micro-size capsules. In recent times, nano spray-driers have been evolved as an efficient method for the creation of NPs through an effective breakdown [21].

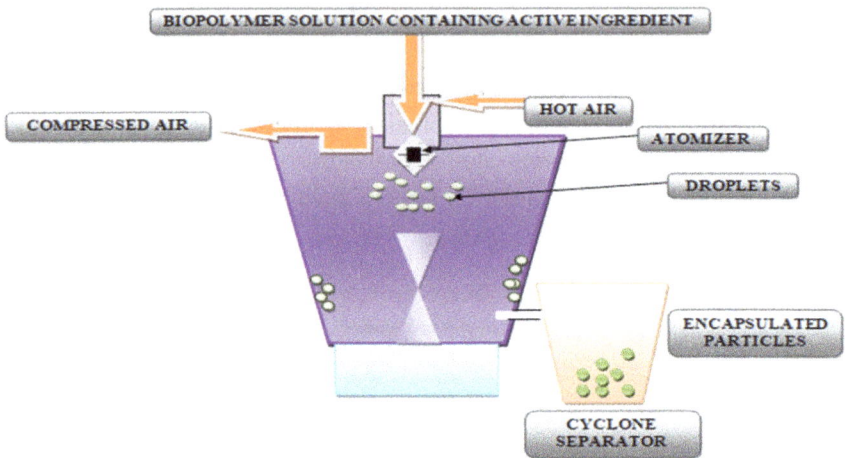

FIGURE 6.9   Diagrammatic representation of a spray dryer (for nanoencapsulation).

6.4.2.4   ELECTRO-HYDRODYNAMIC PROCESS

Electro-hydrodynamic processes utilize an electrical charge to give very lean fibers or minute particles [15]. Electro-spinning and electro-spraying have emerged as the two most widely used electro-hydrodynamic processes used for the fabrication of nanosized fibers and particles when a biopolymer solution is subjected to high-intensity electric fields (kV/cm range). Electrospinning is a technique that gives continuous fibers having size reduced to nano dimension which could be effectively used for carrying bioactive constituents [9, 84]. This technique produces polymer fibers utilizing electrostatic forces.

Electrospray is another advance nanoencapsulation technique that works analogous to electrospinning but varies in giving output as NPs instead of nanofibers as in the case of electrospinning (Figure 6.10). This method utilizes electrostatic force brought by high voltage atomization of the liquid into fine droplets, which facilitate evaporation of the solvent. Electrospray technique has high encapsulation efficiency [88]. Various structural and functional advantages are offered by electrospun and electrosprayed products such as enhanced stability, induces porosity, reduced denaturation, continuous, and regulated release, etc.

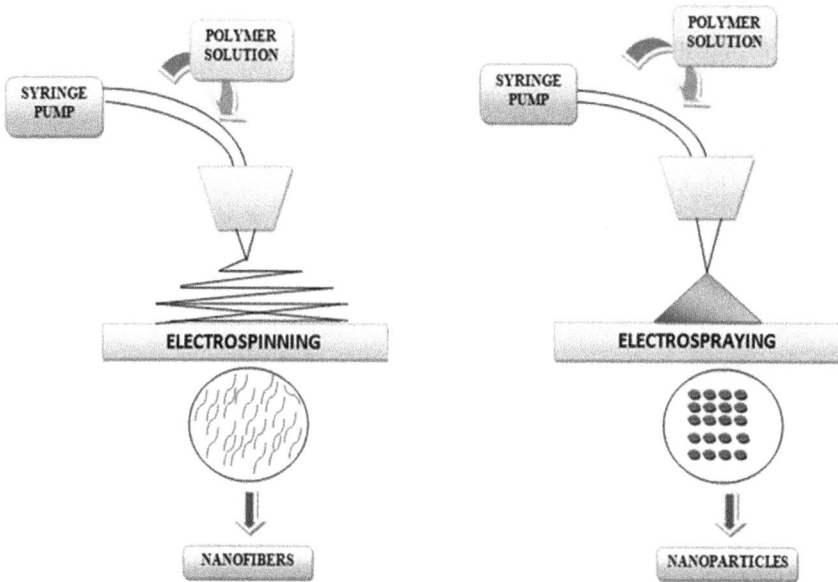

FIGURE 6.10 Comparison of electrospinning and electrospraying technique.

From an application point of view, electrospinning and electrospraying are less explored for food applications. Literature studies carried out till now reveal the tremendous adequacy of these techniques for encapsulation of bioactive compounds within a micro or nano-dimension carrier matrix prior to incorporation into food products. In addition, these techniques can be engaged in the fabrication of functional materials for the active food packaging system.

6.4.3 APPLICATIONS IN FOOD PROCESSING AND PRESERVATION

6.4.3.1 FOOD PACKAGING

Packaging performs an essential part in the protection and containment of food during the product cycle starting from processing to the ultimate consumption of the product. A food package may be considered as an active system when it performs a specific function apart from its basic barrier function. Encapsulation of various active ingredients into the packaging matrix enhances the efficiency of an active packaging system.

Various encapsulation techniques can be employed to serve this purpose, among which the most recent one is electrospinning. Electrospinning is a promising fabrication technique that gives nanostructured layers an output for the development of an active system for food packaging that facilitates food preservation. Electrospun nanofibers act as active packaging materials because of high area to volume ratio, nanoscale dimension, and high suitability for encapsulation of sensitive-constituents. A range of studies has been performed to optimize various parameters for the encapsulation of active components using the electrospinning technique. A study has evaluated the adequacy of electrospinning as an encapsulation technique for the humidity prompt release of aroma compounds [45].

6.4.3.2   IMMOBILIZATION OF ENZYMES

Enzymes are biocatalysts that exhibit elevated specificity and efficacy that can be utilized for various industrial applications. However, their higher cost, recovery difficulties, high susceptibility to heat, and pH limits their application in certain cases [35]. To surpass these constraints, the concept of enzyme immobilization came as a solution that enables its confinement of these enzymes within physical or solid support for the retention of their catalytic activities. Herein, various advanced nanoencapsulation techniques can be used to serve the purpose of enzyme immobilization. Biocomposite fibers fabricated using Chitosan and PVA using the electrospinning technique have shown to enhance the stability of glucose oxidase (GOD) used for the fabrication of biosensors for glucose detection [85].

6.4.3.3   ENCAPSULATION OF BIOACTIVE COMPOUNDS

Encapsulation serves as a booming area of investigation to facilitate targeted and safe conveyance of biologically active ingredients confined within a wall material that is generally regarded as safe (GRAS) for food use [9]. Targeted release of these bioactive components has been associated to confer various health-promoting effects on the host and is used for the formulation of various nutraceutical and functional foods. Spray drying has been most widely used for the development of various nutraceutical and functional foods using encapsulated bioactive ingredients. Spray drying technique was used to the fabrication of pectin coated SLNs composed of sodium caseinate (protein) and Compritol ATO 888 (glyceryl behenate) showed greater stability [80]. Amongst the most recent technologies for nanoencapsulation, electrospinning, and electrospraying hold an important potential. Ultrafine fibers have been fabricated from pectin and pullulan using an electrospinning technique that showed retained bioactivity of the compound [43].

6.4.3.4   ENCAPSULATION OF ANTIOXIDANTS

Safeguarding nutrients, as well as the oxidizable substrate from oxidative breakdown, have been widely studied using various microencapsulation methodologies. Conversely, to facilitate their targeted and restricted liberation, efficient and more effective techniques are needed, such as nanoencapsulation. For encapsulation of antioxidants that are susceptible to oxidative changes, lipid-based nanoencapsulation systems are preferred as they shield effectively against reactive radicals, metal species, pH, and high temperature that may result in degradation of the oxidizable food component [51].

6.5   ISSUES, CHALLENGES, AND FUTURE PROSPECTS

Nanotechnology enables the creation and manipulation of materials to nano-sized dimension inducing changes in physicochemical properties distinct from their bulk analog. Precisely, the reduction in dimension is accompanied by greater surface area with a consecutive increase in reactivity [57]. These nanotechnological interventions had created a lot of potential applications with an incredible lift in the research publications and technology and process linked patents in food domain and promoted fascinating nanotechnological approach for advancements in processing of food, food packaging system, efficient nutraceutical deliverance, efficient control of product quality, and development of novel functional foods [67].

The key allied areas of food where nanotechnologies offer inventive frontiers are:

- As a food processing aid through nanofiltration;
- Efficient food packaging systems through nanocomposites active and smart packaging;
- Health-promoting foods incorporated with nutraceutical component formulated using advanced encapsulation methodologies that deal with nano-dimension;
- To ensure food safety and quality using e-nose, biosensors, etc.

However, for the efficacious establishment of nanotechnological intervention in the food industry, there are several factors related to the advent of this inspiring technology that needs to be investigated critically. A number of technical, communal, and regulatory obstacles need to be addressed for the effective establishment of this technology. Safety issues linked with the use of these nanomaterials in nano processes or nano package food products cannot be ignored as there may be a chance of possible migration of these nanomaterials from the package to surface of food in case of an active and smart packaging system which may exert a detrimental effect on consumers health.

The limpidity of safety aspects and ecological influence should be of prime importance when dealing with the development of a food system using nanotechnology and therefore require obligatory testing of nano processed foods prior to their release in the market for consumer access. Although, use of nano-based packaging systems has shown a rapidly growing pattern in the food industry, the associated safety and regulatory issues limits their use to a great extent. Hence, it is obligatory to obtain adequate knowledge regarding the associated risks with the application of these NPs and nanoformulations for food purposes.

Apart from this, consumer insight towards a product plays a crucial role in the profitable success of that product. Consumer acceptability towards nanotechnological application in the manufacturing and packaging of food is to a degree affected by consumers' thoughts associated with the appliance of nanotechnology to the food system [27]. A large part of the population is unaware of the prospects of this novel technology. Therefore, it is imperative to study consumers' perspectives towards nanotechnological interventions in food for the expansion and booming promotion of these foodstuffs. Faith and awareness serve as two crucial elements for making a successful nano-technological intervention. Presently, nano processed products are costly, but after commercialization and an increase in demand, the market price can be reduced.

6.6 SUMMARY

Green nanotechnological interventions have the capability to improve food quality and boost food safety along with providing tasty, healthy, and more nutritious food products with innovative packaging systems. Nanotechnology can be applied to improve the flavor as well as the texture of food or to encapsulate nutrients, such as vitamins that illustrate increased bioavailability and stability throughout the product shelf-life. Additionally, nano-based packaging systems can be used as efficient technology for prolonging the freshness of food for an extended period. As a whole, nanotechnology offers tremendous benefits to the food industry. Further, consumer acceptances of nano-processed food are important and require communication of potential benefits and probable safety risks associated with the product that may facilitate their marketing.

KEYWORDS

- **energy efficient**
- **food preservation**
- **generally regarded as safe**
- **green nanotechnology**
- **nanoparticles**
- **packaging**

REFERENCES

1. Ahad, A., Raish, M., Al-Jenoobi, F. I., & Al-Mohizea, A. M., (2018). Sorbitane monostearate and cholesterol-based niosomes for oral delivery of telmisartan. *Current Drug Delivery, 15*(2), 260–266.
2. Ahmed, S., Ahmad, M., Swami, B. L., & Ikram, S., (2016). Review on plants extract mediated synthesis of silver nanoparticles for antimicrobial applications: A green expertise. *Journal of Advanced Research, 7*(1), 17–28.
3. Alfuraydi, A. A., Devanesan, S., Al-Ansari, M., & AlSalhi, M. S., (2019). Eco-friendly green synthesis of silver nanoparticles from the sesame oil cake and its potential anticancer and antimicrobial activities. *Journal of Photochemistry and Photobiology B: Biology, 192*, 83–89.
4. Alting, A. C., De Jongh, H. H. J., Visschers, R. W., & Simons, J. W. F. A., (2002). Physical and chemical interactions in cold gelation of food proteins. *Journal of Agricultural and Food Chemistry, 50*(16), 4682–4689.
5. Awad, T. S., Helgason, T., Kristbergsson, K., Decker, E. A., Weiss, J., & McClements, D. J., (2008). Effect of cooling and heating rates on polymorphic transformations and gelation of tripalmitin solid lipid nanoparticle (SLN) suspensions. *Food Biophysics, 3*, 155–162.

6. Balasooriya, E. R., Jayasinghe, C. D., Jayawardena, U. A., Ruwanthika, R. W. D., Mendis, D. S. R., & Udagama, P. V., (2017). Honey mediated green synthesis of nanoparticles: New era of safe nanotechnology. *Journal of Nanomaterials, 2017*, 1–10.

7. Bazana, M. T., Codevilla, C. F., & De Menezes, C. R., (2019). Nanoencapsulation of bioactive compounds: Challenges and perspectives. *Current Opinion in Food Science, 26*, 47–56.

8. Berekaa, M. M., (2015). Nanotechnology in food industry: Advances in food processing, packaging and food safety. International Journal *of* Current *Microbiology and* Applied *Sciences, 4*(5), 345–357.

9. Bhushani, J. A., & Anandharamakrishnan, C., (2014). Electrospinning and electrospraying techniques: Potential food-based applications. *Trends in Food Science and Technology, 38*(1), 21–33.

10. Castro, L., Blazquez, M. L., Munoz, J. A., Gonzalez, F., Garcia-Balboa, C., & Ballester, A., (2011). Biosynthesis of gold nanowires using sugar beet pulp. *Process Biochemistry, 46*, 1076–1082.

11. Chaudhry, Q., Castle, L., & Watkins, R., (2010). Nanotechnologies in the food arena: New opportunities, new questions, new concerns. In: Chaudhry, Q., Watkins, R., & Castle, L., (eds.), *Nanotechnologies in Food* (pp. 1–17). Cambridge - UK: Royal Society of Chemistry Publishers.

12. Chaudhry, Q., Scotter, M., Blackburn, J., Ross, B., Boxall, A., Castle, L., Aitken, R., & Watkins, R., (2008). Applications and implications of nanotechnologies for the food sector. *Food Additives and Contaminants: Part A, 25*, 241–258.

13. Cherpinski, A., Gozutok, M., Sasmazel, H., Torres-Giner, S., & Lagaron, J., (2018). Electrospun oxygen scavenging films of poly (3-Hydroxybutyrate) containing palladium nanoparticles for active packaging applications. *Nanomaterials, 8*(7), 469.

14. Deshmukh, A. R., Gupta, A., & Kim, B. S., (2019). Ultrasound-Assisted green synthesis of silver and iron oxide nanoparticles using fenugreek seed extract and their enhanced antibacterial and antioxidant activities. *BioMed Research International, 2019*, 1–14.

15. Drosou, C. G., Krokida, M. K., & Biliaderis, C. G., (2016). Encapsulation of bioactive compounds through electrospinning/electrospraying and spray drying: A comparative assessment of food-related applications. *Drying Technology, 35*(2), 139–162.

16. Dwivedi, A. D., & Gopal, K., (2010). Biosynthesis of silver and gold nanoparticles using *Chenopodium album* leaf extract. Colloids and Surfaces *A: Physicochemical and Engineering Aspects, 369*, 27–33.

17. Eliasson, A. C., (2006). *Carbohydrates in Food* (2nd edn., p. 521). London: Taylor & Francis Group.

18. Espitia, P. J. P., Soares, N. D. F. F., Dos, R. C. J. S., De Andrade, N. J., Cruz, R. S., & Medeiros, E. A. A., (2012). Zinc oxide nanoparticles: Synthesis, antimicrobial activity and food packaging applications. *Food and Bioprocess Technology, 5*(5), 1447–1464.

19. Fang, Z., & Bhandari, B., (2010). Encapsulation of polyphenols: A review. *Trends in Food Science and Technology, 21*, 510–523.

20. Fathi, M., Donsi, F., & McClements, D. J., (2018). Protein-based delivery systems for the nanoencapsulation of food ingredients. *Comprehensive Reviews in Food Science and Food Safety, 17*(4), 920–936.

21. Fathi, M., Martin, A., & McClements, D. J., (2014). Nanoencapsulation of food ingredients using carbohydrate-based delivery systems. *Trends in Food Science and Technology, 39*(1), 18–39.

22. Fathi, M., Mozafari, M. R., & Mohebbi, M., (2012). Nanoencapsulation of food ingredients using lipid-based delivery systems. *Trends in Food Science and Technology, 23*(1), 13–27.

23. Feng, J., Lin, C., Wang, H., & Liu, S., (2017). Gemini dodecyl O-glucoside-based vesicles as nanocarriers for catechin laurate. *Journal of Functional Foods, 32,* 256–265.

24. Francis, S., Joseph, S., Koshy, E. P., & Mathew, B., (2018). Microwave-assisted green synthesis of silver nanoparticles using leaf extract of *Elephantopus scaber* and its environmental and biological applications. *Artificial Cells, Nanomedicine, and Biotechnology, 46*(4), 795–804.

25. Gallarate, M., Trotta, M., Battaglia, L., & Chirio, D., (2009). Preparation of solid lipid nanoparticles from W/O/W emulsions: Preliminary studies on insulin encapsulation. *Journal of Microencapsulation, 26*(5), 394–402.

26. Garg, S., (2013). Microwave-assisted rapid green synthesis of silver nanoparticles using *Saraca indica* leaf extract and their antibacterial potential. *International Journal of Pharmaceutical Sciences and Research, 4*(9), 3615–3620.

27. Gaskell, G., Ten, E. T., & Jackson, J., (2005). Imagining nanotechnology: Cultural support for the technological innovation in Europe and United States. *Public Understanding of Science, 14*(1), 81–90.

28. Ghosh, A., Mandal, A. K., Sarkar, S., Panda, S., & Das, N., (2009). Nanoencapsulation of quercetin enhances its dietary efficacy in combating arsenic-induced oxidative damage in the liver and brain of rats. *Life Sciences, 84,* 75–80.

29. Gowramma, B., Keerthi, U., Mokula, R., & Rao, D. M., (2015). Biogenic silver nanoparticles production and characterization from native stain of *Corynebacterium* species and its antimicrobial activity. *Biotechnology, 5,* 195–201.

30. Hu, A. W., & Fu, Z. H., (2003). Nanotechnology and its application in packaging and packaging machinery. *Packaging Engineering, 24,* 22–24.

31. Hussain, I., Singh, N. B., Singh, A., Singh, H., & Singh, S. C., (2016). Green synthesis of nanoparticles and its potential application. *Biotechnology Letters, 38*(4), 545–560.

32. Jafari, S. M., (2017). Introduction to nanoencapsulation techniques for the food bioactive ingredients. In: Jafari, S. M., (ed.), *Nanoencapsulation of Food Bioactive Ingredients* (pp. 1–62). Cambridge, MA-USA: Academic Press.

33. Katouzian, I., & Jafari, S. M., (2016). Nano-encapsulation as a promising approach for targeted delivery and controlled release of vitamins. *Trends in Food Science and Technology, 53,* 34–48.

34. Kaushik, N., Thakkar, M. S., Snehit, S., Mhatre, M. S., Rasesh, Y., & Parikh, M. S., (2010). Biological synthesis of metallic nanoparticles. *Nanomedicine: Nanotechnology, Biology, and Medicine, 6,* 257–262.

35. Keyes, M. H., & Saraswathi, S., (1985). Immobilized enzymes. In: Gebelein, C. G., (ed.), *Bioactive Polymeric Systems* (pp. 249–278). Boston - USA: Springer.

36. Koupantsis, T., Pavlidou, E., & Paraskevopoulou, A., (2014). Flavor encapsulation in milk proteins -CMC coacervate-type complexes. *Food Hydrocolloids, 37,* 134–142.

37. Kowsalya, E., Mosa-Christas, K., Balashanmugam, P., & Rani, J. C., (2019). Biocompatible silver nanoparticles/poly(vinyl alcohol) electrospun nanofibers for potential antimicrobial food packaging applications. *Food Packaging and Shelf-Life, 21,* 1–8.

38. Krishnaswamy, K., & Orsat, V., (2017). Sustainable delivery systems through green nanotechnology. In: Grumezescu, A. M., (ed.), *Nano-and Microscale Drug Delivery Systems* (pp. 17–32). London-U.K.: Elsevier.

39. Kumar, I., Mondal, M., & Sakthivel, N., (2019). Green synthesis of phytogenic nanoparticles. In: Shukla, A., & Iravani, S., (eds.), *Green Synthesis, Characterization, and Applications of Nanoparticles* (pp. 37–73). London - UK: Elsevier.

40. Kuswandi, B., (2017). Environmental friendly food nano-packaging. *Environmental Chemistry Letters, 15*(2), 205–221.

41. Lakkis, J. M., (2007). Introduction. In: Lakkis, J. M., (ed.), *Encapsulation and Controlled Release: Technologies in Food Systems* (pp. 1–12). Iowa - USA: Blackwell Publishing.

42. Li, X., Xu, H., Chen, Z. S., & Chen, G., (2011). Biosynthesis of nanoparticles by microorganisms and their applications. Journal of Nanomaterials, *2011*, 1–16. Article ID 270974.

43. Liu, S. C., Li, R., Tomasula, P. M., Sousa, A. M., & Liu, L., (2016). Electrospun food-grade ultrafine fibers from pectin and pullulan blends. *Food and Nutrition Sciences, 7*(7), 636–646.

44. Livney, Y. D., (2015). Nanostructured delivery systems in food: Latest developments and potential future directions. *Current Opinion in Food Science, 3*, 125–135.

45. Mascheroni, E., Fuenmayor, C. A., Cosio, M. S., & Di Silvestro, G., (2013). Encapsulation of volatiles in nanofibrous polysaccharide membranes for humidity triggered release. *Carbohydrate Polymers, 98*(1), 17–25.

46. McClements, D. J., (2015). Nanoscale nutrient delivery systems for food applications: Improving bioactive dispersibility, stability, and bioavailability. *Journal of Food Science, 80*(7), 1602–1611.

47. McClements, D. J., & Gumus, C. E., (2016). Natural emulsifiers-biosurfactants, phospholipids, biopolymers, and colloidal particles: Molecular and physicochemical basis of functional performance. *Advances in Colloid and Interface Science, 234*, 3–26.

48. McClements, D. J., Li, F., & Xiao, H., (2015). The nutraceutical bioavailability classification scheme: Classifying nutraceuticals according to factors limiting their oral bioavailability. *Annual Review of Food Science and Technology, 6*, 299–327.

49. Mezzenga, R., & Fischer, P., (2013). The self-assembly, aggregation and phase transitions of food protein systems in one, two, and three dimensions. *Reports on Progress in Physics, 76*(4), 1–43.

50. Momin, J. K., Jayakumar, C., & Prajapati, J. B., (2013). Potential of nanotechnology in functional foods. Emirates Journal of Food *and* Agriculture, *25*(1), 10–19.

51. Mozafari, M. R., Flanagan, J., MatiaMerino, L., & Awati, A., (2006). Recent trends in the lipid-based nanoencapsulation of antioxidants and their role in foods. *Journal of the Science of Food and Agriculture, 86*(13), 2038–2045.

52. Ortega, F., Giannuzzi, L., Arce, V. B., & García, M. A., (2017). Active composite starch films containing green synthesized silver nanoparticles. *Food Hydrocolloids, 70*, 152–162.

53. Patel, A. R., & Velikov, K. P., (2011). Colloidal delivery systems in foods: A general comparison with oral drug delivery. *LWT-Food Science and Technology, 44*(9), 1958–1964.

54. Patel, A., Patra, F., Shah, N., & Khedkar, C., (2018). Application of nanotechnology in the food industry: Present status and future prospects. In: Grumezescu, A., & Holban,

A. M., (eds.), *Impact of Nanoscience in the Food Industry* (Vol. 12, pp. 1–27). The Handbook of Food Bioengineering series; New York-USA: Academic Press.

55. Peralta-Videa, J. R., Huang, Y., Parsons, J. G., Zhao, L., Lopez-Moreno, L., Hernandez-Viezcas, J. A., & Gardea-Torresdey, J. L., (2016). Plant-based green synthesis of metallic nanoparticles: Scientific curiosity or a realistic alternative to chemical synthesis? *Nanotechnology for Environmental Engineering, 1*(1), 4–10.

56. Pérez-López, B., & Merkoçi, A., (2011). Nanomaterials based biosensors for food analysis applications. *Trends in Food Science and Technology, 22*(11), 625–639.

57. Peters, R. J. B., Bouwmeester, H., & Gottardo, S., (2016). Nanomaterials for food products and applications in agriculture, feed, and food. *Trends in Food Science and Technology, 54*, 155–164.

58. Philip, D., (2009). Honey mediated green synthesis of gold nanoparticles. *Spectrochimica Acta Part A, 73*(4), 650–653.

59. Pimtong-Ngam, Y., Jiemsirilers, S., & Supothina, S., (2007). Preparation of tungsten oxide-tin oxide nanocomposites and their ethylene sensing characteristics. Sensors *and* Actuators *A: Physical, 139*, 7–11.

60. Qidwai, A., Kumar, R., Shukla, S. K., & Dikshit, A., (2018). Advances in biogenic nanoparticles and the mechanisms of antimicrobial effects. *Indian Journal of Pharmaceutical Sciences, 80*(4), 592–603.

61. Rai, M., Ingle, A. P., Gupta, I., Pandit, R., Paralikar, P., Gade, A., Dos, S. C. A., (2019). Smart nano packaging for the enhancement of food shelf-life. *Environmental Chemistry Letters, 17*(1), 277–290.

62. Rajeshkumar, S., (2016). Synthesis of silver nanoparticles using fresh bark of *Pongamia pinnata* and characterization of its antibacterial activity against gram-positive and gram-negative pathogens. *Resource-Efficient Technologies, 2*(1), 30–35.

63. Rajput, J. K., (2018). Bio-polyphenols promoted green synthesis of silver nanoparticles for facile and ultrasensitive colorimetric detection of melamine in milk. *Biosensors and Bioelectronics, 120*, 153–159.

64. Ravichandran, R., (2010). Nanotechnology applications in food and food processing: Innovative green approaches, opportunities and uncertainties for global market. *International Journal of Green Nanotechnology: Physics and Chemistry, 1*(2), 72–96.

65. Restuccia, D., Spizziri, U. G., Parisi, O. I., Cirillo, G., Curio, M., Iemma, F., Puoci, F., Vinci, G., & Picci, N., (2010). New EU regulation aspects and global market of active and intelligent packaging for food industry applications. *Food Control, 21*, 1425–1435.

66. Rezvani, M., Hesari, J., Peighambardoust, S. H., Manconi, M., Hamishehkar, H., & Escribano-Ferrer, E., (2019). Potential application of nanovesicles (niosomes and liposomes) for fortification of functional beverages with isoleucine-proline-proline: A comparative study with central composite design approach. *Food Chemistry, 293*, 368–377.

67. Rubio, A. L., Gómez-Mascaraque, L. G., Fabra, M. J., & Sanz, M. M., (2019). Nanomaterials for food applications: General introduction and overview of the book: Chapter 1. In: Amparo, L. R., Rovira, M., Sanz, M., & Gomez-Mascaraque, L. G., (eds.), *Nanomaterials for Food Applications* (pp. 1–9). London - UK: Elsevier.

68. Saha, S. K., Chowdhury, P., Saini, P., Babu, S. P. S., (2014). Ultrasound-assisted green synthesis of poly(vinyl alcohol) capped silver nanoparticles for the study of its antifilarial efficacy. *Applied Surface Science, 288*, 625–632.

69. Said, M. I., & Othman, A. A., (2019). Fast green synthesis of silver nanoparticles using grape leaves extract. *Materials Research Express, 6*(5), 1–22.

70. Samaranayaka, A. G. P., Li-Chan, E. C. Y., (2011). Food-derived peptidic antioxidants: A review of their production, assessment, and potential applications. *Journal of Functional Foods, 3*(4), 229–254.

71. Scheafer-Korting, M., & Mehnert, W., (2005). Delivery of lipophilic compounds with lipid nanoparticles-applications in dermatics and for transdermal therapy. In: *Lipospheres in Drug Targets and Delivery* (pp. 170–184). Boca Raton - FL: CRC Press.

72. Scroccarello, A., Della, P. F., Neri, L., Pittia, P., & Compagnone, D., (2019). Silver and gold nanoparticles based colorimetric assays for the determination of sugars and polyphenols in apples. *Food Research International, 119*, 359–368.

73. Sercombe, L., Veerati, T., Moheimani, F., Wu, S. Y., Sood, A. K., & Hua, S., (2015). Advances and challenges of liposome assisted drug delivery. *Frontiers in Pharmacology, 6*, 286–299.

74. Sharma, P. K., Saxena, P., Jaswanth, A., Chalamaiah, M., Tekade, K. R., & Balasubramaniam, A., (2016). Novel encapsulation of lycopene in niosomes and assessment of its anticancer activity. *Journal of Bioequivalence and Bioavailability, 8*(5), 224–32.

75. Silva, L. P., Pereira, T. M., & Bonatto, C. C., (2019). Frontiers and perspectives in the green synthesis of silver nanoparticles. In: Iravani, S., (ed.), *Green Synthesis, Characterization and Applications of Nanoparticles* (pp. 137–164). London - UK: Elsevier.

76. Taccola, S., Greco, F., Zucca, A., Innocenti, C., De JuliánFernández, C., Campo, G., & Mattoli, V., (2013). Characterization of freestanding PEDOT: PSS/iron oxide nanoparticle composite thin films and application as conformable humidity sensors. *ACS Applied Materials and Interfaces, 5*(13), 6324–6332.

77. Timilsena, Y. P., Wang, B., Adhikari, R., & Adhikari, B., (2017). Advances In microencapsulation of polyunsaturated fatty acids (PUFAs)-rich plant oils using complex coacervation: A review. *Food Hydrocolloids, 69*(SC), 369–381.

78. Varshosaz, J., Ghaffari, S., Khoshayand, M. R., Atyabi, F., Azarmi, S., & Kobarfard, F., (2010). Development and optimization of solid lipid nanoparticles of amikacin by central composite design. *Journal of Liposome Research, 20*, 97–104.

79. Vijayan, R., Joseph, S., & Mathew, B., (2018). *Indigo feratinctoria* leaf extract mediated green synthesis of silver and gold nanoparticles and assessment of their anticancer, antimicrobial, antioxidant, and catalytic properties. *Artificial Cells, Nanomedicine, and Biotechnology, 46*(4), 861–871.

80. Wang, T., Hu, Q., Zhou, M., Xia, Y., Nieh, M. P., & Luo, Y., (2016). Development of all-natural layer-by-layer redispersible solid lipid nanoparticles by nanospray drying technology. *European Journal of Pharmaceutics and Biopharmaceutics, 107*, 273–285.

81. Washington, C., (1998). Drug release and interfacial structure in emulsions. In: Miller, R. H., Benita, S., & Bohm, B., (eds.), *Emulsions and Nanosuspensions for the Formulation of Poorly Soluble Drugs* (pp. 101–117). Berkin Waldstraße-Stuttgart, Germany: MedPharm Scientific Publishers.

82. Weiss, J., Gaysinsky, S., Davidson, M., & McClements, J., (2009). Nanostructured encapsulation systems: Food antimicrobials. In: *Global Issues in Food Science and Technology* (pp. 425–479). New York: Academic Press.

83. Wong, S., & Karn, B., (2012). Ensuring sustainability with green nanotechnology. *Nanotechnology, 23*(29), 1, 2.
84. Wongsasulak, S., Patapeejumruswong, M., Weiss, J., Supaphol, P., & Yoovidhya, T., (2010). Electrospinning of food-grade nanofibers from cellulose acetate and egg albumen blends. *Journal of Food Engineering, 98*(3), 370–376.
85. Wu, J., & Yin, F., (2013). Sensitive enzymatic glucose biosensor fabricated by electrospinning composite nanofibers and electrodepositing Prussian blue film. *Journal of Electroanalytical Chemistry, 694*, 1–5.
86. Yan, C., Zhang, W., Gaonkar, A. G., & Vasisht, N., (2014). Coacervation processes. In: *Microencapsulation in the Food Industry* (pp. 125–137). San Diego, CA - USA: Academic Press.
87. Yeo, Y., Baek, N., & Park, K., (2001). Microencapsulation methods for delivery of protein drugs. *Biotechnology and Bioprocess Engineering, 6*, 213–230.
88. Zhang, S., & Kawakami, K., (2010). One-step preparation of chitosan solid nanoparticles by electrospray deposition. *International Journal of Pharmaceutics, 397*, 211–217.
89. Zimet, P., & Livney, Y. D., (2009). Beta-lactoglobulin and its nanocomplexes with pectin as vehicles for omega-3 polyunsaturated fatty acids. *Food Hydrocolloids, 23*, 1120–1126.

CHAPTER 7

FOOD ENCAPSULATION: PRINCIPLES, NOVEL METHODS, AND APPLICATIONS

NIKUNJ SHARMA, SYED MANSHA RAFIQ, and SYED INSHA RAFIQ

ABSTRACT

The bioactive compounds in foods are highly unstable and prone to oxidation. The encapsulation methods (such as spray drying, fluid bed coating, spray chilling, spray bed drying, microwave drying, freeze-drying, melt injection, melt extrusion, electrospinning, and electrospraying) employ a coating (encapsulating material) to cover these bioactive components (such as vitamins, minerals, peptides, fatty acids, antioxidants, lipids, and probiotics, etc.), to protect them from migration, evaporation or degradation of volatiles (aroma), etc., and to maintain their stability. The encapsulating material may be of carbohydrate/protein/fat origin. This chapter gives an insight to various encapsulation techniques, their advantages, and applications in the food industry.

7.1 INTRODUCTION

Encapsulation is a technique to provide a barrier around a substance to protect it from the outside environment [103]. It is one of the useful tools to cause less harm to the food components that may have physiological effects on our body. The encapsulated substances include bioactive compounds, food ingredients, enzymes, cells, lipids, acids, bases, probiotics, flavoring agents, sweeteners, colorants, leavening agents, preservatives, etc., [7]. This chapter focuses on the use of various encapsulation technologies to prevent the bioactive components from deterioration by contact with the external environment. It also deals with various techniques, their advantages, and applications in the food industry.

7.2   ENCAPSULATION TECHNOLOGY AND ITS ADVANTAGES

According to the size of the capsule, terms microcapsule (1–5000 μm) or nanocapsule (<1 μm) may be used [79]. The outer cover may be called carrier, wall-material, membrane, shell, coating, capsule, external phase or matrix, whereas the encapsulated material may be called fill, core material, active agent, internal phase, payload, etc., which should be food grade and able to form barrier for the core material [75]. This technique can be used to release the core material slowly or when desired. The encapsulating materials may be derived from fats, alginates, protein, starches, etc., [31]. Two representative forms are produced by the encapsulation technique (Figure 7.1).

The product formed after the encapsulation process is called encapsulated product or microcapsules or microspheres. The microcapsules are also known as reservoir, and the microspheres are also known as matrix [99]. The reservoir type encapsulation involves a shell surrounding the active agents. Other names include mono core or single core, or capsules. In case of matrix type encapsulation, the actives are well distributed within or over the coating material. Unlike the reservoir type, the matrix type may occur on the surface of the shell. These may be of varying shapes, such as spherical, oval, or any irregular shape [121].

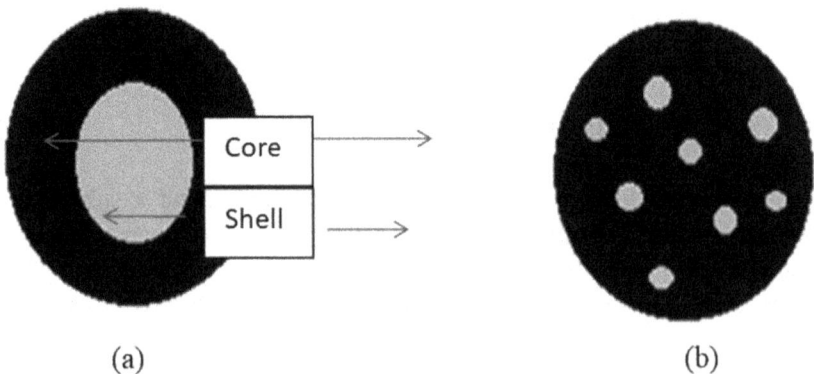

(a)   (b)

FIGURE 7.1   (A) Microcapsules (reservoir); and (B) microspheres (matrix).

The encapsulation finds wide application in areas including textile, pharmaceutical, cosmetics, drugs, biomedical, biotechnology, electronics, agriculture, and food [72]. Further, the encapsulation technique is employed in various segments within the food industries. These include oil, beverage, baked goods, probiotics, meat, poultry, dairy, and fermented foods [83].

The encapsulation technique provides various advantages [35, 71, 76], such as:

- Can be used to extract important components from a mixture and preserved;
- Control the release of internal phase at the right time when desired;
- Conversion of liquid to solid form;
- Easier handling of core material (due to changed physical characteristics);
- Enhance the shelf life, handling, and flow properties;
- Improve solubility and stability of sensitive compounds;
- It also helps in immobilization of enzymes and cells;
- Maintain viability or increase bioavailability of specific ingredients;
- Mask unwanted smell or odor or taste or flavor;
- Protecting the ingredients or flavors from the external environment (oxygen, moisture, heat, light, free radicals or several other extreme conditions, etc.);
- The transfer rate of active agents to the external environment is reduced.

7.3   ENCAPSULATION AND IMMOBILIZATION

The terms encapsulation and immobilization are used interchangeably, but they differ based on various criteria (Table 7.1).

TABLE 7.1   Encapsulation versus Immobilization

Points of Difference	Encapsulation	Immobilization
Definition	Process of entrapping one material (core material or active agent) into another material (wall material or shell) thereby producing micro or nanosized particles [117].	Physical confinement or localization in a specific region of space that helps in retaining the catalytic properties and viability of the cells which may be used repeatedly and continuously [64].

TABLE 7.1   *(Continued)*

Points of Difference	Encapsulation	Immobilization
Other definition	In other words, it is the process of formation of continuous wall surrounding the core which is fully contained within the coating material [43].	It is the process of entrapment of material within or throughout a matrix [43].
Methods	Spray drying, spray chilling/cooling, fluid bed coating, melt injection, melt extrusion, coacervation, liposome entrapment, emulsification, Inclusion complexation, freeze-drying [121].	Adsorption, covalent bonding, encapsulation, entrapment, and cross-linking [17].
Exposure	No leaks at the surface [109].	Leaks may be present [109].

7.4   CHARACTERISTICS OF GOOD COATING MATERIALS

The wall material must possess certain characteristics to develop a stable capsule with desired functions. The characteristics for production of an idea microcapsule are as follows [89]:

- Capable of providing a safe and protective environment for the active agent against the outside environment such as oxygen, heat, free radicals, etc.
- Capable to retain the core materials in its boundary during the processing and storage.
- Good flow properties at higher amounts or easier to handle with better stability.
- Must be able to release the active agents under desired conditions.
- Must be chemically inert with the actives during processing or storage.
- Must be food grade and able to emulsify the core materials to form a stable emulsion.
- Upon drying, the wall material must be capable of releasing the solvent employed during such processes.

7.5 MATERIALS USED IN ENCAPSULATION

The major components of encapsulation include core material and shell. Different materials can be employed for the preparation of microcapsules (Table 7.2).

TABLE 7.2 Commonly Used Core and Wall Materials for Preparation of Microcapsules [53, 87]

Core Material	Wall Material
Flavors: essential oils, d-limonene, resins	Starches: modified starch, maltodextrins, cellulose esters (carboxy methylcellulose, cellulose acetate phthalate), chitosan
Oils and lipids: butter, fat, fish oils or liver oils	Proteins: Casein, gluten, soy protein, gelatin, milk protein, albumin, peptides
Food additives: preservatives, colorants, antioxidants, vitamins, and minerals	Gums: acacia gum, gum Arabic, sodium alginate, gum tragacanth, carrageenan
Bioactive ingredients: enzymes, probiotics, lactic acid bacteria	Other materials: cyclodextrins, liposomes, sucrose, dextrose, lipids (wax, beeswax, paraffin)

7.6 METHODS OF CONTROLLED RELEASE OF THE CORE MATERIAL

Encapsulation provides one of the advantages of release of core material at a desired rate. In other words, the controlled release may be defined as the way of releasing one or more active agents at a desired site at the right time of the desired rate [67]. The following four methods can be used to release the active agents in a controlled manner [47]:

- Fracturation (such as pressure/shear).
- Diffusion (wall acts as semi-permeable membrane).
- Melting (by heat) or Dissolution (in a solvent).
- Biodegradation (through enzymes).

7.7 METHODS OF ENCAPSULATION

Different techniques have been employed for the encapsulation of food materials. These techniques may be classified into three broad categories [1,

53, 73, 109]: Physical, chemical, and physicochemical (as shown in Figure 7.2).

7.7.1 PHYSICAL METHODS

7.7.1.1 SPRAY DRYING

It is one of the oldest and technologically efficient techniques to produce encapsulated materials. It works on the principle of atomization of liquid product into a dry powder using hot air or inert gas such as nitrogen [37]. This unit operation is also used as a dehydration technique which finds wide application in food industry, such as vitamins (ascorbic acid, β-carotene), antioxidant's colors (lycopene, bixin, anthocyanins), flavors (limonene, citrus peel oil, cardamom essential oil), essential fatty acids (omega-3 fatty acids), minerals (Fe, Na, P, K, I, Ca) [73] and probiotics [110].

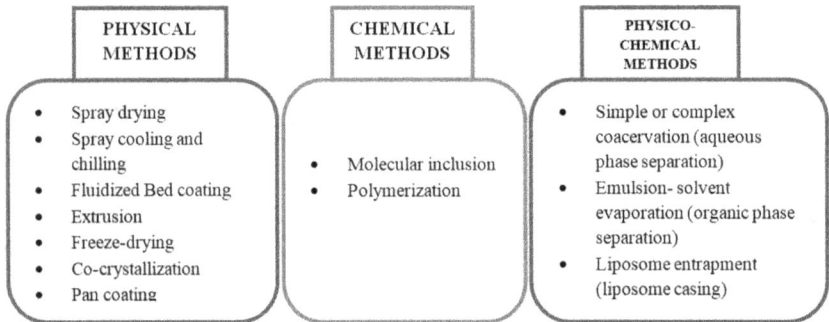

FIGURE 7.2 Methods of encapsulation.

In this method, following three major steps are involved [45, 68, 86, 121]:

1. **Dissolving Active Agent in Hydrated Wall Material:** This involves the preparation of emulsion by dispersing the core material with the wall material (1:4) with or without the addition of emulsifiers.
2. **Atomization of Emulsion:** The fine and stable emulsion thus formed is then heated and homogenized to form smaller droplets to undergo atomization process using the spray dryer. The homogenization step is important as the initial diameter of oil droplets must range between 1 and 100 μm. In addition, the solution must be kept undisturbed and

stable for a certain period of time. The fine emulsion thus prepared is now passed through the nozzle or atomizer to atomize the solution.

3. **Dehydration of Atomized Particles:** The hot air or steam or inert gas (N_2) flowing in the same or opposite direction then comes in contact with the atomized particles. This causes evaporation of water and formation of spherical droplets with core material enclosed in the wall material. The dried encapsulated products are collected at the bottom of the equipment in the collection chamber.

This unit operation has the following advantages [8, 37]:

- Simple to perform;
- Processing cost is low;
- Equipment (spray dryer) is readily available;
- Better protection to the active agents;
- Efficient.

The spray drying technique has also the following disadvantages [93]:

- Production of very fine particles that need further processing such as agglomeration (Fluid bed coating is employed to form agglomerated products);
- Not suitable for heat-sensitive materials.

7.7.1.2 SPRAY CHILLING AND COOLING

Spray chilling and cooling are similar to spray drying with the exception of the presence of a cold chamber, where the particles are formed by chilling in the presence of cold air [56, 78, 93]. It works on the principle of atomization of liquid product, which is passed through the vessel containing CO_2 ice bath, where the atomized particles stick to the wall material and form microcapsules. This technique finds application in dry soup mixes, bakery foods, and foods with high level of fat [67]. It finds many applications, such as flavors, vitamins, and minerals, prebiotics, probiotic microorganisms, etc., [28]. The wall material used may include fats, lipids, waxes, etc.

The core material (hydrophilic in nature) is mixed with the wall material (hydrophobic in nature) [96] and emulsified to form a stable emulsion, which is then homogenized and passed through the pneumatic nozzle for atomization. The atomized particles reach the vessel containing a cool chamber where these droplets adhere on the surface of particles, thus solidifying to form a protective coat [67].

7.7.1.3   FLUIDIZED BED COATING

Fluid bed coating is also known as air suspension method [65]. It works on the principle of spraying of shell through the nozzle into the fluidized bed containing active agents in hot chamber [103]. It finds wide application in encapsulation of food ingredients such as enzymes, preservatives, flavors, spices, leavening agents, fortifiers, etc., [33].

In this method [8, 52], the first stage involves the fluidization of active agents through the hot air present in the coating chamber, which is followed by the spraying of the wall material through the nozzle to form a protective coating layer surrounding the core material. The final stage involves the alternate wetting and drying processes [97]. It has following advantages [50, 105]:

- Requires less processing time and energy;
- No particle shrinkage;
- Cost-effective and flexible;
- No involvement of solvent and additional drying step;
- Higher rate of mass transfer, high heat and proper particles mixing.

The disadvantages of this unit operation include [59, 68]:

- Less production due to smaller size of batch (50 g in one cycle);
- Require additional processing time and equipment for shell preparation;
- It degrades the highly heat-sensitive materials.

7.7.1.4   EXTRUSION

The extrusion method is also considered as a true encapsulation method as it surrounds the core materials completely [93]. Extrusion method works on the principle of passing the active agents in the molten shell mass through small openings followed by the bath of desiccant solution that results in hardening of the shell thus entrapping the actives [60, 103]. Extrusion is a method which, utilizes a combination of high pressure and temperature The technique finds application in preservation of volatile and unstable flavors in glassy carbohydrate matrices [14], preparation of liposomes by filter extrusion [15], encapsulation of canola oil with alginate [112], encapsulation of probiotics [110], polyphenolic antioxidants from medicinal plant extracts in alginate-chitosan system [13] and preparation of immobilized lactic cultures in gels [75].

The first step in extrusion involves the preparation of molten polymer matrix in the presence of high temperature (110°C). The next step involves the incorporation of active agents into the polymer matrix through mixing or high shear [20, 65, 121]. The final step involves the passing of the emulsion through the small opening called die. The pressure usually employed is <100 psi and temperature reach above 115°C [90]. The strands obtained after passing through the die is cooled and dehydrated in desiccant liquid. The strands are further broken down into smaller pieces (1 mm), separated from the IPA and dried (maybe vacuum dried). The final dried microcapsules are produced [67]. The desiccant liquid generally used is IPA (isopropyl alcohol), which has following uses [67]:

1. Dehydration and cooling of strands (drying agent);
2. Hardens or solidifies the mass (transition to glassy state);
3. Surrounding the active agents;
4. Excess residual oil or active agents are removed;
5. Provide longer shelf life.

The advantages of extrusion method include [24, 29, 67]:

• Simple to perform;
• Inexpensive and flexible;
• Increased viability of probiotic cells;
• Production of stable flavors;
• Prevention of oxidation of volatile and unstable flavors or oils;
• Glassy carbohydrate matrices possess good barrier properties;
• Encapsulation of microbes can be performed in aerobic or anaerobic environment with no involvement of hazardous solvents.

The extrusion method the disadvantage of lesser production rate [70]. Although for the scaling up issue, the multiple nozzles are introduced and employed in industries [65].

7.7.1.5   FREEZE DRYING

Freeze-drying/cryo-desiccation/lyophilization is an expensive method for the production of heat-sensitive compounds [39]. It works on the principle of entrapping the emulsion (solution prepared by mixing core and shell) by lyophilization [103]. It finds wide applications in encapsulation of DHA [61], chemicals such as limonene [62], pigments such as β-carotene [30], garcinia fruit extract [38]. The basic method of formation of encapsulated product includes three main steps [56, 79, 92, 121]:

1. The preparation of core material involves the formation of dispersion of core material in the shell (using water) which is followed by the freeze-drying step.
2. Freeze-drying involves the freezing step, wherein the conversion of the majority of water to ice takes place in several hours, which is followed by the primary drying step or sublimation of the ice formed in the presence of low temperature and pressure, which results in removal of ice.
3. The final step is the secondary drying or desorption, which involves the removal of water (unfrozen or bound) which is performed until the desired moisture content is attained.

Advantages of freeze-drying [61, 65, 92] are:

- Avoid particles shrinking partly;
- Enhanced stability of active agents;
- Good encapsulation efficiency;
- Help maintain the particle size;
- Highly porous particles result;
- Offer better stability to nano-capsules;
- Used for encapsulation of heat-sensitive actives.

Disadvantages of freeze-drying [41, 61] are:

- High operating costs;
- Time consuming;

7.7.1.6 CO-CRYSTALLIZATION

Co-crystallization is defined as a technique involving concentration of sucrose solution by crystallization, which causes supersaturation of solution, thus forming the microcrystals of sucrose. It is then followed by the incorporation of active agents in the porous microcrystals. Further, it is subjected to an intense agitation that causes nucleation and agglomeration of the final product [6]. In a recent study, the incorporation of bio-macromolecules via this method is performed successfully using hemacyanin crystals [48].

It works on the principle of formation of an irregular shaped crystal, which is known as conglomerate from a perfect crystal via spontaneous crystallization at higher temperatures (above 120°C) that creates huge space or voids inside the crystals; and the core materials can be incorporated or fused into the spaces resulting in the encapsulated products [39].
Advantages are [16, 85]:

- Economic and flexible in nature;
- Relatively simple to perform;
- Enhanced flowability, dispersing power, stability, solubility, wettability, hydration, anticaking, and homogeneity of the microcapsules.

7.7.1.7 PAN COATING

Pan coating is used in the pharmaceutical industries [109]. It works on the principle of tumbling of particles in a pan followed by wall material being applied slowly to form coated tablets/particles of small sizes [2]. In this method, the first step involves the mixing of solid particles and dry coating material followed by an increase in temperature in order to meet the shell and surround the active agents. The final step involves the hardening through cooling [54].

An alternative method may also be utilized, which includes the spraying of the wall material (in atomized form) on the solid particles (active agents) in the coating pans. In addition, for effective coating, the shell employed for this purpose must be greater than 600 μm [103]. In practice, the hot air is passed over the coated mass while applying the wall materials in order to remove the coating solvent or may be dried in an oven [9].

7.7.2   CHEMICAL METHODS

7.7.2.1   MOLECULAR INCLUSION

Molecular inclusion in the formation of a special complex of one molecule, referred to as guest molecule, incorporated partially or completely in the second component's cavity structure, referred to as host molecule. The host molecules generally employed is cyclodextrins or their derivatives and guest molecules include certain hydrophobic drugs with low molecular weight [49]. The cavity is hydrophobic in nature, and the walls are hydrophilic in nature. Applications of these methods are: inclusion of essential vitamins (nicotinic acid, ascorbic acid) into β-cyclodextrins [95], rifampicin as an inclusion complex of hydroxypropyl-cyclodextrin [49], flavors as helical inclusion complexes of amylose [114], herbicide terbuthylazine in native and modified β-cyclodextrin [42].

It works on the principle of entrapping the polar molecule via hydrophobic interactions taking place in the cavity of β-cyclodextrin, thus replacing water molecules [103, 121]. Cyclodextrins are most widely used as the host molecule for molecular inclusion due to the non-toxicity, physical, and chemical stability. The guest molecule (cyclodextrins) consists of an internal non-polar hole and hydroxyl (-OH) groups on the surface. The process of molecular inclusion of hydrophobic compounds includes hydrophobic interactions between guest molecules and the walls of the cyclodextrin cavity. Many other forces may also be involved in such interactions, such as weak van der walls forces and dipole-dipole interactions [22]. The methods for preparation of inclusion complex can include [21, 44]:

1. Co-evaporation;
2. Drying techniques (freeze-drying and spray drying);
3. Complex formation in aqueous phase;
4. Kneading.

Advantages [66, 88] are:

- Decreased volatility and toxicity of core materials;
- Enhance retention of active agents (due to increase in size and increased hydrophilicity);
- improved bioavailability of components, odor masking and easier handling;
- Improves solubility;

- Simple;
- Stable complexes are formed.

Disadvantages are [21]:

- Cost of preparation is high;
- Difficulty in incorporating into formulation of dosage forms;
- High manufacturing cost;
- Laborious work;
- Only minute concentrations of drugs form complexes.

7.7.2.2   INTERFACIAL POLYMERIZATION

This method uses the polymerization technique to produce microcapsules with protective coverings, which generally include the interaction of single units located at the interface, i.e., between the active agents and a wall supporting phase in which the core material is dispersed. The core material supporting phase is also termed as continuous phase, which mostly consists of a liquid or a gas; and the polymerization reaction occurs at a liquid-liquid, liquid-gas, solid-liquid, or solid-gas interface [54, 80 84]. It finds application in the formation of polymer nanoparticles (NPs) [27, 74, 101].

This method involves the reaction at the interface of two liquids. It is also considered as a step polymerization process of two reactive monomeric units, which are dissolved in two phases, respectively (continuous and dispersed phase) [74]. In other words, the two reactants on interacting at an interface react at a faster rate. The basis of this method is the reaction between an acid chloride (ROCl) and a compound containing an active hydrogen atom, such as an amine (polyamides) or alcohol, polyesters, polyurea, polyurethane. Under optimum conditions, thin flexible walls form rapidly at the interface [9]. It finds many applications, such as formation of thin-film composite membranes employed in reverse osmosis and nano-filtration [101], ammonium sulfate with polyurethane shell [34], production of polypyrrole nanocapsules [119], fabrication of polyaniline composites [51], etc.

7.7.2.3   IN-SITU POLYMERIZATION

In-situ polymerization is a process of chemical reaction between two immiscible liquids (oil and water) in a continuous phase. The methods involved in these polymerizations are as follows [46]:

- **Emulsion Poly-Condensation:** In this method, the O/W (Oil-in-water) emulsion is obtained by the dispersion of an oil phase (containing oil (monomer) and an active agent) in the water phase containing an adequate amount of surfactant carried out by fast stirring and ultra-sonication [46].
- **Interfacial Poly-Condensation:** It involves the formation of a polymer wall at the interface of the two phases with each of them containing some monomer [102].
- **Suspension Poly-Condensation:** It involves the dissolution of water phase (monomer) into organic phase (core materials) followed by the formation of oil/water (O/W) emulsion that eventually leads to separation and precipitation of the monomer molecules from core materials and formation of solid shell [102].

7.7.3   PHYSICOCHEMICAL METHODS

7.7.3.1   COACERVATION OR PHASE SEPARATION

This encapsulation method is the phase separation of one or more hydro-colloids from an initial solution followed by deposition of the coacervates formed around the core material [32, 39, 58, 94]. It works on the principle of entrapping due to deposition of a liquid wall material surrounding the active agents by the electrostatic force of attraction [103]. It finds wide applications, such as incorporation of wheat proteins (gliadins) by simple coacervation technique to produce nanospheres [91], preparation of fatty acid lipid nanoparticle dispersions [25] and biofunctional hydrophobic compounds [111] and probiotics [10]. This method involves the following steps:

1. First step involves the formation of an emulsion (core material and coating material) by the processes, such as high shear, mechanical agitation, micro-fluidization, etc.
2. Second step involves the process of coacervation [111]: simple and complex (Table 7.3).
3. The third step is shell formation or hardening by cooling to low temperature to form the outer coating.
4. Fourth step, the cross-linking is done to form a stable 3-D network of microcapsules using crosslinkers such as glutaraldehyde [57, 111, 115].

TABLE 7.3 Differences between Simple and Complex Coacervation Techniques

Points of Difference	Simple Coacervation	Complex Coacervation	References
Number of polymers involved	One	Two or more (of opposite charges)	[32, 121]
Types of polymers involved	Gelatin, carboxymethyl cellulose or polyvinyl alcohol	Gum Arabic and gelatin; Gelatin and sodium carboxymethyl cellulose, gelatin, and alginate	[32, 115]
Driving force	Force-induced between charged food components and an oppositely charged capsule wall material	Electrostatic force of attraction between molecules with opposite charges	[104]
Method	Addition of non-solvent, ions, change in pH, ion concentration, temperature, etc.	Complex formation between polyelectrolyte polymers with opposite charges	[57, 65]
Nature of resultant complexes	Dehydrated molecules	Viscous and insoluble complexes	[65]

Advantages are [39, 54]:

- Bio-functional ingredients with higher value are obtained;
- Encapsulation of heat-labile polyphenols can be performed; and
- Involve its versatility and efficiency of controlling the particles size.

Disadvantages [4, 54, 67] are:

- Complicated;
- Expensive;
- Removal of volatiles by evaporation;
- Oxidation;
- Formation of irregular structure;
- Unstable in nature;
- Chemical agents involved are toxic in nature;
- The difficulty in selection of shells; and
- Difficult to scale up and agglomeration issues.

7.7.3.2   EMULSION-SOLVENT EVAPORATION (ORGANIC PHASE SEPARATION)

Solvent evaporation is a method to emulsify the polymer-organic solvent (volatile in nature) in water followed by removal of solvent [54]. It works on the principle of entrapping drug within the microspheres, which depend on the rapid precipitation rate of the polymer from the organic solvent phase, less hydrophilicity of the drug in the aqueous phase; and greater concentration of the polymer in the organic phase [19]. It finds many applications, such as encapsulation of probiotics [110], encapsulation of hydrophobic drugs in polymeric micelles of methoxy poly(ethylene oxide)-block-poly(-caprolactone) (MePEO-b-PCL) [3]. Encapsulation of insulin in PLGA (poly(lactide-co-glycolide)) microspheres [11], and development of SS-AG20 loaded polymeric microparticles [23], etc. It generally involves two main steps [120]:

1. The first step involves the dissolution of active agent and the polymer in an organic solvent (water-immiscible and volatile in nature) [103]. The solution thus obtained is mixed in an aqueous medium using stirrer to form O/W (oil in water) emulsion; and

2. The second step includes the evaporation of the solvent from the solution, which may be carried out by heating or continuous stirring (if necessary). This step induces precipitation of polymer followed by filtration and final drying to form the microcapsules [2]. In the case, where the active agent is dispersed in the polymer solution, the shrinkage of the polymer around the core takes place. In the case where active agent is dissolved in the coating polymer solution, a matrix-type microcapsule results [9].

Advantages [54, 58] are:

•   Cost-effective; and
•   Facilitate controlled release of drugs.

Disadvantages are:

•   Scaling up issues.

7.7.3.3   LIPOSOME ENTRAPMENT

It is an encapsulation of certain food ingredients. It works on the principle of entrapment of core material into the liposome via hydrophobic or hydrophilic interactions [103]. It finds wide application in pharmaceutical and cosmetics industry, such as preparation of encapsulated lead [63], encapsulated anticancer drugs [100], encapsulated antibiotic drugs such as griseofulvin [77], encapsulation of cheese ripening enzymes and antioxidants [71].

The formation of liposomes involves the interaction of hydrophobic (phospholipids) and hydrophilic (water) components [4]. Liposomes are biodegradable, non-toxic, and non-immunogenic, providing several advantages, such as carriers of the encapsulated molecules by improving their drug delivery along with target selective properties. Due to their structure, they can incorporate hydrophobic, hydrophilic, and amphiphilic molecules [98]. The components can be incorporated in the liposome into two ways:

1. In the active drug loading technique, the active agents are coated after liposome formation; and
2. In the passive method, the active agents are encapsulated during the liposome formation. The passive loading techniques involve three main methods [2]: (a) Mechanical dispersion (sonication, extrusion, freeze-thawing, lipid film hydration by freeze-drying or micro-emulsification); (b) Solvent dispersion (ether injection, ethanol injection and reverse-phase evaporation); and (c) Detergent removal or un-encapsulated material removal method (dialysis, absorption, gel permeation chromatography and dilution).

Advantages [2, 26] are:

* Increased drug activity against intracellular bacteria;
* Less toxic;
* Better target selectivity;
* Controlled release rate of incorporated compounds;
* Enhanced drugs solubility (lipophilic and amphiphilic); and
* Improved tissue penetration.

Disadvantages [26, 108] are:

- Expensive;
- Scaling up issues;
- Difficulty in liposome sterilization;
- shorter shelf-life;
- Low stability; and
- possibility of introduction of lower amounts of bioactives.

7.8 ENCAPSULATION METHODS OF FOOD COMPONENTS

7.8.1 ELECTROSPINNING

Electrospinning is an encapsulation technology for the production of nano-size polymer fibers. It involves the application of electric field for the conversion of polymer solution into fibrous structure followed by their continuous deposition on the grounded collector [113]. It finds wide application in areas of enzyme immobilization, biomedical science, textiles, piezoelectric, catalyst, electronics, sensors, energy devices, filtration [82], for the production of antioxidants (such as carotene, retinyl acetate, epigallocatechin gallate, gallic acid), and probiotics (such as *Bifidobacterium,* flavor such as menthol, vanillin) [76]. Electrospinning is classified as [36]: co-axial (separate solution of core material and polymer are introduced); and co-solubilization or emulsion method (prior mixing of core material and polymer is done).

Electrospinning consists of five basic components [82]: Source of high voltage; Grounded collector (flat pump or rotated); Syringe; and Stainless-steel needle or capillary. Cycle of events occurring during the electrospinning process [5] includes:

1. Melt polymer is introduced through the syringe pump that reaches the needle or capillary.
2. The source of high voltage potential provides free charges to the polymer solution.
3. These free charges are responsible for the generation of electrostatic forces within the particles of the solution.
4. These forces namely: electrostatic repulsion of like charges and the coulombic force of the external electric field, are responsible for the

conversion of initial hemispherical shape of the droplets to the shape of cone termed as 'Taylor cone.'

5. Ejection of a charged jet polymer from the Taylor cone (due to counteraction of surface tension with the electrostatic forces).
6. Polymer jet reaches the collector in whipping or bending motion due to an uneven charge distribution.
7. Elongation of jet (rate is generally 1000 sec^{-1}) [122].
8. Quick evaporation of the solvent.
9. Deposition of thin fibers (in a random manner) on the grounded collector.
10. Formation of non-woven mat.

Advantages [76, 113] are:

• Generation of organic or inorganic fibers (diameter <100 nm);
• Simple and cheap;
• No application of heat;
• Ease in incorporation of bioactive materials (vitamins, drugs, metals) in nanosize fibers; and
• Applicable for heat-sensitive materials.

Disadvantages [5, 40, 76] are:

• Low throughput;
• Low scaling up;
• Forms beads or porous tissues that affect the release into the target system.

7.8.2   ELECTROSPRAYING

Electrospinning is a technique for atomization of solution by means of an electrical force that helps to overcome the surface tension [18, 55]. It finds application in food industry, such as preparation of nutraceuticals (omega-3 fatty acids) [106], biomedical sciences [118], formation of biopolymer particles, antioxidants (β-carotene, lycopene), folic acid, DHA, α-linolenic acid [36], food coating, preparation of material for filtration and active packaging of food [5].

Tables 7.4 and 7.5 indicate some differences and similarities between electrospinning and electrospraying.

TABLE 7.4   Differences between Electrospinning and Electrospraying

Point of Difference	Electrospinning	Electrospraying	References
Product	Ultrafine threads or fibers	Ultrafine droplets or particles	[107]
Jet	Stabilized	Destabilized	[5]
Mechanism	Whipping instability	Varicose instability	
Polymer solution concentration	High	Low	[36]
Viscosity of solution used	High	Low	
Cohesion force	High	Low	
Types	Co-axial, emulsion, and co-solubilization	Colliding droplets of opposite polarity, co-axial, emulsion, chemical or ionic crosslinker and evaporation	[1, 36, 92]
Surfactant or plasticizer	Require surfactant or plasticizer	No surfactant or plasticizer required	[92]

TABLE 7.5   Similarities between Electrospinning and Electrospraying

Point of Similarity	Electrospinning and Electrospraying	References
Driving force	Electrostatic force	[36]
Components	Source for high voltage, needle or capillary, syringe pump and collector	
Type of treatment	Non-thermal	[92]
Core material incorporation techniques	Blending, co-axial process	[116]
Characteristics of structures obtained	Larger surface area and trapping efficiency	[106]
Application	Encapsulation of bioactive materials, such as antioxidants, immobilization of enzymes, also employed in mass spectrometry, electronics, nano-industry, tissue engineering and textile industry	[5, 36, 55, 76]
Advantages	Nano-sized threads or particles, porous in nature, high ratio of surface to volume, controlled release, non-thermal products, improved stability, food-grade polymers	[5]
Other advantages	Low cost, easy, versatile, and emerging technologies, No droplet agglomeration and coagulation complexity	[92]

7.9 SUMMARY

Encapsulation preserve the bioactive components of foods (such as vitamins, minerals, antioxidants, fatty acids, probiotics, etc.), by covering the active agent by the shells to provide an effective protection for core materials against oxidation, evaporation or migration in foods. Microencapsulation plays a major role in the development of high-quality functional food ingredients with improved physical and functional properties. It can be performed by spray drying, freeze-drying, fluid bed coating, co-crystallization, pan-coating, extrusion, molecular inclusion, polymerization, coacervation or liposome entrapment. The novel approaches include electro-hydrodynamic methods, i.e., electrospinning and electrospraying.

KEYWORDS

- **bioactive compound**
- **electrospinning**
- **electrospraying**
- **fluid bed coating**
- **melt extrusion**
- **spray chilling**

REFERENCES

1. Abbas, S., Da Wei, C., Hayat, K., & Xiaoming, Z., (2012). Ascorbic acid: Microencapsulation techniques and trends: A review. *Food Reviews International, 28*(4), 343–374.
2. Agnihotri, N., Mishra, R., Goda, C., & Arora, M., (2012). Microencapsulation: A novel approach in drug delivery: A review. *Indo Global Journal of Pharmaceutical Sciences, 2*(1), 1–20.
3. Akbarzadeh, A., Rezaei-Sadabady, R., Davaran, S., Joo, S. W., Zarghami, N., Hanifehpour, Y., & Nejati-Koshki, K., (2013). Liposome: Classification, preparation, and applications. *Nanoscale Research Letters, 8*(1), 1–9.
4. Aliabadi, H. M., Elhasi, S., Mahmud, A., Gulamhusein, R., Mahdipoor, P., & Lavasanifar, A., (2007). Encapsulation of hydrophobic drugs in polymeric micelles through co-solvent evaporation: The effect of solvent composition on micellar properties and drug loading. *International Journal of Pharmaceutics, 329*(1/2), 158–165.

5. Anubhushani, J., & Anandharamakrishnan, C., (2014). Electrospinning and electrospraying techniques: Potential food-based applications. *Trends in Food Science and Technology, 38*(1), 1–13.

6. Astolfi-Filho, Z., Souza, A. C., Reipert, E. C. D., Telis, V. R. N., (2005). Encapsulation of passion fruit juice by co-crystallization with sucrose: Crystallization kinetics and physical properties. *Science and Technology of Food, 25*(4), 795–801.

7. Augustin, M. A., & Hemar, Y., (2009). Nano- and micro-structured assemblies for encapsulation of food ingredients. *Chemical Society Reviews, 38*(4), 902–912.

8. Bakry, A. M., Abbas, S., Ali, B., Majeed, H., Abouelwafa, M. Y., Mousa, A., & Liang, L., (2015). Microencapsulation of oils: A comprehensive review of benefits, techniques, and applications. *Comprehensive Reviews in Food Science and Food Safety, 15*(1), 143–182.

9. Bansode, S. S., Banarjee, S. K., Gaikwad, D. D., Jadhav, S. L., & Thorat, R. M., (2010). Microencapsulation: A review. *International Journal of Pharmaceutical Sciences Review and Research, 1*(2), 38–43.

10. Bao, C., Jiang, P., Chai, J., Jiang, Y., Li, D., Bao, W., & Li, Y., (2019). The delivery of sensitive food bioactive ingredients: Absorption mechanisms, influencing factors, encapsulation techniques, and evaluation models. *Food Research International, 120*, 130–140.

11. Bao, W., Zhou, J., Luo, J., & Wu, D., (2006). PLGA microspheres with high drug loading and high encapsulation efficiency prepared by a novel solvent evaporation technique. *Journal of Microencapsulation, 23*(5), 471–479.

12. Becerril, R., Nerín, C., & Silva, F., (2020). Encapsulation systems for antimicrobial food packaging components: An update. *Molecules, 25*(5), 1–40.

13. Belscak-Cvitanovic, A., Stojanovic, R., Manojlovic, V., Komes, D., Cindric, I. J., Nedovic, V., & Bugarski, B., (2011). Encapsulation of polyphenolic antioxidants from medicinal plant extracts in alginate-chitosan system enhanced with ascorbic acid by electrostatic extrusion. *Food Research International, 44*(4), 1094–1101.

14. Benczedi, D., & Blake, A., (1999). Encapsulation and the controlled release of flavors. *Leatherhead Food RA Industry Journal, 2*, 36–48.

15. Berger, N., Sachse, A., Bender, J., Schubert, R., & Brandl, M., (2001). Filter extrusion of liposomes using different devices: Comparison of liposome size, encapsulation efficiency, and process characteristics. *International Journal of Pharmaceutics, 223*(1/2), 55–68.

16. Beristain, C. I., Vazquez, A., Garcia, H. S., & Vernon-Carter, E. J., (1996). Encapsulation of orange peel oil by co-crystallization. *LWT-Food Science and Technology, 29*(7), 645–647.

17. Bickerstaff, G., & Bickerstaff, G. F., (2003). Immobilization of enzymes and cells: Some practical considerations. *Immobilization of Enzymes and Cells, 1*, 1–12.

18. Bock, N., Woodruff, M. A., Hutmacher, D. W., & Dargaville, T. R., (2011). Electrospraying, a reproducible method for the production of polymeric microspheres for biomedical applications. *Polymers, 3*, 131–149.

19. Bodmeier, R., & McGinity, J. W., (1988). Solvent selection in the preparation of poly(dl-lactide) microspheres prepared by the solvent evaporation method. *International Journal of Pharmaceutics, 43*(1/2), 179–186.

20. Castro, N., Durrieu, V., Raynaud, C., Rouilly, A., Rigal, L., & Quellet, C., (2016). Melt extrusion encapsulation of flavors: A review. *Polymer Reviews, 56*(1), 137–186.

21. Chaudhary, V. B., & Patel, J. K., (2013). Cyclodextrin inclusion complex to enhance solubility of poorly water-soluble drugs: A review. *International Journal of Pharmaceutical Sciences and Research, 4*(1), 68–76.

22. Cheirsilp, B., & Rakmai, J., (2016). Inclusion complex formation of cyclodextrin with its guest and their applications. *Biology, Engineering, and Medicine, 2*(1), 1–6.

23. Choi, E., Bai, C. Z., Hong, A., & Park, J., (2012). Development of SS-AG20-loaded polymeric microparticles by oil-in-water (O/W) emulsion solvent evaporation and spray drying methods for sustained drug delivery. *Bulletin of the Korean Chemical Society, 33*(10), 3208–3212.

24. Choinska-Pulit, A., Mituła, P., Sliwka, P., Łaba, W., & Skaradzinska, A., (2015). Bacteriophage encapsulation: Trends and potential applications. *Trends in Food Science and Technology, 45*(2), 212–221.

25. Chougule, M. B., Patel, S. G., & Patel, M. D., (2018). Solid lipid nanoparticles for targeted brain drug delivery. In: Kesharwani, P., & Gupta, U., (eds.), *Nanotechnology-Based Targeted Drug Delivery Systems for Brain Tumors* (pp. 191–244). London - UK: Academic Press.

26. Clark, A. P., (1998). Liposomes as drug delivery systems. *Cancer Practice, 6*(4), 251–253.

27. Cocero, M. J., & Martin, A., (2009). Encapsulation and coprecipitation processes with supercritical fluids: fundamentals and applications. *J Supercrit Fluid, 47*, 546–555.

28. Consoli, L., Grimaldi, R., Sartori, T., Menegalli, F. C., & Hubinger, M. D., (2016). Gallic acid microparticles produced by spray chilling technique: Production and characterization. *LWT-Food Science and Technology, 65*, 79–87.

29. De Vos, P., Faas, M. M., Spasojevic, M., & Sikkema, J., (2010). Encapsulation for preservation of functionality and targeted delivery of bioactive food components. *International Dairy Journal, 20*(4), 292–302.

30. Desai, K. G. H., & Jin, P. H., (2005). Recent developments in microencapsulation of food ingredients. *Drying Technology, 23*(7), 1361–1394.

31. Desobry, S. A., Netto, F. M., & Labuza, T. P., (1997). Comparison of spray-drying, drum drying and freeze-drying for β-carotene encapsulation and preservation. *Journal of Food Science, 62*(6), 1158–1162.

32. Devi, N., Sarmah, M., Khatun, B., & Maji, T. K., (2017). Encapsulation of active ingredients in polysaccharide-protein complex coacervates. *Advances in Colloid and Interface Science, 239*, 136–145.

33. Dewettinck, K., & Huyghebaert, A., (1999). Fluidized bed coating in food technology. *Trends in Food Science and Technology, 10*(4/5), 163–168.

34. Dhouha, S., Isabelle, V., Giraud, S., & Bourbigot, S., (2006). Microencapsulation of ammonium phosphate with a polyurethane shell. Part II: Interfacial polymerization technique. *Reactive and Functional Polymers, 66*, 1118–1125.

35. Dias, M. I., Ferreira, I. C. F. R., & Barreiro, M. F., (2015). Microencapsulation of bioactives for food applications. *Food and Function, 6*(4), 1035–1052.

36. Drosou, C. G., Krokida, M. K., & Biliaderis, C. G., (2016). Encapsulation of bioactive compounds through electrospinning/electrospraying and spray drying: A comparative assessment of food-related applications. *Drying Technology, 35*(2), 139–162.

37. Estevinho, B. N., Rocha, F., Santos, L., & Alves, A., (2013). Microencapsulation with chitosan by spray drying for industry applications: A review. *Trends in Food Science and Technology, 31*(2), 138–155.

38. Ezhilarasi, P. N., Indrani, D., Jena, B. S., & Anandharamakrishnan, C., (2013). Freeze drying technique for microencapsulation of garcinia fruit extract and its effect on bread quality. *Journal of Food Engineering, 117*(4), 513–520.
39. Fang, Z., & Bhandari, B., (2010). Encapsulation of polyphenols: A review. *Trends in Food Science and Technology, 21*(10), 510–523.
40. Faridi-Esfanjani, A., & Jafari, S. M., (2016). Biopolymer nanoparticles and natural nano-carriers for nano-encapsulation of phenolic compounds. *Colloids and Surfaces B: Biointerfaces, 146*, 532–543.
41. Fucinos, C., Miguez, M., Pastrana, L. M., Rúa, M. L., & Vicente, A. A., (2017). Creating functional nanostructures: Encapsulation of caffeine into α-lactalbumin nanotubes. *Innovative Food Science and Emerging Technologies, 40*, 10–17.
42. Garrido, E. M., Rodrigues, D., Milhazes, N., Borges, F., & Garrido, J., (2017). Molecular encapsulation of herbicide terbuthylazine in native and modified β-cyclodextrin. *Journal of Chemistry*, 1–9.
43. Gbassi, G. K., & Vandamme, T., (2012). Probiotic encapsulation technology: From microencapsulation to release into the gut. *Pharmaceutics, 4*(1), 149–163.
44. Gharib, R., Greige-Gerges, H., Fourmentin, S., Charcosset, C., & Auezova, L., (2015). Liposomes incorporating cyclodextrin-drug inclusion complexes: Current state of knowledge. *Carbohydrate Polymers, 129*, 175–186.
45. Gharsallaoui, A., Roudaut, G., Chambin, O., Voilley, A., & Saurel, R., (2007). Applications of spray-drying in microencapsulation of food ingredients: An overview. *Food Research International, 40*(9), 1107–1121.
46. Ghosh, S. K., (2006). *Functional Coatings by Polymer Microencapsulation* (p. 369). Weinheim: Wiley-VCH Verlag GmbH & Co.
47. Barbosa-Canovas, G. V., & Ortega-Rivas, E., (2005). Encapsulation processes. *Food Powders, 2005*, 199–219.
48. Hashimoto, T., Ye, Y., Matsuno, A., Ohnishi, Y., Kitamura, A., Kinjo, M., & Tanaka, Y., (2019). Biochemical and biophysical research communications encapsulation of biomacromolecules by soaking and co-crystallization into porous protein crystals of hemocyanin. *Biochemical and Biophysical Research Communications, 509*(2), 577–584.
49. He, D., Deng, P., Yang, L., Tan, Q., Liu, J., Yang, M., & Zhang, J., (2013). Molecular encapsulation of rifampicin as an inclusion complex of hydroxypropyl-β-cyclodextrin: Design; characterization and *in vitro* dissolution. *Colloids and Surfaces B: Biointerfaces, 103*, 580–585.
50. Hemati, M., Cherif, R., Saleh, K., & Pont, V., (2003). Fluidized bed coating and granulation: Influence of process-related variables and physicochemical properties on the growth kinetics. *Powder Technology, 130*(1–3), 18–34.
51. Hyun-Ji, L., Taek-Jun, C., Hueck-Jin, K., Hyun-Joong, K., & Tze, W. T. Y., (2012). Fabrication and evaluation of bacterial cellulose-polyaniline composites by interfacial polymerization. *Cellulose, 19*(4), 1251–1258.
52. Jacquot, M., & Pernetti, M., (2004). Spray coating and drying processes. In: Nedovic, V., & Willaert, R., (eds.), *Fundamentals of Cell Immobilization Biotechnology* (pp. 343–356). Dordrecht-The Netherlands: Kluwer Academic Publishers.
53. Jafari, S. M., Assadpoor, E., He, Y., & Bhandari, B., (2008). Encapsulation efficiency of food flavors and oils during spray drying. *Drying Technology, 26*(7), 816–835.
54. Jamekhorshid, A., Sadrameli, S. M., & Farid, M., (2014). A review of microencapsulation methods of phase change materials (PCMs) as a thermal energy storage (TES) medium. *Renewable and Sustainable Energy Reviews, 31*, 531–542.

55. Jaworek, A., (2008). Electrostatic micro- and nanoencapsulation and electro emulsification: A brief review. *Journal of Microencapsulation, 25*(7), 443–468.
56. Jeyakumari, A., Zynudheen, A. A., & Parvathy, U., (2016). Microencapsulation of bioactive food ingredients and controlled release: A review. *MOJ Food Processing and Technology, 2*(6), 214–224.
57. Jyothi, N. V. N., Prasanna, P. M., Sakarkar, S. N., Prabha, K. S., Ramaiah, P. S., & Srawan, G. Y., (2010). Microencapsulation techniques, factors influencing encapsulation efficiency. *Journal of Microencapsulation, 27*(3), 187–197.
58. Jyothi, S. S., Seethadevi, A., Suria, P. K., Muthuprasanna, P., & Pavitra, P., (2012). Microencapsulation: A review. *International Journal of Pharma and Bio Sciences, 3*(1), 509–531.
59. Kablitz, C. D., Harder, K., & Urbanetz, N. A., (2006). Dry coating in a rotary fluid bed. *European Journal of Pharmaceutical Sciences, 27*(2/3), 212–219.
60. Kailasapathy, K., (2002). Microencapsulation of probiotic bacteria: Technology and potential applications. *Current Issues in Intestinal Microbiology, 3*, 39–48.
61. Karthik, P., & Anandharamakrishnan, C., (2012). Microencapsulation of docosahexaenoic acid by spray-freeze-drying method and comparison of its stability with spray drying and freeze-drying methods. *In Food and Bioprocess Technology, 6*(10), 2780–2790.
62. Kaushik, V., & Roos, Y. H., (2007). Limonene encapsulation in freeze-drying of gum Arabic-sucrose-gelatin systems. *LWT-Food Science and Technology, 40*(8), 1381–1391.
63. Kensova, R., Blazkova, I., Konecna, M., Kopel, P., Chudobova, D., Zitka, O., & Kizek, R., (2013). Lead ions encapsulated in liposomes and their effect on *Staphylococcus aureus*. *International Journal of Environmental Research and Public Health, 10*(12), 6687–6700.
64. Klein, J., & Wagner, F., (1983). Methods for the immobilization of microbial cells. In: *Immobilized Microbial Cells: Applied Biochemistry and Bioengineering* (Vol. 4).
65. Lengyel, M., Kallai-Szabo, N., Antal, V., Laki, A. J., & Antal, I., (2019). Microparticles, microspheres, and microcapsules for advanced drug delivery. *Scientia Pharmaceutica, 87*(3), 1–31.
66. Ma-Domingues, P., & Santos, L., (2019). Essential oil of pennyroyal (*Mentha pulegium*): Composition and applications as alternatives to pesticides: New tendencies. *Industrial Crops and Products, 139*, 1–20.
67. Madene, A., Jacquot, M., Scher, J., & Desobry, S., (2006). Flavor encapsulation and controlled release: A review. *International Journal of Food Science and Technology, 41*(1), 1–21.
68. Mahdavi, S. A., Jafari, S. M., Ghorbani, M., & Assadpoor, E., (2014). Spray-drying microencapsulation of anthocyanins by natural biopolymers: A review. *Drying Technology, 32*(5), 509–518.
69. Milian, Y. E., Gutierrez, A., Grageda, M., & Ushak, S., (2017). a review on encapsulation techniques for inorganic phase change materials and the influence on their thermophysical properties. *Renewable and Sustainable Energy Reviews, 73*, 983–999.
70. Mortazavian, A., Razavi, S. H., Ehsani, M. R., & Sohrabvandi, S., (2007). Principles and methods of microencapsulation of probiotic microorganisms. *Iranian Journal of Biotechnology, 5*(1), 1–18.
71. Mozafari, M. R., Khosravi- Darani, K., Borazan, G. G., Cui, J., Pardakhty, A., & Yurdugul, S., (2008). Encapsulation of food ingredients using nanoliposome technology. *International Journal of Food Properties, 11*(4), 833–844.

72. Muhamad, I. I., Jusoh, Y. M. M., Nawi, N. M., Aziz, A. A., Padzil, A. M., & Lian, H. L., (2018). Advanced natural food colorant encapsulation methods: Anthocyanin plant pigment. *Natural and Artificial Flavoring Agents and Food Dyes, 2018*, 495–526.

73. Murugesan, R., & Orsat, V., (2011). Spray drying for the production of nutraceutical ingredients: A review. *Food and Bioprocess Technology, 5*(1)*, 3–14.

74. Nagavarma, B. V. N., Yadav, H. K. S., Ayaz, A., Vasudha, L. S., & Shivakumar, H. G., (2012). Different techniques for preparation of polymeric nanoparticles: A review. *Asian Journal of Pharmaceutical and Clinical Research, 5*(3), 16–23.

75. Nedovic, V., Kalusevic, A., Manojlovic, V., Levic, S., & Bugarski, B., (2011). An overview of encapsulation technologies for food applications. *Procedia Food Science, 1*, 1806–1815.

76. Noruzi, M., (2016). Electrospun nanofibers in agriculture and the food industry: A review. *Journal of the Science of Food and Agriculture, 96*(14), 4663–4678.

77. Ong, S., Ming, L., Lee, K., & Yuen, K., (2016). Influence of the encapsulation efficiency and size of liposome on the oral bioavailability of griseofulvin-loaded liposomes. *Pharmaceutics, 8*(3), 25.

78. Oxley, J. D., (2012). Spray cooling and spray chilling for food ingredient and nutraceutical encapsulation. *Encapsulation Technologies and Delivery Systems for Food Ingredients and Nutraceuticals,* 110–130.

79. Ozbek, Z. A., & Ergonu, P. G., (2017). A Review on encapsulation of oils. *Celal Bayar University Journal of Science, 13*(2), 293–309.

80. Venkatesan, P., Manavalan, R., & Valliappan, K., (2009). Microencapsulation: Vital technique in novel drug delivery system. *Journal of Pharmaceutical Sciences and Research, 1*(4), 26–35.

81. Pandey, P., Turton, R., Joshi, N., Hammerman, E., & Ergun, J., (2006). Scale-up of a pan-coating process. *AAPS PharmSciTech., 7*(4), E1–E8.

82. Panthi, G., Park, M., Kim, H. Y., & Park, S. J., (2015). Electrospun polymeric nanofibers encapsulated with nanostructured materials and their applications: A review. *Journal of Industrial and Engineering Chemistry, 24*, 1–13.

83. Peanparkdee, M., Iwamoto, S., & Yamauchi, R., (2016). Microencapsulation: A review of applications in the food and pharmaceutical industries. *Reviews in Agricultural Science, 4*(1), 56–65.

84. Perignon, C., Ongmayeb, G., Neufeld, R., Frere, Y., & Poncelet, D., (2014). Microencapsulation by interfacial polymerization: Membrane formation and structure. *Journal of Microencapsulation, 32*(1), 1–15.

85. Phanindra, P., Ramesh, P., & Poshadri, A., (2018). Microencapsulation of bioactive compounds through co-crystallization. *International Journal of Pure and Applied Bioscience, 6*(2), 1366–1371.

86. Poornima, K., & Sinthya, R., (2017). Application of various encapsulation techniques in food industries. *International Journal of Latest Engineering Research and Applications (IJLERA), 2*(10), 37–41.

87. Poshadri, A., & Aparna, K., ***(2010). Microencapsulation technology: A review.*** *Journal of Research ANGRAU, 38*(1), 86–102.

88. Rajauria, G., & Tiwari, B. K., (2017). *Fruits Juices: Extraction, Composition, Quality, and Analysis* (1st edn., p. 910). Boston-USA: Academic Press.

89. Ray, S., Raychaudhuri, U., & Chakraborty, R., (2016). An overview of encapsulation of active compounds used in food products by drying technology. *Food Bioscience, 13*, 76–83.

90. Reineccius, G. A., (2009). Edible films and coatings for flavor encapsulation. *Edible Films and Coatings for Food Applications, 269*–294.

91. Renard, D., Robert, P., Lavenant, L., Melcion, D., Popineau, Y., Gue, J., & Schmitt, C., (2002). *Biopolymeric Colloidal Carriers for Encapsulation or Controlled Release Applications, 242*, 163–166.

92. Rhishir, M. R., Xie, L., Sun, C., Zheng, X., & Chen, W., (2018). Advances in micro and nano-encapsulation of bioactive compounds using biopolymer and lipid-based transporters. *Trends in Food Science and Technology, 78,* 34–60.

93. Risch, R., (1995). Encapsulation: Overview of uses and techniques: Chapter 1. In: *Encapsulation and Controlled Release of Food Ingredients ACS Symposium Series* (Vol. 590, pp. 1–7). Washington, DC-USA: American Chemical Society.

94. Sachan, N. K., (2006). Controlled drug delivery through microencapsulation. *Malaysian Journal of Pharmaceutical Sciences, 4*(1), 65–81.

95. Saha, S., Roy, A., Roy, K., & Roy, M. N., (2016). Study to explore the mechanism to form inclusion complexes of β-cyclodextrin with vitamin molecules. *Scientific Reports, 6*(1), 1–12.

96. Sartori, T., Consoli, L., Hubinger, M. D., & Menegalli, F. C., (2015). Ascorbic acid microencapsulation by spray chilling: Production and characterization. *LWT-Food Science and Technology, 63*(1), 353–360.

97. Schoebitz, M., Lopez, M. D., & Roldan, A., (2013). Bioencapsulation of microbial inoculants for better soil-plant fertilization: A review. *Agronomy for Sustainable Development, 33*(4), 751–765.

98. Sherry, M., Charcosset, C., Fessi, H., & Greige-Gerges, H., (2013). Essential oils encapsulated in liposomes: A review. *Journal of Liposome Research, 23*(4), 268–275.

99. Shishir, M. R. I., Xie, L., Sun, C., Zheng, X., & Chen, W., (2018). Advances in micro and nano-encapsulation of bioactive compounds using biopolymer and lipid-based transporters. *In Trends in Food Science and Technology, 78*, 34–60.

100. Skalickova, S., Nejdl, L., Kudr, J., & Ruttkay-Nedecky, B., (2016). Fluorescence characterization of gold modified liposomes with antisense N-myc DNA bound to the magnetizable particles with encapsulated anticancer drugs (doxorubicin, ellipticine, and etoposide). *Sensors, 16*(3), 290–299.

101. Song, Y., Sun, P., Henry, L. L., & Sun, B., (2005). Mechanisms of structure and performance controlled thin film composite membrane formation via interfacial polymerization process. *Journal of Membrane Science, 251*, 67–79.

102. Su, W., Darkwa, J., & Kokogiannakis, G., (2015). Review of solid-liquid phase change materials and their encapsulation technologies. *Renewable and Sustainable Energy Reviews, 48*, 373–391.

103. Suganya, V., & Anuradha, V., (2017). Microencapsulation and nanoencapsulation: A review. *International Journal of Pharmaceutical and Clinical Research, 9*(3), 233–239.

104. Talegaonkar, S., Pandey, S., Rai, N., Rawat, P., Sharma, H., & Kumari, N., (2016). Exploring nanoencapsulation of aroma and flavors as new frontier in food technology. *Encapsulations, 2016*, 47–88.

105. Teunou, E., & Poncelet, D., (2002). Batch and continuous fluid bed coating: Review and state of the art. *Journal of Food Engineering, 53*(4), 325–340.

106. Torres-giner, S., Martinez-abad, A., Ocio, M. J., & Lagaron, J. M., (2010). Stabilization of a nutraceutical omega-3 fatty acid by encapsulation in ultrathin electrosprayed zein prolamine. *Journal of Food Science, 75*(6), 69–79.

107. Townsend-Nicholson, A., & Jayasinghe, S. N., (2006). Cell electrospinning: A unique biotechnique for encapsulating living organisms for generating active biological microthreads/scaffolds. *Biomacromolecules, 7*(12), 3364–3369.

108. Trifković, K., Đorđević, V., Balanč, B., Kalušević, A., Lević, S., Bugarski, B., & Nedović, V., (2016). Novel approaches in nanoencapsulation of aromas and flavors: Chapter 9. In: Grumezescu, A. M., (ed.), *Encapsulations Nanotechnology in the Agri-Food Industry* (Vol. 2, pp. 363–419). Series. London- England: Academic Press.

109. Vidhyalakshmi, R., Bhakyaraj, R., & Subhasree, R. S., (2009). Encapsulation: The future of probiotics: A review. *In Advances in Biological Research, 3*(3/4), 96–103.

110. Vivek, K. B., (2013). Use of encapsulated probiotics in dairy-based foods. *International Journal of Food, Agriculture and Veterinary Sciences, 3*(1), 188–199.

111. Wang, B., Akanbi, T. O., Agyei, D., Holland, B. J., & Barrow, C. J., (2018). Coacervation technique as an encapsulation and delivery tool for hydrophobic biofunctional compounds. *Role of Materials Science in Food Bioengineering, 2018*, 235–261.

112. Wang, W., Waterhouse, G. I. N., & Sun-Waterhouse, D., (2013). Co-extrusion encapsulation of canola oil with alginate: Effect of quercetin addition to oil core and pectin addition to alginate shell on oil stability. *In Food Research International, 54*(1), 837–851.

113. Wen, P., Zong, M., Linhardt, R. J., Feng, K., & Wu, H., (2017). Trends in food science and technology electrospinning: A novel nano-encapsulation approach for bioactive compounds. *Trends in Food Science and Technology, 70*, 56–68.

114. Wulff, G., Avgenaki, G., Guzmann, M. S. P., (2005). Molecular encapsulation of flavors as helical inclusion complexes of amylose. *Journal of Cereal Science, 41*(3), 239–249.

115. Yeo, Y., & Park, K., (2005). Recent advances in microencapsulation technology. *Encyclopedia of Pharmaceutical Technology, 2005*, 1–15.

116. Zamani, M., Prabhakaran, M. P., & Ramakrishna, S., (2013). Advances in drug delivery via electrospun and electrosprayed nanomaterials. *International Journal of Nanomedicine, 8*, 2997–3017.

117. Zanetti, M., Carniel, T. K., Dalcanton, F., Dos, A. R. S., Gracher, R. H., De Araújo, P. H. H., & Antônio, F. M., (2018). Use of encapsulated natural compounds as antimicrobial additives in food packaging: A brief review. *Trends in Food Science and Technology, 81*, 51–60.

118. Zhang, L., Huang, J., Si, T., & Xu, R. X., (2012). Coaxial electrospray of microparticles and nanoparticles for biomedical applications. *Expert Review of Medical Devices, 9*(6), 595–612.

119. Zhang, L., Liu, P., Ju, L., Wang, L., & Zhao, S., (2010). Polypyrrole nanocapsules via interfacial polymerization. *Macromolecular Research, 18*(7), 648–652.

120. Zohri, M., (2009). Polymeric nanoparticles : Production, applications, and advantage. *The Internet Journal of Nanotechnology, 3*(1), 1–14.

121. Zuidam, N. J., & Shimoni, E., (2010). Overview of microencapsulates for use in food products or processes and methods to make them: Chapter 2. In: Zuidam, N. J., & Nedovic, V., (eds.), *Encapsulation Technologies for Active Food Ingredients and Food Processing* (Vol. 12, pp. 3–29). New York- USA: Springer.

122. Zussman, E., (2010). Encapsulation of cells within electrospun fibers. *Polymers for Advanced Technologies, 22*(3), 366–371.

PART II

Non-Destructive Techniques for Food Quality and Safety

CHAPTER 8

FOOD AUTHENTICATION: BASICS AND DETECTION METHODS

SANGITA BANSAL, KANIKA SHARMA, and ERA V. MALHOTRA

ABSTRACT

Food frauds have become a growing concern globally for food processors, distributors, and consumers. To mitigate the challenges of food frauds and ensure wholesome food and nutrition for the people, authenticity testing of food and its ingredients is necessary through authentication techniques. Research studies to authenticate several plants and animal-based food products suggest that the food frauds have reached an alarming level. This chapter provides a basic understanding of food frauds, food authentication, and methods to detect food frauds.

8.1 INTRODUCTION

Food authentication is used to analyze a product, if it is true to its description or not. Food safety being a global concern makes food authenticity one of its major issues. Frauds in food is centuries old issue and has made its way till today because of weak laws and lack of awareness among the consumers. Earliest food authenticity survey done by Arthur Hill Hassall in 1861 using microscopy on expensive product coffee showed adulteration in 31 out of 34 samples. Another study conducted in 1995 by the Ministry of Agriculture Fisheries and Food in the United Kingdom revealed 15% unauthentic samples of instant coffee, suggesting that adulteration issues are far more complex.

Foods with high nutritional value or high demand or high cost are often the target of frauds, including adulteration and/or false labeling often misleading and harmful to health. To overcome this problem, the methods of detecting adulterants and other authenticity issues have been developed for accurately detecting undesirable parts. Food authentication is a prevalent research area in both food sciences and biotechnology, thus, comprising many techniques viz. visual analysis, biochemical analysis, molecular-based approaches, etc. This field is in the phase of exponential growth and is attracting a high level of attention from authorities and media around the world. Food authentication methods must be scientifically proven, accurate, and reliable so that bonafide products can be protected and clearly distinguished from any illegal substitutes. This chapter explores the potential of food frauds and the food authentication methods.

8.2   TECHNICAL TERMS IN FOOD AUTHENTICATION

- **Chromatography:** A method of separating components of a mixture by their distribution between two phases.
- **Deoxyribonucleic Acid (DNA):** A genetic material that is capable of self-replication.
- **Electrophoresis:** A technique for the separation and analysis of charged substances, especially biopolymers under the influence of electric field.
- **Food Adulteration**: Inclusion or extraction of any food ingredient, either intentional or unintentional, that affects its natural composition and quality.
- **Food Authentication:** Testing or verifying the food products for its originality and genuineness.
- **Food Contamination:** It can be defined as the presence of harmful products, chemicals, and microorganisms in food, which can cause a variety of illness in consumers.
- **Food fraud:** Refers to intentional addition, substitution, tampering, or false/misleading claims of food or any ingredient for economic gain.
- **Food Safety:** Refers to handle, prepare, process, and storage of food/food products in order to prevent the foodborne illnesses.
- **Genetic Engineering (GE):** Artificial manipulation or modification of genetic material (gene) of an organism to alter its character(s).
- **Genetically Modified Organisms (GMOS):** Living organisms with artificially modified genetic makeup.

- **Peptide Nucleic Acid (PNA):** It is a synthetic polymer similar to DNA or RNA.
- **Polymerase Chain Reaction (PCR):** A technique used to make multiple copies of a specific segment of DNA in the laboratory.
- **Ribonucleic Acid (RNA):** It is another genetic material and is one of the three essential biological macromolecules in living organisms.
- **Spectroscopy:** A technique used to study the absorption and emission of light and other radiation by matter, as related to the dependence of these processes on the wavelength of the radiation.

8.3   WHY IS FOOD AUTHENTICATION NECESSARY?

Although food adulteration is practiced since the inception of commerce itself, yet statistics show that it is on the rise [39]. Food has also been associated with crimes resulting into the food frauds. One of the worst cases of food frauds is in China, where 6 babies died because of the addition of melamine in baby formula [37]. Although food authentication is challenging due to complex food matrix and increasing types of adulterants further make their detection difficult, yet food authentication is important to avoid the adverse impact of food adulteration on human nutrition and health, and to gain consumers' confidence [1].

8.4   FOOD FRAUD

Food fraud is a broader term that includes intentional addition, substitution, tampering or misrepresentation or false/misleading claims of food and its ingredients, for economic gain. Substitution of cheaper for expensive food and other commodities is not a new problem [42]. Though food frauds are generally done with economic motives, yet it may involve the public health and can cause serious illness or even death, thus becoming a food safety issue. Few infamous examples of food frauds induced illness are:

- **Spanish Olive Oil:** In 1981, industrial rapeseed oil after removing aniline was sold as olive oil to the street vendors across Spain that resulted in 1000 deaths due to severe allergic reaction called 'Toxic Oil Syndrome.
- **Chinese Milk:** In 2008, few Chinese companies adulterated the diluted milk with melamine (to allow the milk to pass the protein test)

resulting in illness of more than 300,000 people and hospitalization of 50,000 children.

- **American Peanuts:** In 2009, salmonella-contaminated peanut butter was knowingly shipped across 46 states by Peanut Corporation of America, resulting in 7 deaths and illness of hundreds of people.

However, in general, quantification of the impact of food fraud is difficult because the health impacts are usually mild, and it is difficult to correlate an illness with consumption of fraudulent food.

8.4.1   TYPES OF FOOD FRAUDS

Distinct types of food frauds have been observed [61]. These frauds may be food adulteration, tampering, over-run, theft, diversion, simulation, and counterfeiting (Table 8.1).

TABLE 8.1   Terms Related to Food Frauds, Their Definitions and Examples

Term	Example
Adulteration: It is the addition or substitution of any food component	Addition of sugar to honey
Counterfeiting: It is an infringement of IPR to completely replicate a product	Marketing replicas of a branded product
Dilution: It is mixing an inferior quality ingredient to high-quality food	Dilution of milk by water
Diversion: It is to sell and distribute a product beyond designated places/purpose	To sell relief food in markets
Grey Market: This production is to sell an unreported food	
Overrun: It is excess production of a legitimate product beyond approved limits	Stocking a product by under-reporting
Simulation: It is copying the appearance of a legitimate product to sell an illegitimate product	Selling low-quality product as a standard product
Substitution: It is the replacement of a food component with the inferior component	Palm oil substitution with Babassu oil
Tampering: It is to use a genuine product or its packaging in a fraudulent manner	To tamper expiry date of a product
Theft: It is selling a stolen product as legitimately procured	–

8.4.2   RISKS IN FOOD

- **Direct Risk:** It includes the immediate and direct risk to consumers due to fraudulent food which can also be lethal.
- **Indirect Risk:** When the risk is due to long-term exposure to fraudulent food, such as accumulation of chronically toxic contaminant in the body. Indirect risk also includes the removal of important nutrients or vitamins.
- **Technical Risk:** It does not include any ingredient but is related to the packaging. An example includes mislabeling of product content or other information like country-of-origin, etc.

8.4.3   OVERCOMING FOOD FRAUDS

- Always ensure to check the health claims made by the company while choosing healthier options. For example, several artificial sweeteners possess ingredients that are more harmful than natural sugars.
- Ensure to buy food with FSSAI license number, proper label, the complete list of ingredients with quantities and expiry date.
- Mostly in the case of freshly cooked/prepared food, the ingredients like frying oil quality, food colors, etc., are generally overlooked that can be equally harmful. Therefore, caution must be taken to avoid such foods that seem stale, unhealthy, or adulterated.
- Water quality, which is a major source of food contamination, is generally. Use water after properly purifying it through boiling.

8.5   FOOD ADULTERATION

Food adulteration has become a global concern as it affects almost all food commodities. Food Safety and Standards Authority of India (FSSAI, 2006) [24] defines *"adulterant as any material which is or could be employed for making the food unsafe or sub-standard or misbranded or contain foreign matter."* Federal food, drug, and cosmetic act (FFDCA) declare a food adulterated if:

- A nutritious/important component from food is extracted;
- It appears more valuable than actual;
- It contains a substance harmful to health;

- It has a sub-standard food quality;
- It includes a cheaper or inferior quality item;
- It is added with any substance to increase bulk or weight.

Adulteration not only results in severe economic problems but may also distress the nutritional and health status of an individual, as adulterated food may be toxic, allergic, or deprived of nutrients [4].

Prevention of food adulteration (PFA) is difficult and multifaceted. Food companies, authorities, and academics have put the best efforts not only in devising effective testing methods but also in acquiring the information regarding the supply chain in terms of stages, intricacy, and susceptibility. For prevention of food fraud, transparency of food chain and full raw material traceability is vital and substantial efforts have been made recently on this aspect.

Seeing the greater risks and threats to the health of the citizens, governments have also framed different policies and acts to overcome these problems; and one such act in India is the PFA Act of 1954. The act was further amended to incorporate PFA Rules (1955). Food standards, sampling procedures and analysis, nature of penalties, and other parameters related to food are covered under PFA act. It also includes the provisions related to food additives and coloring issues; packaging and food labeling; and marketing regulations, etc.

Food adulteration may either be intentional or unintentional (incidental). Intentional adulteration is deliberate addition or substitution with inferior materials to gain more profits [33], whereas incidental adulteration results due to ignorance, negligence, or improper (unclean) facilities.

8.5.1   WAYS OF FOOD ADULTERATION

- **Addition of Toxicants:** Addition of inedible toxic substances like synthetic dyes, low-quality preservatives, etc.
- **Decomposed Food:** Mixing stale decomposed food with fresh ones mainly in fruits and vegetables.
- **Hiding Quality:** Labeling a low-quality food as qualitative food.
- **Misbranding/False Labels:** Copy of a quality/branded food, tampering expiry date/nutritional composition.
- **Mixing:** Addition of cheaper substances like clay, sand, marbles, etc.
- **Substitution:** Replacing a food component with cheaper/inferior substances either completely or partially.

Adulteration has become common in different foods, food grains and other essential commodities that affect the health of the consumers.

8.5.2 OVERCOMING FOOD ADULTERATION

Different food laws have been adopted by the government agencies, but there must be proper surveillance because even after their implementation, food crimes are committed at a greater pace. The food adulteration can be mitigated effectively only after detection of the type of adulterant. Various detection methods ranging from preliminary and quick detection methods to sophisticated specific analytical/molecular methods (such as chromatography, enzyme-linked immunosorbent assay (ELISA), spectroscopy, PCR, etc.), are available to avoid the consumption of adulterated foods. Quick detection tests are done to preliminary analyzes the adulteration, some of which can easily be done by the common person in their households.

8.6 FOOD CONTAMINATION

Food contamination is generally defined as the presence of either microorganisms or harmful substances that make the food unfit for consumption. Consumption of contaminated foods may lead to severe consequences: for example, *Shigella dysenteriae* type-2 in the United States that was a result of consumption of muffins or doughnuts placed in the staff break room that got contaminated from *Shigella* stock culture maintained in the laboratory [34]. Contamination of food is a serious issue and has also been recognized by the World Health Organization (WHO) as a global challenge. The list of food contaminants is continuously growing, including pesticide residues in fruits/vegetables, antibiotics in animal products and contamination of foods, etc., [31]. Broadly, food contamination can be of three types:

- **Biological Contamination:** It usually occurs due to ignorance, improper infrastructure, and poor handling by people who do not practice good hygiene. Biological contamination is most common and cannot be detected in the early stages and generally results in spoiled food, foodborne illness, and even death.
- **Chemical Contamination:** There are various chemical contaminants, ranging from agrochemicals used and industrial and environmental

contaminants to emerging chemical hazards, such as pesticides sprayed on fruits and vegetables, heavy metals, refrigerants, food additives, and other chemicals like melamine, which may show adverse effects on human and animal health.

- **Physical Contamination:** Hair, glass, papers, scabs, etc., can be classified as physical contamination. These can be added intentionally or sometimes unintentionally, by mistake. Care should be taken while consuming physically contaminated food, as this can choke the person, leading to death.

8.6.1   COMMON CAUSES OF CONTAMINATION

- **Cross-Contamination:** It occurs due to transfer of microorganisms from a surface or food to another. Maintenance of proper hygiene during processing and storage is effective in the prevention of cross-contamination.
- **Time-Temperature Abuse:** It occurs when the food is exposed to the temperature danger zone of 41°F to 140°F, for more than 4 hours, which allows the pathogens to multiply rapidly.
- **Poor Hygiene:** Food handlers are the carriers of contamination, and if they do not keep good hygiene practices like washing hands, clean utensils, surfaces, environment, etc., they would contaminate the foods by their own habits and handling. Thus, the ones dealing with food need to be very much particular about their good hygiene.

8.7   GENETICALLY MODIFIED (GM) FOODS

Genetically modified (GM) or genetically engineered or bioengineered foods (GM/GE foods) are basically those foods that are derived from organisms having DNA or genetic material modified using genetic engineering (GE) techniques, not by conventional breeding. Flavr Savr, delayed-ripening tomato was first but unsuccessful GM crop commercialized in 1994 by Calgene. Soybean, corn, canola, and cotton are some other examples of GM crops that have been modified either for resistance to pathogens/herbicides or for better nutrient profiles. Specific systems are in place for an extensive evaluation of GM organisms or foods in relation to human health and the environment, both. The potential concerns about GM foods are: the potential to elicit allergic responses, horizontal gene transfer from GM organisms to other non-GM organisms.

8.8 METHODS OF FOOD AUTHENTICATION

Customers are entitled to know what is in the food they eat, and manufacturers are obligated to educate and not intentionally mislead consumers about the food product's origin or content. Nevertheless, there is a higher risk that adulteration may occur throughout the food chain due to the increasing demand for food and the increased globalization of trade [13]. A number of methods ranging from basic, analytical to sophisticated quantitative methods are available for the detection of chemical contaminants. The advantages of these methods are high sensitivity and good accuracy, and their disadvantages are that they are complicated and time-consuming and need expensive instruments and trained personnel with specific skills, which strongly limit their application. Nanosensors are other efficient and effective tools in the detection of chemical contaminants [36]. For biological adulterants, various biochemical, immunological, and molecular methods have been developed for food authentication.

8.8.1 IMMUNOLOGICAL TECHNIQUES

The immunological techniques are based on antigen and antibody reactions. Antibodies specific for an antigenic component of a food product is generated in experimental animals in the laboratory [43]. Immunoassays being fast, cheap, sensitive, and specific, have become a popular tool for authenticating quality standards of various types of food and food ingredients. In addition, they are user-friendly, have a high throughput, and are amenable to field-testing. In the food industry, antibodies are developed against specific antigens (allergens, toxins, pathogens, etc.), [25], and are then used as capture molecules to trap their target antigens. Some of the methods that include this type of interaction are discussed in this section.

8.8.1.1 ENZYME-LINKED IMMUNO-SORBENT ASSAY (ELISA)

Enzyme-linked immunosorbent assay (ELISA) is a widely used technique of food authentication. The basic procedure involves the coating of antigens or antibodies to the wells of 96-well polystyrene plates. Then the test serum along with a positive control serum and a negative control serum is incubated in different wells. After some time, the serum is removed, and the plate is washed with wash buffer to remove unbound antibodies or antigens. Bound antigens or antibodies are then detected with the help of secondary antibodies labeled with an enzyme, such as peroxidase or alkaline phosphatase, that produce color after reaction with a suitable substrate. On the basis of sequence of antigen and antibody addition, there are three main types of ELISA test:

- **Indirect ELISA:** Antigen is coated on the microtiter well first, then primary antibody is added and incubated. Unbound primary antibody is washed away using wash buffer and enzyme-labeled secondary antibody is used to detect the bound antibody. Unbound secondary antibody is removed by washing. Enzyme hydrolyzes the substrate to form colored products. The end product is measured by recording the absorbance of all the wells of the plate through spectrophotometric plate readers.
- Sandwich ELISA: the antibody is coated on the microtiter well first then the antigen is added and incubated to allow antigen-antibody interaction and formation of an antigen-antibody complex. After washing to remove any unbound antigen, a second enzyme-linked antibody specific for a different epitope on the antigen is added and allowed to interact with the bound antigen. Enzyme specific substrate is added that gives colored products after removing unbound antibody by washing.
- **Competitive ELISA: Both direct and indirect ELISA** can be adapted to competitive format. This technique utilizes a reference antigen that competes with sample antigen for binding to the known quantity of enzyme-labeled antibody. This technique is also known as inhibition ELISA or competitive immunoassay. First, the reference antigen is coated to the well of microtiter plate. Then labeled antibody pre-incubated with the test sample is added to the wells. Few or more free antibodies will be left to interact with the reference antigen that will depend on the quantity of antigen in the test sample. If more antigens are present in the test sample, only a few antibodies will be available to bind reference antigen, thus resulting in weak signals. Another variant of competitive ELISA uses labeled reference antigen instead of the antibody.

8.8.1.2   WESTERN BLOT

Western blotting, also known as protein immunoblotting, is a technique to visualize an individual specific protein in a given sample. The technique uses sodium dodecyl sulfate (SDS)-polyacrylamide gel electrophoresis (SDS-PAGE) to separate different proteins present in a sample. The separated proteins are then transferred on to a nitrocellulose or polyvinylidene difluoride membrane (PVDF), and blotted by their interaction with specific labeled antibodies. Western blotting is highly specific and sensitive to detect even 1 ng concentration of target protein. There are three basic steps in this technique: (a) extraction and separation of protein according to size by electrophoresis; (b) transfer of separated proteins to membranes; and (c) visualization of proteins on membrane with the help of labeled antibodies [41]. Western blot assays are performed to obtain reliable quantitative data using standardization procedures coupled with the updated reagents and detection methods [62].

8.9   SPECTROSCOPY TECHNIQUES

Spectroscopic methods (Table 8.2) are successful in evaluating the food products, as they require minimum sample preparation, provide rapid analysis, and single sample can be tested multiple times. UV-Vis spectroscopy has extensively been used by various researchers for assessment of the quality of vegetable oils, alcoholic beverages; coffee extracts, etc., [45, 46]. In a study, the NMR metabolite fingerprinting was done to evaluate saffron adulteration with cheaper agents (i.e., saffron stamens, safflower petals, and corn silk, etc.), [53]. In a recent study in the UK on 78 commercial oregano samples, researchers used a combination of LC-MS and FTIR to find 24% adulteration [21]. Specific Natural Isotopic Fractionation-nuclear magnetic resonance (NMR) is another recent technique used in food authentication [56].

TABLE 8.2   Spectroscopic Methods Used for Food Authentication of Different Products

Item	Authenticity Factor	Test Method	References
Basmati rice	Geographic location	SNIF-NMR (specific natural isotopic fractionation-nuclear magnetic resonance)	[66]
Black pepper	Papaya seeds	Near infrared-hyperspectral imaging (NIR-HSI), magnetic resonance imaging (MRI)	[16, 51]

TABLE 8.2   *(Continued)*

Item	Authenticity Factor	Test Method	References
Cheese	Detection of rennet whey	HPLC	[14]
Chili powder	Brick powder, salt powder or talc powder. water-soluble coal tar color	NIR, Raman spectroscopy, paper chromatography	[16, 27]
Cloves	Volatile oil	GC-MS	[32]
Coffee	Chicory	SPME-GC-MS	[50]
Honey	Water-soluble honey proteins	MALDI-TOF MS protein profiling	[60]
Milk	Species identification and adulterant detection	Indirect competitive ELISA	[30]
Olive oil	Lead from olive presses, Soybean oil	NIR, FTIR, UV-Vis Spectroscopy	[46, 68]
Orange juice	Grapefruit juice	NMR spectroscopy	[11]
Paprika	Lead oxide, Sudan I, Sudan IV, Lead chromate	FT-MIR	[28]
Porter and stout	Water, Brown sugar, source of picrotoxin, a poisonous crystalline alkaloid, $C_{30}H_{34}O_{13}$, $FeSO_4$, cream of tartar (Potassium bitartrate, ($KC_4H_5O_6$), hartshorn shavings (horns of the male Red Deer)	MALDI-TOF	[21]
Saffron	Synthetic dyes, other chemicals	NMR, Metabolite fingerprinting	[53]
Vinegar	Sulfuric acid (H_2SO_4), and dissolved tin and lead when boiled vessels	SNIF-NMR; site-specific natural isotopic fractionation-nuclear magnetic resonance	[29]
Wheat	L-cysteine	Raman spectroscopy	[8]

8.10 CHROMATOGRAPHY

Chromatography is a technique for separation and analysis of components of a mixture on the basis of their different solubility. The mixture dissolved in the mobile phase is passed through a column containing stationary phase. Different components of a mixture get separated while passing through the column due to their different speeds that depends on their relative solubility in the mobile and stationary phase. Chromatography alone or in combination with mass spectroscopy has been extensively used in food authentication and quality analysis including composition [38], detection of mycotoxins (fungal toxins), food additives, vitamins, amino acids, food preservatives, antibiotic residues, etc.

8.11 NUCLEIC ACID-BASED FOOD AUTHENTICATION

During recent years, molecular methods have proven to be a valuable option for the detection of food authenticity and adulteration detection [22]. The analytical techniques generally use protein as a marker for species identification and food authenticity testing. However, in processed foods, the detection becomes difficult as proteins are thermolabile and get denatured during processing. Typical examples include, wines produced from defined grape varieties, monocultivar olive oils, and so on (Table 8.4). DNA based techniques or DNA markers are another powerful tools that can be used for species identification and food authentication as DNA is more thermostable.

DNA based methods (Table 8.3) are hybridization-based markers such as restriction fragment length polymorphisms (RFLP), or PCR-based markers that include Random Amplified Polymorphic DNA (RAPD), multiplex PCR, quantitative PCR (qPCR), simple sequence repeat (SSR), also known as microsatellite, amplified fragment length polymorphisms (AFLPs), RFLPs, sequence-characterized amplified regions (SCAR), high-resolution melting (HRM) PCR, loop-mediated isothermal amplification (LAMP), and DNA bar-coding, single nucleotide polymorphism (SNP).

8.11.1 DNA MARKERS

A molecular marker is a specific DNA sequence in the genome. Variations in DNA sequences are the result of insertion, deletion, the substitution of bases due to mutation. These differences, collectively called polymorphisms can be mapped and identified.

TABLE 8.3 Various Molecular Markers Used in Food Authentication

Molecular Marker	Phenotype Expression	Type of Primer Used
Hybridization Based		
Restriction fragment length polymorphism (RFLP)	Codominant	cDNA/gDNA sequence

TABLE 8.3   *(Continued)*

Molecular Marker	Phenotype Expression	Type of Primer Used
Double fluorescence protein (DFP) marker	Codominant	Minisatellite synthetic oligos
PCR-Based		
Polymerase chain reaction-restriction fragment length polymorphic DNA (PCR-RFLP)	Codominant	Sequence-specific primers
Random amplified polymorphic DNA (RAPD)	Codominant/ Dominant	Arbitrarily designed primers
Inter simple sequence repeat (ISSR)	Dominant	Repeated DNA motifs (2–4 bp each) meant to be complementary to microsatellite regions in the genome
Simple sequence repeats (SSR)	Codominant	Sequence-specific primers
Amplified fragment length polymorphic DNA (AFLP)	Dominant	The sequence corresponding to flanking regions of target DNA
Start codon targeted polymorphism (SCoT)	Dominant	The sequence corresponding to the short conserved region near start codon (ATG)
Expressed sequence tags (EST)	Codominant	Single-pass (from 5' or 3' end) sequencing of a clone randomly selected from cDNA libraries
Sequence characterized amplified region (SCAR)	Codominant	Specific DNA sequence

TABLE 8.4   DNA-based methods for Food Authentication in Plant-based Foods

Food Product	Adulterant	Technique Used for Authentication	References
Black pepper	Papaya seeds, Chili powder	SCAR	[17, 51]
Chill powder	Red beet pulp, Jujube (*Ziziphus nummularia*)	RAPD, SCAR	[16, 18]
Cinnamon	Cassia; beechnut husk; hazelnut; almond shell dust	Sequencing	[56]
Food allergens	Hazelnut, peanut	Species-specific PCR PCR-ELISA PCR/ PNA-HPLC, Real-time PCR (Cor a gene)	[40]
Kala zeera	Safed zeera	DNA barcoding	[5]

TABLE 8.4 *(Continued)*

Food Product	Adulterant	Technique Used for Authentication	References
Mustard seed	Argemone seeds (*Argemone mexicana*); rapeseed	Real time-PCR	[57]
Olive oil	Oil from plants other than *Oleaeuropeae*	AFLP, NIR, FTIR, SSR/Real-time PCR for identification of microsatellite DNA	[6]
Potato	Commercial variety	DNA technology	[3]
Rice	Other seeds, damaged grains, infected grain	Real-time PCR method, SSR Microsatellite DNA, multiplex SSR, microsatellite DNA	[38, 65]
Saffron	Biological adulterants (safflower/calendula petals)	SCAR	[44]
Turmeric powder	Adulteration with other inferior Curcuma spp.	RAPD, PCR, DNA barcoding	[19, 59]
Gluten	Wheat	Species-specific PCR for genes-gliadin identification	[58]
	Barley	Real-time PCR for hordey identification	
	Oats	Real-time PCR for avenin identification	
Adulteration	Mustard	Real-time PCR for 2S albumin identification	[47]
	Sesame	Real-time PCR for sinA identification	
Bread	From durum wheat cv.	Duplex PCR for puroindoline b identification	[2]
		SSR for microsatellite DNA	[52]

8.11.1.1 MARKERS BASED ON DNA HYBRIDIZATION

RFLP was the first DNA marker used for the assessment of genetic diversity within and among species. RFLP markers have been extensively used for DNA fingerprinting, construction of linkage maps in several species, as they are reproducible and co-dominant. In RFLP, the genomic DNA is cut into smaller fragments using restriction enzymes, and then specific fragments are visualized by hybridizing labeled probe to the separated fragments transferred on blotting membranes. However, due to its tedious, expensive, and time-consuming detection procedure, RFLP is considered outdated at present.

8.11.1.2   MARKERS BASED ON PCR AMPLIFICATION

Polymerase chain reaction (PCR) is a technique for artificial amplification of specific DNA segments from small DNA concentrations. PCR based markers involves use of primers (specific or random) to amplify a specific or random regions of DNA. PCR-based markers may be Locus specific or Locus non-specific markers.

> **Random Amplified Polymorphic DNA (RAPD):** These marker uses single random primer to amplify multiple regions in target DNA. The marker is useful when no sequence information is available; however, the reproducibility of this marker is poor. In RAPD, the template DNA molecules are denatured and annealed with random primers to amplify by PCR. Single short oligonucleotide primer (usually a 10-base primer) can be arbitrarily selected and used for the amplification of DNA segments of the genome (which may be in distributed throughout the genome). Different sized bands are obtained after gel electrophoresis due to difference in the nucleotide sequence. Genomic DNA from two different plants often results in different amplification patterns. This represents polymorphism and can be used as a molecular marker of a particular species.

> **Amplified Fragment Length Polymorphism (AFLP):** This technique combines the usefulness of restriction digestion and PCR. In AFLP, the DNA of the genome is extracted and subjected to restriction digestion by two enzymes (a rare cutter, e.g., Msel; a frequent cutter, e.g., EcoRI). The cut ends on both sides are then ligated to known sequences of oligonucleotides. PCR is now performed for the pre-selection of a fragment of DNA which has a single specific nucleotide. By this approach of pre-selective amplification, the pool of fragments can be reduced from the original mixture. In the second round of amplification by PCR, three-nucleotide sequences are amplified. This further reduces the pool of DNA fragments to a

manageable level (<100). Autoradiography can be performed for the detection of DNA fragments. The use of radiolabeled primers and fluorescently labeled fragments quickens AFLP.

AFLP analysis is tedious and requires the involvement of skilled technical personnel. Hence some people are not in favor of this technique. In recent years, commercial kits are available for AFLP analysis. AFLP is very sensitive and reproducible. It does not require prior knowledge of sequence information. By AFLP, a large number of polymorphic bands can be produced and detected.

8.11.1.3   SEQUENCE TAGGED SITES (STS)

Sequence tagged sites (STS) represent unique simple copy segments of genomes, whose DNA sequences are known, and which can be amplified by using PCR. STS markers are based on the polymorphism of simple nucleotide repeats, e.g., $(GA)_n$, $(GT)_n$, $(CAA)_n$, etc., on the genome. STS have been recently developed in plants. When the STS loci contain simple sequence length polymorphisms (SSLPs), they are highly valuable as molecular markers. STS loci have been analyzed and studied in a number of plant species.

> ➢ **Microsatellites:** Or SSR are extensively used in varietal identification and fingerprinting makes use of 2–8 base pairs simple repeats that are repeated several times in a genome and are highly polymorphic markers. In food authentication, SSR markers have been used to accurately match olive oil and its leaf profiles [54].
> ➢ **Sequence Characterized Amplified Regions (SCARS):** These are sequence-specific markers developed from RAPD profile of a given plant species. SCARs are the modified forms of STS markers. They are developed by PCR primers that are made for the ends of RAPD fragment. SCARs are useful for the rapid development of STS markers.

8.11.1.4   POLYMERASE CHAIN REACTION (PCR)

PCR is a routine molecular biology technique used to make multiple copies of a specific region of DNA *in vitro*. PCR employs a thermostable DNA polymerase (Taq DNA Polymerase obtained from a bacterium *Thermus aquaticus*) that can withstand high temperature during denaturation. Other ingredients for a PCR reaction are DNA primers (short oligonucleotide sequences 10–20 bp), dNTPs (deoxyribonucleotide triphosphates), template DNA and PCR buffer (contains $MgCl_2$, the cofactor for DNA polymerase).

A standard PCR reaction involves 3 steps: denaturation at 94°C to open DNA double-strand; annealing at a temperature between 40 and 63°C to allow primers to bind specific sequences; extension at 72°C to allow polymerization of dNTPs to synthesize new complementary strand. These steps are repeatedly cycled to synthesize multiple copies of target DNA.

PCR technique has extensively been used for food authentication and GMO detection [23]. Lack of quantitative information with normal PCR has led to the development of quantitative competition PCR (QC-PCR) and Real-time PCR that are being extensively utilized for quantification and food authentication, particularly in the detection of allergens and GMOs. Simultaneous detection of multiple fragments is possible by using specific probes or labeled primers. Specific and reliable methods of food authentication [40] and species identification need to be developed [63].

8.11.1.5   DNA BARCODING

DNA barcode is also a sequence-specific marker that amplifies standard regions of DNA using universal primers that can be amplified in different species. The variation in amplified sequences generates a different barcode for different genotypes. Barcoding has been extensively used in species identification and food authentication, for example, differentiating edible from inedible species of Solanum (*Solanum tuberosum, Solanum lycopersicum* group) and Prunus (*Prunus armeniaca, Prunus avium, Prunus cerasus, Prunus domestica*) [7] and adulteration detection in tea [20]. Similarly, DNA barcoding has been used for traceability of different aromatic spices of Lamiaceae family [15, 26].

8.11.1.6   MICROARRAY TECHNOLOGY

The microarray technology, also known as chip technology, is known for an orderly arrangement of probes to detect the base pairs in DNA sequences. The technology uses microplates or blotting membranes, and known and unknown DNA samples are matched based on base-pairing rule.

The sample spot sizes generally about two hundred microns in diameter may contain thousands of spots. The probes (DNA with well-known identity) are immobilized on a solid support, usually a silicon chip, microscopic slide, or nylon membrane that will bind to the complementary sequence of the unknown DNA templates, therefore, permitting parallel analysis for target sequence and gene discovery. An experiment with one silicon chip can provide information on thousands of genes at the same time. The orderly arrangement of the probes on the support is vital because the location of every spot on the array is employed for the identification of a specific sequence.

8.11.2   USE OF PNAS IN THE FIELD OF FOOD ANALYSIS

Owing to their unique properties, peptide nucleic acids (PNAs, a type of molecular probes) have been considered very promising. PNAs are imitator macromolecules having a neutral polyamide backbone composed of N-(2-aminoethyl) glycine that is similar to the sugar-phosphate backbone of DNA/RNA. Since the length and distance of bases from the backbone remain unchanged, therefore, PNAs can bind their complementary DNA or RNA sequences similarly to molecular probes. Due to the neutral charge of the PNA backbone, the thermostability of PNA/DNA or PNA/RNA complexes is better, thus increasing the sensitivity to detect even minute quantities. PNA in combination with PCR has been successfully used to detect hazelnut in complex food matrices. Surface plasmon enhanced fluorescent spectroscopy (SPFS) is another advanced technique that was used to detect GMO samples even at 0.1% concentrations without using PCR [55, 64] and oligonucleotide sequences matching with PNA probe.

8.11.3   LOOP-MEDIATED ISOTHERMAL AMPLIFICATION (LAMP)

Loop mediated isothermal amplification (LAMP) is an alternative technique to PCR that amplifies the DNA under isothermal condition. This method makes use of; (a) a DNA polymerase that possesses dual activity of displacement the DNA strands and its amplification and (b) set of inner and outer primer sets to specifically amplify the target DNA [35, 49]. Being accurate, sensitive, and simple, LAMP has been used in food authentication such as verifying medicinal plant species such as ginger [10] and detecting various contaminants of food [9, 12, 48, 67].

8.12 COMMERCIAL KITS FOR FOOD AUTHENTICATION

With the advances in food authentication and the related research areas, many commercial kits have been developed to ease the process of food analysis in different aspects. Some of the commercial kits based on molecular methods are listed in Table 8.5.

TABLE 8.5 Various Commercial Food Authentication Kits

Name of Kit	Assay	Target Product	Firm
EHEC (VT gene) PCR Screening Set	PCR based	*E. coli*	TaKaRa
ENLITEN® ATP Assay System	Quantitative ATP based	Indirect Detection of Biocontamination	ENLITEN, Promega
ENLITEN® rLuciferase/ Luciferin Reagent	Quantitative ATP based	Detection of contamination in liquid samples	ENLITEN, Promega
Food Control™ qPCR	Real-time PCR	Bacterial contaminations detection	Minerva biolabs
Food detection kit	Real-time PCR based Kit	Salmonella	Himedia
Food safety testing kits	Real-time PCR	Bacteria and GMO detection	Mericon, Qiagen
Foodproof® Celery Detection Kit	real-time PCR	Celery adulteration detection	BIOTECON Diagnostics
Foodproof® Gluten Detection Kit	real-time PCR	Gluten detection in mislabeled foods	BIOTECON Diagnostics
Gene TrakTM Campylobacter	Gene probe	*E. coli*; *Listeria*; Salmonella	Gene Trak, neogen, MI
GENE-UP® Real-time PCR kits	Real-Time PCR	Pathogen detection	Biomerieux-India
GoTaq® Probe qPCR and RT-qPCR Systems	Q PCR based	Seed admixtures	GoTaq, Promega
Maxwell® RSC PureFood GMO and Authentication Kit	PCR based	GM food ingredient	Promega
Maxwell® RSC PureFood Pathogen Kit	Bacterial DNA based	Bacterial contamination	Promega

8.13 SUMMARY

Food authentication, labeling requirements, and engagement in authenticity testing varies among different countries. Food authentication involves composition analysis, adulteration detection, species identification, and traceability of origin. Various techniques like ELISA and agar-gel immuno-diffusion have been employed widely for the authentication-related analysis followed by DNA based methods. In addition, spectroscopic techniques including IR spectroscopy, NMR spectroscopy, and spectral imaging have great potentials in authentication of various food products. Spectral imaging (such as VIS/NIR hyperspectral imaging, FTIR imaging, and Raman imaging) could capture large image data within spectral ranges, the imaging spectroscopy has now been acknowledged as an advanced technique that can meet the speedy and rising demand in the food industry.

KEYWORDS

- detection methods
- food adulteration
- food authentication
- food contamination
- food fraud
- loop-mediated isothermal amplification
- polymerase chain reaction

REFERENCES

1. Abbas, O., Zadravec, M., Baeten, V., Mikus, T., Lesic, T., Vulic, A., Prpic, J., Jemersic, L., & Pleadin, J., (2017). Analytical methods used for the authentication of food of animal origin. *Food Chemistry, 25*, 6–17.
2. Arlorio, M., Coïsson, J. D., Cereti, E., Travaglia, F., Capasso, M., & Martelli, A., (2003). Polymerase chain reaction (PCR) of puroindoline b and ribosomal/puroindoline b multiplex PCR for the detection of common wheat in Italian pasta. *European Journal of Food Research Technology, 216*, 253–258.
3. Ashkenazi, V., Chani, E., Lavi, U., & Veilleux, R., (2001). Development of microsatellite markers in potato and their use in phylogenetic and fingerprinting analyses. *Genome, 44*, 50–62.

4. Bansal, S., Singh, A., Mangal, M., Mangal, A. K., & Kumar, S., (2015). Food adulteration: Sources, health risks and detection methods. *Critical Reviews in Food Science and Nutrition, 57*(6), 1–80.

5. Bansal, S., Thakur, S., Mangal, M., Mangal, A. K., & Gupta, R. K., (2018). DNA barcoding for specific and sensitive detection of *Cuminum cyminum* adulteration in *Bunium persicum*. *Phytomedicine, 50*, 178–183.

6. Breton, C., Claux, D., Metton, I., Skorski, G., & Berville, A., (2004). Comparative study of methods for DNA preparation from olive oil samples to identify cultivar SSR alleles in commercial oil samples: Possible forensic applications. *Journal of Agriculture and Food Chemistry, 52*, 531–537.

7. Brunei, I., De Mattia, F., Galimberti, A., Galasso, G., Banfi, E., Casiraghi, M., & Labra, M., (2010). Identification of poisonous plants by DNA barcoding approach. *International Journal of Legal Medicine, 124*, 595–603.

8. Cebi, N., Dogan, C. E., Develioglu, A., Yayla, M. E. A., & Sagdic, O., (2017). Detection of L-cysteine in wheat flour by Raman microspectroscopy combined chemometrics of HCA and PCA. *Food Chemistry, 696*(1), 84–93.

9. Chaudhary, A. A., (2014). Rapid and easy molecular authentication of medicinal plant *Zingiber officinale* roscoe by loop-mediated isothermal amplification (LAMP)-based marker. *Journal of Medicinal Plant Research, 8*(20), 756–762.

10. Chaudhary, A., Ahmad, A., & Iqbal, M., (2014). Molecular biology techniques for authentication of medicinal plants. In: Iqbal, M., & Ahmad, A., (eds.), *Current Trends in Medicinal Botany* (pp. 313–347). Karnataka- India: I.K International Publishing House Pvt. Ltd.

11. Cuny, M., Vigneau, E., & Le Gall, G., (2008). Fruit juice authentication by ¹H NMR spectroscopy in combination with different chemometrics tools. *Analytical and Bioanalytical Chemistry, 390*, 419–427.

12. Curtis, K. A., Rudolph, D. L., & Owen, S. M., (2008). Rapid detection of HIV-1 by reverse-transcription, loop-mediated isothermal amplification (RT-LAMP). *Journal of Virological Methods, 151*, 264–270.

13. Danezis, G., Tsagkaris, A. S., Brusic, V., & Georgiou, C. A., (2016). Food authentication: State of the art and prospects. *Current Opinions in Food Science, 10*, 8. doi: 10.1016/j.cofs.2016.07.003; Accessed on August 31, 2020.

14. De La Fuente, M. A., & Juarez, M., (2005). Authenticity Assessment of dairy products. *Critical Reviews in Food Science and Nutrition, 45*(7/8), 563–585.

15. De Mattia, F., Bruni, I., Galimberti, A., Cattaneo, F., Casiraghi, M., & Labra, M., (2011). A Comparative study of different DNA barcoding markers for the identification of some members of *Lamiaceae*. *Food Research International, 44*, 693–702.

16. Dhanya, K., & Sasikumar, B., (2010). Molecular marker-based adulteration detection in traded food and agricultural commodities of plant origin with special reference to spices. *Current Trends in Biotechnology and Pharmacy*, 454–489.

17. Dhanya, K., Siv, P. S., & Bhas, S., (2009). Development and application of SCAR Marker for the detection of papaya seed adulteration in traded black pepper powder. *Food Biotechnology, 23*, 97–106.

18. Dhanya, K., Syamkumar, S., Siju, S., & Sasikumar, B., (2011). SCAR markers for adulterant detection in ground chili. *British Food Journal, 113*(5), 656–668.

19. Dhanya, K. P., Syamkumar, S. A., Jaleel, K. U., & Sasikumar, B., (2008). Random amplified polymorphic DNA technique for detection of plant-based adulterants in chili powder (*Capsicum annuum*). *Journal of Spices and Aromatic Crops, 17*(2), 75–81.

20. Dhiman, B., & Singh, M., (2003). Molecular detection of cashew husk (*Anacardium occidentale*) adulteration in market samples of dry tea (*Camellia sinensis*). *Planta Medica, 69*(9), 882–884.

21. Ellis, D. I., Brewster, V. L., Dunn, W. B., Allwood, W., Golovanov, A. P., & Goodacre, R., (2012). Fingerprinting of food: Current technologies for the detection of food adulteration and contamination. *Chemistry Society Reviews, 41*, 5706–5727.

22. Ellis, D. I., Muhammadali, H., Allen, D. P., Elliot, C. T., & Goodacre, R., (2016). A flavor of omics approaches for the detection of food fraud. *Current Opinion in Food Science, 10*, 7–15.

23. Erlich, H., (2015). *PCR Technology: Principles and Applications for DNA Amplification* (p. 246). New York: Springer.

24. FSSAI, (2015). Manual of Methods of Analysis of Foods (Spices and Condiments). Lab Manual 10; New Delhi – India: Food Safety and Standards Authority of India (Ministry of Health and Family Welfare), Government of India; p. 50.

25. Gajewski, K., (2008). *Characterization of Monoclonal Antibody Specific to Fish Major Allergen-Parvalbumin* (p. 137). MSc Thesis; Gainesville-FL: Florida State University.

26. Galimberti, A., Mattia, F. D., Losa, A., Bruni, I., Federici, S., Casiraghi, M., Martellos, S., & Labra, M., (2013). DNA barcoding as a new tool for traceability. *Food Research International, 50*(1), 55–63.

27. Haughey, S., Galvin-King, P., Ho, Y. C., & Bell, S., (2015). The feasibility of using near-infrared and Raman spectroscopic techniques to detect fraudulent adulteration of chili powders with Sudan dye. *Food Control, 48*, 75–83.

28. Horn, B., Esslinger, S., Pfister, M., Faul-Hassek, C., & Riedl, J., (2018). Nontargeted detection of paprika adulteration using mid-infrared spectroscopy and one-class classification—is it data preprocessing that makes the performance? *Food Chemistry, 257*, 112–119.

29. Hsieh, C., Li, P. H., Cheng, J. Y., & Ma, J. T., (2013). Using SNIF-NMR method to identify the adulteration of molasses spirit vinegar by synthetic acetic acid in rice vinegar. *Industrial Crops and Products, 50*, 904–908.

30. Hurley, I. P., Coleman, R. C., Ireland, H. E., & Williams, J. H., (2004). Measurement of bovine IgG by indirect competitive ELISA as a means of detecting milk adulteration. *Journal of Dairy Science, 87*(3), 543–549.

31. Hussain, M. A., (2016). Food Contamination: Major challenges of the future. *Foods, 5*, 21–28.

32. Jentzch, P. V., Gualpa, F., Ramos, L. A., & Ciobota, V., (2017). Adulteration of clove essential oil: Detection using a handheld Raman spectrometer. *Flower and Fragrance Journal, 33*(2), 184–190.

33. Jha, S. N., (2016). Spectroscopy and chemometrics. In: Jha, S. N., (ed.), *Rapid Detection of Food Adulterants and Contaminants* (Vol. 1, pp. 147–214). San Diego-CA, USA: Academic Press.

34. Kolavic, S. A., Kimura, A., Simons, S. L., Slutsker, L., Barth, S., & Haley, C. E., (1997). An outbreak of *Shigella dysenteriae* type-2 among laboratory workers due to intentional food contamination. *JAMA, 278*(5), 396–398.

35. Kumar, Y., Bansal, S., & Jaiswal, P., (2017). Loop-mediated isothermal amplification (LAMP): A rapid and sensitive tool for quality assessment of meat products. *Reviews in Food Science and Food Safety, 16*, 1359–1378.

36. Kuswandi, B., Futra, D., & Heng, L. Y., (2017). Nanosensors for the detection of food contaminants. In: Oprea, A. E., & Grumezescu, A. M., (eds.), *Nanotechnology Applications in Food* (pp. 307–333). San Diego - USA: Academic Press.

37. Liu, Y. (2009). *Control Engineering China, Dairy Safety in Control in China*. Available at: https://www.controleng.com/articles/dairy-safety-in-control-in-china/ (accessed on 27 January 2021).

38. Lopez, M. D., Jordan, M., & Pascual, M. J., (2008). Toxic compounds in essential oils of coriander, caraway, and basil active against stored rice pests. *Journal of Stored Products Research, 44*, 273–278.

39. Lotta, F., & Bogue, J., (2015). Defining food fraud in the modern supply chain. *European Food and Feed Law Review, 10*, 114–122.

40. Mafra, I., Ferreira, I. M. P. L. V., Oliveira, M. B. P. P., (2007). Food authentication by PCR-based methods. *European Food Research and Technology, 227*, 649–665.

41. Mahmood, T., & Yang, P. C., (2012). Western blot: Technique, theory, and troubleshooting. *North American Journal of Medical Sciences, 4*(9), 429–434.

42. Makie, I. M., (1996). Authenticity of fish. In: Ashurt, P. R., & Dennis, M. J., (eds.), *Food Authentication* (pp. 140–170). London: Blackie Academic and Professional.

43. Mangal, M., Bansal, S., Sharma, S. K., & Gupta, R. K., (2016). Molecular detection of foodborne pathogens: A rapid and accurate answer to food safety. *Critical Reviews in Food Science and Nutrition, 56*(9), 1568–1584.

44. Marieschi, M., Torelli, A., & Bruni, R., (2017). Quality control of saffron (*Crocus sativus* L.): Development of SCAR markers for the detection of plant adulterants used as bulking agents. *Journal of Agriculture and Food Chemistry, 44*, 10998–11004.

45. Martins, A. R., Talhavini, M., Vieira, M. L., Zacca, J. J., & Braga, J. W. B., (2015). Discrimination of whisky brands and counterfeit identification by UV-Vis spectroscopy and multivariate data analysis. *Food Chemistry, 229*, 142–151.

46. Milanez, K. D. T. M., Nobrega, T. C. A., Nascimento, D. S., Insausti, M., & Band, B. S. F., (2017). Multivariate modeling for detecting adulteration of extra virgin olive oil with soybean oil using fluorescence and UV-Vis spectroscopies: A preliminary approach. *Food Science and Technology, 85*, 9–15.

47. Mustorp, S., Engdahl, C., Svensson, U., & Holck, A., (2008). Detection of celery (*Apium graveolens*), mustard (*Sinapis alba, Brassica juncea, Brassica nigra*) and sesame (*Sesamum indicum*) in food by real-time PCR. *European Journal of Food Research Technology, 226*(4), 771–778.

48. Nagdev, K. J., Kashyap, R. S., Parida, M. M., Kapgate, R. C., Purohit, H. J., & Taori, G. M., (2011). Loop-mediated isothermal amplification for rapid and reliable diagnosis of *Tuberculous meningitis. Journal of Clinical Microbiology, 49*, 1861–1865.

49. Notomi, T., Okayama, H., Masubuchi, H., Yonekawa, T., Watanabe, K., Amino, N., & Hase, T., (2000). Loop-mediated isothermal amplification of DNA. *Nucleic Acids Research, 2000*, E63–E69.

50. Oliveira, R. C. S., Oliveira, L. S., Franca, A. S., & Augusti, R., (2009). Evaluation of the potential of SPME-GC-MS and chemometrics to detect adulteration of ground roasted coffee with roasted barley. *Journal of Food Composition and Analysis, 22*, 257–261.

51. Parvathy, V. A., Swetha, V. P., Sheeja, T. E., Leela, N. K., Chempakam, B., & Sasikumar, B., (2014). DNA barcoding to detect chili adulteration in traded black pepper powder. *Food Biotechnology, 28*, 25–40.

52. Pasqualone, A., Lotti, C., & Blanco, A., (1999). Identification of durum wheat cultivars and monovarietal semolinas by analysis of DNA microsatellites. *European Journal of Food Research Technology, 210*, 144–147.

53. Petrakis, E. A., Cagliani, L. R., Polissiou, M. G., & Consonni, R., (2015). Evaluation of saffron (*Crocus sativus* L.) adulteration with plant adulterants by [(1)]H NMR metabolite fingerprinting. *Food Chemistry, 217*, 418–424.

54. Rabiei, Z., Tahmasebi, E. S., Saidi, A., Patui, S., & Vannozzi, G. P., (1999). SSRs (simple sequence repeats) amplification: A tool to survey the genetic background of olive oils. *Iranian Journal of Biotechnology, 8*, 24–31.

55. Rant, U., Arinaga, K., Scherer, S., Pringsheim, E., Fujita, S., Yokoyama, N., Tornow, M., & Abstreiter, G., (2007). Switchable DNA interfaces for the highly sensitive detection of label-free DNA targets. *PNAS, 104*(44), 17364–17369.

56. Remaud, G., Debon, A. A., Martin, Y. L., & Martin, G. G., (1997). Authentication of bitter almond oil and cinnamon oil: Application of the SNIF-NMR method to benzaldehyde. *Journal of Agriculture and Food Chem*istry, *45*, 4042–4048.

57. Ropelewska, E., & Jankowski, K. J., (2019). Classification of the seeds of traditional and double-low cultivars of white mustard based on texture features. *Journal of Food Process Engineering, 2019*, 11. doi: 10.1111/jfpe.13077; Accessed on August 31, 2020.

58. Sandberg, M., Lundberg, L., Ferm, M., & Yman, I., (2003). Real-time PCR for the detection and discrimination of cereal contamination in gluten-free foods. *European Food Research and Technology, 10*, 344–349.

59. Sasikumar, B., Siv, P. S., Remya, R., & Zachariah, J., (2004). PCR based detection of adulteration in the market samples of turmeric powder. *Food Biotechnology, 18*(3), 299–306.

60. Sivakesava, S., & Induraj, J., (2001). A rapid spectroscopic technique for determining honey adulteration with corn syrup. *Journal of Food Science, 66*, 46–64.

61. Spink, J., & Moyer, D. C., (2011). *Defining the Public Health Threat of Food Fraud* (p. 79). East Lansing-MI: Michigan State University.

62. Taylor, S. C., Berkelman, T., Yadav, G., & Hammond, M., (2013). A defined methodology for reliable quantification of western blot data. *Molecular Biotechnology, 55*(3), 217–226.

63. Teletchea, F., Maudet, C., & Hanni, C., (2005). Food and forensic molecular identification: Update and challenges. *Trends in Biotechnology, 10*, 359–366.

64. Toma, M., Toma, K., Adam, P., Vala, M., Homola, J., Knoll, W., & Dostalek, J., (2012). Surface plasmon-coupled emission on plasmonic Bragg gratings. *Optical Express, 20*(13), 14042–14053.

65. Vemireddy, L. R., Archak, S., & Nagaraju, J., (2007). Capillary electrophoresis is essential for microsatellite marker-based detection and quantification of adulteration of basmati rice (*Oryza sativa*). *Journal of Agriculture and Food Chemistry, 55*, 8112–8117.

66. Verma, S. K., Khanna, V., & Singh, N., (1999). Random amplified polymorphic DNA analysis of Indian scented basmati rice (*Oryza sativa*) germplasm for identification for variability and duplicate accessions, if any. *Electrophoresis, 20*(8), 1786–1789.

67. Wang, L., Shi, L., Alam, M. J., Geng, Y., & Li, L., (2008). Specific and rapid detection of foodborne *Salmonella* by loop-mediated isothermal amplification method. *Food Res. International, 41*, 69–74.

68. Yang, H., & Irudayaraj, J., (2001). Comparison of near-infrared, Fourier transform-infrared, and Fourier transform-Raman methods for determining olive pomace oil adulteration in extra virgin olive oil. *Journal of American Oil Chemists' Society, 78*, 889–895.

CHAPTER 9

IMAGING TECHNIQUES FOR QUALITY ASSESSMENT OF SPICES AND NUTS

LEENA KUMARI, MONIKA SHARMA, and GAJANAN DESHMUKH

ABSTRACT

Many of the imaging techniques (such as X-ray and MRI) have led to several novel applications in different areas of the agri-food industry (including spices and nuts), particularly handling the food quality and food safety-related issues. Imaging techniques can be utilized to identify the adulterants and contaminants and internal or external defects in spices and nuts, which are difficult to observe by visual inspection. This book chapter discusses the important imaging techniques along with their applications in the quality inspection of different spices and nuts.

9.1 INTRODUCTION

The quality of food and raw products is complex in nature being affected by hygiene, nutrition, sensory, toxicology, and several technological and processing attributes [126]. The food quality influences the acceptance and consumer satisfaction of food. Therefore, worldwide attention and major social concern is focused on developing novel methods of evaluation of food quality and safety [5]. External quality attributes (viz., color, appearance, shape, size, and texture) of a food are in general perceived by human senses based on touch and sight, which instantly helps the customer to judge its quality [21]. Recently, consumers have shown an inclination towards the greater use of spices and nuts as health foods, owing to the presence of various essential oils, functional ingredients, bioactive compounds (such as antioxidants), etc.

International Organization for Standardization (ISO) describes spices "*as natural plant products either in whole or ground form having a characteristic flavor, aroma, or pungency.*" Maintaining good product quality is very vital in this era of globalization, where products from one nation are being exported to several countries. This can be only possible when there are robust, rapid, and effective quality assessment techniques.

There is a great demand of good quality spices and nuts and people are ready to pay high prices for quality products. However, there is a critical issue of adulterating the spices and nuts with cheaper substances, which may prove hazardous for human health as well as results in economic losses. Mostly in the case of ground spices, certain dyes (both natural and chemical) are added, such as Sudan dye in chili powder, yellow chalk powder in turmeric powder, lead chromate as a bulking agent in turmeric powder, and marble dust in certain spices, etc.

Further, in the case of nuts, some of the nuts can aggravate allergy in some individuals, and the effects could be fatal. Therefore, it becomes important to determine the adulteration in spices and nuts very precisely. Owing to these issues of adulteration, insect, pests, fungal contamination, etc., it becomes vital to adopt methods and techniques of rapid identification of such substances, so that quality standards could be maintained for spices and nuts, thereby delivering good quality and safe products to the consumers.

Numerous steps (*viz.*, handling, grading, sorting, and processing) must be monitored with utmost precision to ensure the wholesomeness in internal and external quality of spices and nuts. Various analytical techniques (such as high performance liquid chromatography (HPLC), Gas chromatography (GC), Polymerase chain reaction (PCR), mass spectrometry (MS), etc.), are commonly used for quality assessment of spices and nuts, which are destructive, time-consuming, require sophisticated instruments and pre-processed samples; whereas visual inspection requires human interpretation that is less reliable due to human errors or inexperienced technicians [158]. Conventional techniques are incapable to provide real-time and online results, which further create a substantial barrier for large-scale quality monitoring system. The current optical and mechanical method are based on monitoring parameters (such as color, size, shape, and mass), which are suitable for assessing the external quality of food [12, 68, 125]. Further, it is necessary to develop faster and accurate non-destructive quality evaluation systems that are able to investigate the internal quality of a food or a food product.

In order to conquer the challenges posed by destructive quality assessment techniques, there is a tremendous shift towards spectroscopic and imaging methods for the quality inspection of spices and nuts. Spectroscopic techniques (*viz.*, Raman spectroscopy, near-infrared (NIR) spectroscopy and nuclear magnetic resonance (NMR) spectroscopy have been well-established techniques for prediction of composition and internal structural evaluation [28, 106]. However, the effectiveness of these spectroscopic techniques can be enhanced through coupling these with image analysis that reveals more spectral and spatial information [128, 140]. Technological developments in image capturing devices (cameras/scanners), potent computer hardware, software system, and analyzing tools have paved the way for successful non-destructive tools for assessing both internal and external quality attributes of an object [5].

The image processing includes: (a) a device (camera/scanner) for capturing images of the product either during processing or of a finished product or during storage of product, under standard illumination conditions; and (b) a computational unit for processing the digitalized image that analyzes the data for defining and characterizing quality attributes [153]. Imaging technique includes numerous steps, such as image acquisition, feature extraction, image segmentation, classification, etc., [36]. Several researchers have used imaging techniques (such as camera-based imaging, X-ray imaging, computed tomography (CT), hyperspectral imaging, thermal imaging (TI), magnetic resonance imaging (MRI), etc.), for non-destructive quality assessment of agricultural and food products [1, 25, 49, 127].

This chapter discusses:

1. Various aspects related to the quality concerns in spices and nuts;
2. Basic concepts of different imaging techniques; and
3. Specific applications of various imaging techniques for quality assessment.

9.2   QUALITY ISSUES IN SPICES AND NUTS

Spices constitute an important group of plants having significant biodiversity with diverse parts, such as roots, fruits, leaves, seeds, peels, buds, flowers, stems, or rhizomes that are being used for distinct taste, flavor, or pungency [6]. The consumers throughout the world are becoming more quality conscious about the spices, nuts, and other food items. Drying and storage of spices at the farm level generally causes contamination with various extraneous matters and may lead to visual and integral defects in the products. There are different worldwide organizations to provide guidelines for maintaining the quality standards of spices and nuts. Usually, the standards vary slightly for whole and ground spices. The limits are given usually for the maximum permissible moisture content, insects count, and minimum quantity of volatile oils, ash content, mammalian excreta, and percentage of infected seeds by insects, etc.

Another cause of great concern in spices is adulteration. Due to high cost of spices, spurious practices are followed by some traders, where adulterants having similar appearances as that of a spice are mixed with ground spices and also in case of whole spices (such as papaya seeds in black pepper, chemicals dyes in chili powder, etc.). Therefore, the major concerns for the quality of spices and nuts are: adulterants, insects, pests, and rodents, which in turn can affect the quality of spices. They are also responsible for the entry of microbes in the spices and thereby leading to an increase in toxins levels secreted by these microbes. In addition, fungal spoilage is one of the major causes of concern for maintaining the quality of spics and nuts (Figure 9.1). Even if the fungal and bacterial loads are inactivated by certain preservation treatments, it becomes very difficult to get rid-off the mycotoxins and other toxins secreted by these microorganisms. Thus, it becomes very important to select a quality inspection technique, which can identify the defects caused by insect and pest infestation, detection of microbial contamination, determination of mycotoxins, etc.

FIGURE 9.1   Fungal growth on varied spices: (A) Red chili, (B) cinnamon, (C) clove, (D) coriander seed, (E) cardamom, and (F) black pepper.
Source: Reprinted with permission from Springer [142].

The term nut is generally used to describe a wide range of seeds having a seed coat or a shell. Nuts are a rich source of several nutrients with significant amounts of proteins, carbohydrates, fats, vitamins, dietary fiber, minerals, and bioactive components (such as phytosterols, antioxidants, and other phytochemicals) [35]. Most of nuts are generally graded based on the visual quality attributes (such as shape, size, density, color, texture, appearance, moisture), pathological, and physical damage, and external defects. The manual and conventional mechanical inspections cannot help us to evaluate internal quality assessment of nuts using these techniques.

Further, the quality of spices and nuts are of prime importance for consumer satisfaction in the market. Their quality assessment is dependent on manual inspection, which is not only time and labor-intensive but also expensive and is influenced by numerous physiological factors and prone to human error [36]. Thus, the non-destructive rapid quality evaluation techniques for spices and nuts have emerged as a major area of interest in the nut industry.

The image analysis has proven as an effective technique for rapid detection of defects and internal quality assessment in nuts [64, 75, 77, 79]. Nuts generally suffer from various defects during production, harvesting, processing, and storage. The main problem in the case of pistachio nut is contamination with aflatoxin, having an adverse effect on the health of consumers and economic profits for sellers [132]. Wang et al. [150, 151] used image pattern recognition for differentiating wholesome and worm-infested nuts based on imaging technique. Donis-González et al. [33] also examined the sorting of chestnut based on its texture, size, color, and geometric orientation using imaging technique, and they indicated an accuracy of >88% in the system.

Jin et al. [59] established a system based on imaging for separating black walnut meat from the shell with overall accuracy of separation of >98%. The system works on the principle of differences in surface characteristics of black walnut meat with its shell. Lee et al. [85] developed a system to evaluate the maturity of dates based on color distribution analysis and color space conversion. In case of fennel seeds, the common adulterants (such as pedestals, top part of seeds and stones) could be detected through capturing the red green blue (RGB) images, followed by their conversion into binary images and application of certain colorization procedures [113]. Abdollahn-ejad et al. [2] applied the imaging technique along with adaptive neural-fuzzy inference system for sorting of pistachio nuts.

Razavi et al. [124] measured physical properties (sphericity, projected area, roundness, aspect ratio, density, angle of repose, angle of friction, funning angles terminal velocity) of wild pistachio nut using image analysis and compared these results with experimental methods. Giraudo et al. [45] developed a rapid detection protocol for identifying flawed hazelnuts using the multivariate analysis of RGB images. Narendra and Hareesha [102] developed the method based on image processing rapid for identification cashew nuts based on their kernels. Ercisli et al. [41] applied the image processing technique for structural characterization of walnut.

9.3 BASICS OF IMAGING PROCESS

Imaging techniques are based on the process of capturing images using suitable devices, and then these images are analyzed using statistical techniques for intended purpose. There are several crucial steps for extracting the useful information from images that start with image acquisition, and then images are processed and analyzed using statistical tools and software. A brief overview of various steps of the imaging process is illustrated in this section.

9.3.1 IMAGE ACQUISITION

In the image acquisition process, the electronic signals received by sensing devices are transferred into numeric digits [94]. Camera is used to capture the images and captured images are stored in the computer for transforming these into digital images for subsequent analysis. Digital images are numeric digits (usually binary numbers), which can be read by computer and modified into pixels or minute dots [94]. A pixel is the smallest single point picture element of a graphic image that cannot be magnified further and it is the very basic element of the image. An image is divided into millions of pixels arranged in a definite pattern of numerous rows with several columns. Arrangement of pixels is in such a way that the pixels look like joined altogether in an image, and in case of color monitors, every single pixel is composed of three dots of RGB (red, blue, green) color.

9.3.2 IMAGE PROCESSING

Image processing refers to the process of applying various operations on images, which results in more meaningful images with improved visibility of different useful features. It facilitates subsequent analysis of images.

There are various steps in image processing that mainly includes image pre-processing, image enhancement, image segmentation, image feature extraction, and image feature classification. In image pre-processing, raw images are primarily processed so that blurring and noise may be removed and geometric distortions, etc., can be corrected [135].

9.3.2.1   IMAGE ENHANCEMENT

Image enhancement is used to correct the issues of low contrast or high noise in the acquired image. The discrepancies produced due to improper illumination can be corrected by various measures, such as filters, morphological, and pixel-to-pixel operations [94]. Image enhancement helps in differentiating the sample of interest from the background. For instance, better contrast of images can be obtained simply by increasing the brightness level. Common methods used for improving the image contrast include: basic point operations (viz.: brightness inversion and brightness scaling through multiplication), and histogram equalization that scale-up intensity of image through flatting the histogram. To minimize the blur and noise in images, different filters (such as average filter, median filter, low pass filter, and Gaussian filter) are utilized [161]. Many times, images of the moving object as if sample moving on the conveyor belt is captured that results in image distortion. Various geometric transforms viz. image rotation, translation, scaling, mirror, and transpose can be used to rectify image distortion [161].

9.3.2.2   IMAGE SEGMENTATION

In the image segmentation process, the image is distributed into multiple segments or regions of interest (ROI) that are more relevant for sample analysis. It is one of the important processes as the accuracy of subsequent analysis is dependent on this step [18]. Some researchers defined the Image segmentation methods based on the properties of discontinuity known as boundary-based methods and based on properties of similarity known as region-based methods.

Generally, image segmentation can be achieved by five approaches that include thresholding, edge-based, region-based segmentation [136, 139], gradient-based segmentation, and classification-based segmentations [51]. In a thresholding technique, image regions are characterized depending on the continuous reflection or absorption of light from surfaces of the sample

(Figure 9.2(a)). In edge-based segmentation technique, edge operators detect dissimilarity in color, texture, and grayness of images and then segment them based on these dissimilarities (Figure 9.2(b)). Region segmentation technique utilizes the clustering of alike pixels together so that ROI are created that depict the particular objects in an image (Figure 9.2(c)). In gradient-based segmentation, the convolute gradient operators such as Laplace operators are used to compute the gradient of images. In classification-based segmentation, every pixel is treated as an independent observer containing variables obtained from various image features viz. color, texture, and shape. A matrix composed of all the pixels is used as input of the classification and then classified into classes utilizing learning model [36].

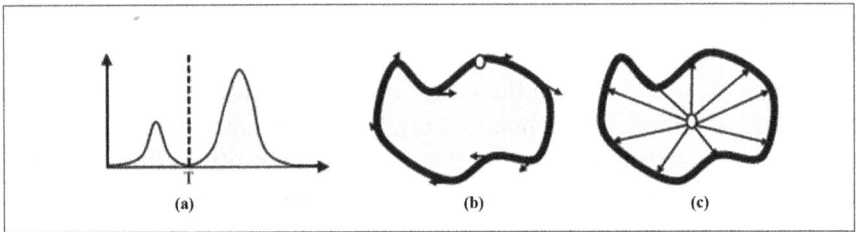

FIGURE 9.2   Different techniques of image segmentation: (A) thresholding; (B) edge-based segmentation; and (C) region-based segmentation.
Source: Reproduced from [139] and reprinted with permission from Elsevier.

It is a supervised learning technique based on training of image set and then classifying the testing image set. However, some unsupervised learning technique like clustering and self-organizing map also exists that classifies the observers (pixels) into different clusters with no need of prior training or knowledge.

Watershed and hybrid-based segmentation are some other segmentation techniques employed for image segmentation. Color image segmentation involves various color spaces, including RGB, HSI, and CIE Lab for segmentation of image based on color features [99].

9.3.2.3 FEATURE EXTRACTION

In case of an image or a picture, the feature is the "interest" part of the image. Feature extraction refers to the extraction of useful features from the image that will be further used for detailed analysis. Different image features (including: color, size, shape, texture, etc.), are beneficial for describing parameters of sample, and are extracted from the image using local feature detector and visual descriptor (such as local binary pattern (LBP), histogram of oriented gradient (HOG)) and speeded up robust features (SURF) according to Naik et al. [100]. Some of the statistical techniques and procedures involved in image analysis (such as mean, standard deviation, variance, and Principle component analysis (PCA), etc.), can also be used to extract important features of images.

9.3.2.4   FEATURE CLASSIFICATION

Feature classification is an important step that refers to the classification of features extracted by various methods. Various techniques (such as neural networks, machine learning algorithms, and fuzzy inference systems) are applied to perform image feature classification. Using a powerful classification function, imaging technique can imitate the human brain for decision-making and able to classify the different objects based on desired parameters with accuracy and consistency [3]. For instance, linear discriminant analysis (LDA), principal component analysis (PCA), KNN (K-nearest neighbor), support vector machine (SVM), artificial neural network (ANN), Fuzzy logic, and Regression methods are some common techniques that are used for classification of features [100].

9.3.3   IMAGE ANALYSIS AND RELATED SOFTWARE

Image analysis is the process of computing the different dimensions and related statistics depending on the data value of image pixels along with their information of spatial location [87] through utilization of features extracted from images. Mainly there are two parts of image analysis, such as vision measurement and pattern classification.

- Vision measurement is used as a method for quantitative analysis and it involves quantification of relevant parameters from the extracted features [112], such as estimation of size, texture, and color of the tested sample; and
- Pattern classification, also known as pattern recognition is the technique used for qualitative analysis of images through statistical tools and techniques such as multivariate analysis and algorithms, etc.

Mainly, there are two types of classification techniques: (a) one is based on supervised methods that build a model or a classifier depending on certain characteristics and mainly involves training and validation of data from images [7]; (b) Other method is unsupervised method, which classifies images using clustering algorithm based on similarity among the selected features. Some popular examples of unsupervised methods are: SVM, KNN and ANN; as described by Zhang et al. [161] and Li et al. [87].

The specific software applications developed in different programming languages, mainly including C++, Visual Basic, MATLAB, LabVIEW, etc., are used for image analysis [161]. Most of the methods required for image processing and analysis are included in this software, which are able to perform spectral processing, multivariate analysis, etc., and sometimes hand-coded specific programs may be developed with the help of this software.

9.3.3.1   COLOR ANALYSIS BASED ON HISTOGRAM

Colors that are perceived by the human eye are a result of the reflection of light from the object surface. Mainly there exist three primary colors including red (R), green (G), and blue (B). The human eye perceives the color of items either as a primary color or in the form of combinations of primary colors.

A color model is used to specify the color in specified coordinate systems so that each color is defined by a single point in the subspace of the coordinate system, thus enabling the standard color system suitable for analysis and comparison. In the RGB color model, black color is at the origin, whereas the farthest corner from origin is for white color [154]. In RGB color model, three primary color R (Red), G (green), B(blue) are depicted at three corners; whereas secondary colors including cyan, magenta, and yellow are depicted at other three corners [55] as shown in Figure 9.3. Different colors can be presented by points in on or inside the cube in terms of vectors from the origin [46, 134].

To analyze the color of an image, usually a color histogram is built, which shows the distribution of colors in the image. There are a number of color spaces or color coordinate systems, which define the color of objects as well as their images, and are used to develop color histogram. For instance, RGB color space, HSI (hue, saturation, and intensity) space and Hunter Lab space are mostly used for color space for image analysis.

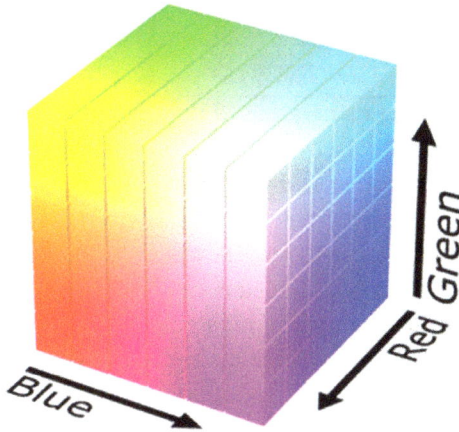

FIGURE 9.3 RGB color model.

Some other color space used for image analysis includes [29]: The CIE L*a*b* space or CIE XYZ as set by International Commission on Illumination. Although RGB color space is most commonly used but HSI color space is considered superior to RGB, as HSI is more imitated for human vision and hue value is unaffected by changes in light [47, 123]. In the case of monochromatic images, intensity histogram is used instead of color histogram; whereas in the case of multispectral images, N-dimensional color histogram is used where N defines the number of measurements [94].

9.4   IMAGING TECHNIQUES FOR QUALITY INSPECTION OF SPICES AND NUTS

Several non-destructive imaging techniques (such as Camera-based imaging, X-ray imaging, Hyperspectral imaging, MRI, CT, NIR (near-infrared) imaging, TI, etc.), have been investigated by researchers for quality assessment of spices and nuts [25, 49, 71]. Many of these techniques (such as X-ray, MRI) have led their novel applications for quality assessment of a variety of nuts and spices [57, 62, 92, 101, 118, 140, 143, 150]. In the subsequent sections of this chapter, various imaging techniques, including their basics and some popular applications for quality inspection of spices and nuts, have been discussed.

9.4.1   COMPUTER VISION OR CAMERA-BASED IMAGING TECHNIQUE

Computer vision or the camera-based imaging technique is a common imaging technique widely used for external quality inspection of a variety of food items. This technique is based on replicating the human eye vision through acquisition of images with a camera. Usually three filters centered at wavelengths corresponding to red (R), green (G) and blue (B) colors are used for capturing images with the camera [93]. The key component of this imaging technique is the camera, which acts like a human eye for investigating the sample in interest. Other main components of computer vision technique include (Figure 9.4): illumination source, image capturing board or frame (or digitizer), software, and computer [146].

Illumination devices are used for illuminating the food samples by generating light of suitable intensity. Mainly different types of lamps (such as incandescent, halogen, fluorescent, LED, and infrared) are used as light sources for illumination [72]. The process of image capturing is greatly influenced by the level, quality, intensity, and type of illumination. Good quality of illumination helps in improving image analysis by noise reduction, shadow removal, reflection correction and enhancing contrast of images; and thus, efficiency, and accuracy of the system is improved [18, 19, 108]. There are generally two ways to apply illumination: (a) one is front lighting that is used to study surface characteristics; and (b) other is backlighting utilized for enhancement of background [102].

An extremely important element of a computer vision technique is the camera that collects light from the surface of the sample object and then captured light is converted in the form of electrical signals. Mainly charged coupled device (CCD) and complementary metal-oxide-semiconductor (CMOS) image sensors-based cameras can also be used for acquisition of digital images [161]. In the case of CCD based camera, mainly two modes (viz. area and line scan) are commonly used for investigation of an object. Computer vision can be used for Monochrome imaging by using a single-chip CCD, whereas color imaging is mostly accomplished by three-chip CCD camera.

Another component is the image capturing board (also known as frame grabber). It is an electronic mechanism, which can receive various types (digital or analog) of images based on the type of camera in use. Main functions of frame grabber include: camera control, acquisition, and storage of raw or compressed digital images for subsequent analysis and pre-processing. In addition, frame grabber divides the image into grids (smaller areas) or pixels. The computer hardware and software perform the functions like a human brain and are used to process the image information for specific analysis. Software may be specifically developed or the existing software (such as MATLAB, etc.), may be used for detailed data analysis of the image. Some of the common steps of image processing are image enhancement, feature extraction from image and feature classification that were discussed in Section 9.3 of this chapter. Various statistical tools and techniques (such as ANN) are used in combination with software for in-depth analysis of image data.

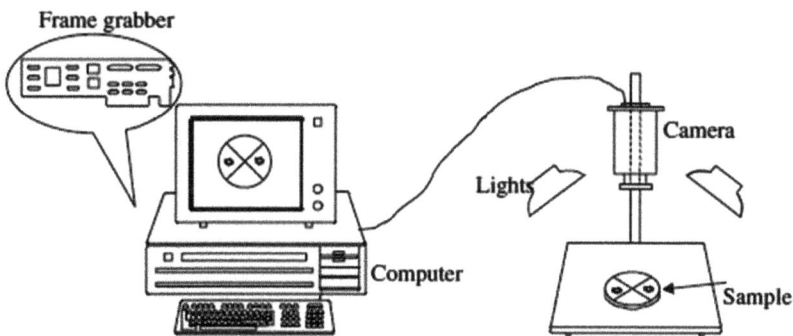

FIGURE 9.4 Different components of a computer vision system.
Source: Reproduced from Ref. [146] and reprinted with permission from John Wiley and Sons.

Computer vision system is extremely useful for inspection of surface defects and determinations of external quality attributes, such as size, shape, and color, etc. In recent years, there has been an increase in popularity of computer vision system for grading and sorting of nuts, spices, and other food items, such as fruits, vegetables, and grains, etc.

9.4.1.1 PREDICTION OF MATURITY

Black Pepper is one of the expensive spices having a strong pungent flavor with many medicinal properties. There are various stages of ripening of peppercorn fruit, such as white, green, and black that defines the color, appearance, and taste of final produce [107]. To predict the maturity stages of black pepper, Mythri et al. [98] developed an android-based application using Java and MVC (model view controller) architecture. During the initial phase of training, sample images of peppers having different maturity stages (4–5 months, 6–7 months, and 8 months) were collected from authorized sources in India, and in the final testing stage, images of pepper were acquired by smartphone camera. Using RGB features extracted from images and SVM classification accuracy of nearly 80% was achieved for classification of peppers into three different classes according to their maturity. The classification was mainly based on the darkening of color during the maturity of peppercorns. Research work by Lim and Gopalai [89] utilized the computer vision system to classify the ripening of peppers based on various features, such as color, intensity, and orientation of images. Vision system was able to spot the regions of importance with a success rate of 91.3%, and contour extraction resulted in a success rate of 84.35% for covering the pepper boundaries.

9.4.1.2 CLASSIFICATION BASED ON QUALITY ATTRIBUTES

Classification is an important process employed in the food industry that involves the segregation of food items into different groups based on certain quality parameters, such as color. Machine vision system was explored by Sajjan et al. [129] for classification of different samples of chili varieties (with stalk and without stalk) from bulk chili collected from Dharwad, India.

A digital camera (resolution of 16 Megapixels) was used for acquiring of images and a total 9 color features (such as R (red), G (green), B (blue) or RGB, H (hue), S (saturation), I (intensity) or HSI and L (brightness), a (red and green chromaticity layer), b (blue and yellow chromaticity layer)) or L*a*b Model were extracted from chili images for classification. The best classification of chili varieties was achieved using L*a*b color features with a recognition rate of 85% and 95%, for chili with stalk and without stalk, respectively. Good classification was achieved using HSI color features with a recognition rate of 80% and 90% chili with stalk and without stalk; whereas using RGB color features recognition rate was 70% and 85% for chili with stalk and without stalk, respectively.

In a research work by Goswami et al. [48], computer vision was used for classification of black pepper and detection of foreign materials. The color image was converted into grayscale images, and different features (such as length of major and minor axis, area, and eccentricity) were used to classify the pepper seeds and quality.

Computer vision system was employed for the classification of betel nuts in a research work by Liu et al. [90]. The CCD camera was used for the acquisition of images in YCbCr color mode, and different features (such as texture, shape, and color) were used for high-speed classification of betel nut in four categories, such as best level, better level, good level and bad level.

Classification of pepper seeds is a difficult task through human inspection as the size of the seed is very small and is similar in shape. A research effort was made by Kurtulmuş et al. [80] for the classification of pepper seeds using machine vision system and neural networks. They used a scanner (office type) having a resolution of 1200 dpi to obtain images of seed sample of eight varieties of pepper (*Capsicum* spp.) in Turkey. Images acquired in RGB color mode were transformed into two color space into YCbCr (luminance-chrominance in blue-chrominance in red) and HSI (hue-saturation intensity) so that a total of eight color channels (such as R, G, B, H, S, I, Cb, and Cr) were obtained for analysis using MATLAB. Using ANN, the accuracy rate of 80% was obtained compared to 85% with MLP topology for classification of pepper seeds for eight varieties.

Sandoval et al. [131] employed a computer vision system for color analysis of Habanero chili pepper during various ripening stages, such as transforming from green (unripe) to yellow and finally to orange (fully ripe stage). Samples of Habanero chili were divided into three groups based on maturity stage (green, yellow, orange), and a total of 900 images of samples were acquired using a CCD camera. Color analysis was carried out with CIELAB color space, and a high color change was observed through average values of hue angle representing transition in ripening stages. Using one-way ANOVA on data, significant color change existed in chili samples with sampling days corresponding to different ripening stages.

In a research work by Nasution and Gusriyan [104], computer vision was used for qualitative grading of nutmeg samples of Indonesian origin. Images of the nutmeg were acquired using Kinect v.2 camera and different features (viz.: color features (RGB)), shape features (such as diameter, perimeter, circularity, area) and texture features (such as entropy) and correlation extracted from images were used for prediction of nutmeg quality. Using LDA for classification, 94.7% of original grouped cases were classified accurately; and for cross-validated groups, 94% of cases were classified accurately. Classification with multilayer perceptron (MLP) proved more effective showing a classification accuracy of 99.2% for training data and

100% accuracy was achieved for classification of testing data. In another work by Paulus and Suryani [115], camera-based vision was used for detection of aflatoxin in the nutmeg and classification was also carried out using various features (such as color, shape), texture features extracted from images.

9.4.1.3   QUALITY EVALUATION USING COLOR

Sharma et al. [133] employed computer vision technique to study the effect of solar drying versus hot air drying on the quality of turmeric slices obtained from Kharagpur, India. CMOS based digital camera was used to acquire images of the turmeric (*Curcuma longa L.*) slices (6 mm), directly during the drying process. RGB images and grayscale image (Figure 9.5) taken during various intervals of drying revealed a smooth decrease in area (83.72%), perimeter (60.69%), roundness (22.34%), solidity (6.12%) and density (83.72%) of slices. Browning determinant (BD) calculated from grayscale images showed a high value of total phenolic content present in dried turmeric as compared to that of fresh one. During drying period of 4 hours, an increase in Lightness Index (L*) by 5.55% was observed; whereas decrease in Redness Index (a*) by 117.06% and Yellowness Index (b*) by 278.26% were observed.

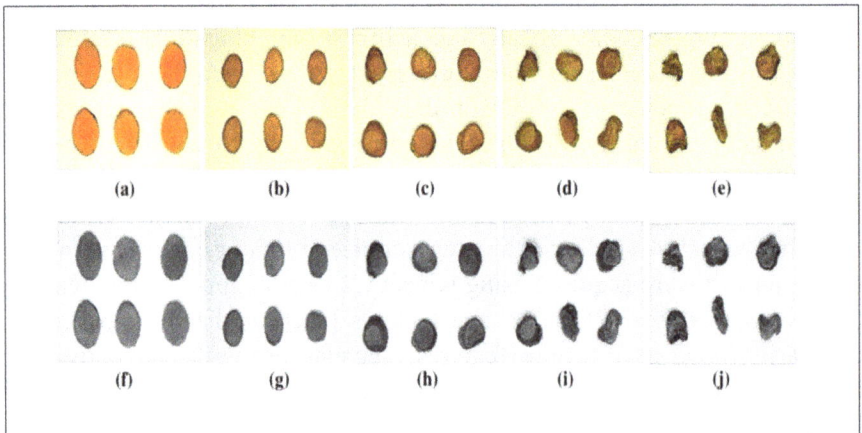

FIGURE 9.5   RGB and grayscale images of dried turmeric slices at different moisture content (wb: wet bases): (A and F) 93.5%; (B and G) 39.39%; (C and H) 16.46%; (D and I) 3.90%; (E and J) 3.46%.
Source: Reproduced from Ref. [133] and reprinted with permission from Springer Nature.

In a research work by Jain et al. [56], a computer vision system was used for the determination of foreign materials in cumin (*Cuminum cyminum L.*) seeds and segmentation of cumin seeds having pedestals attached. A camera having resolution of 6 megapixels was used for the acquisition of images of cumin seeds. Features (such as area and minor axis length) of cumin seeds were extracted from images and used for calculating the number of cumin seeds having long pedestals; and other extraneous elements (stones, etc.), were also determined.

Camera-based imaging technique was also investigated to identify the diseases in chili plants by Husin et al. [55]. The digital camera was used to acquire a total of 107 images of leaf of chili plant, out of which 21 were for healthy leaf sample and 86 were for leaf samples having disease. Using RGB color model and MATLAB for image analysis of leaf samples, they were able to detect the disease in chili plants.

Although camera-based vision technique is quite suitable for the assessment of external quality, yet there is a need to inspect internal quality [3]. Since, the internal structures of the food items are not visible by external appearance, therefore, Hyperspectral imaging, X-ray, MRI, CT, etc., were used to observe internal quality to identify the chemical composition, contaminants, and internal defects, etc., in food items.

9.4.2 HYPERSPECTRAL IMAGING

Optical sensing techniques (viz., imaging, and spectroscopy) are being immensely utilized for the evaluation of food quality and food safety, but these techniques have their own limitations regarding quality assessment of food items. The camera-based imaging techniques as such cannot acquire detailed spectral information of the food sample, while spectroscopic assessment of food has limitations in terms of inability to reveal external features. Traditional camera-based vision technique mimics the human vision by capturing only three monochromatic RGB images at wavelength related to Red (R), Green (G) and blue (B) color, respectively [161]. On the other hand, hyperspectral-imaging technique couples the imaging technique with spectroscopy so that spatial and spectral information of the sample can be obtained [87].

Hyperspectral imaging (HI: also known as spectroscopic or chemical imaging) is a non-destructive imaging technique for quality evaluation of varied food items. Contrary to the conventional RGB images, the hyperspectral images offer numerous monochromatic images in the spectral domain. The data structure of hyperspectral image is a three-dimensional hypercube, or three-dimensional (3D) data cube composed of two spatial and one wavelength dimension as shown in Figure 9.6. It can be viewed as a stack of two-dimensional monochromatic images taken at different wavelengths. As compared to other imaging techniques, Hyperspectral imaging technique provides more valuable information owing to the detailed spectral signature of the image surface being captured by each pixel.

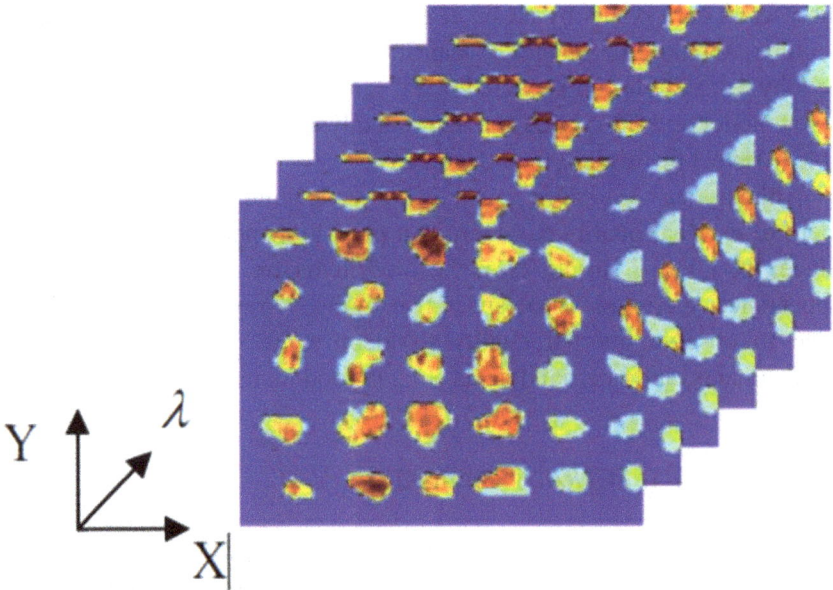

FIGURE 9.6   Hyperspectral image cube with two spatial dimensions and one spectral dimension (X, Y, λ).
Source: Reproduced from Ref. [162] and reprinted with permission from Springer.

9.4.2.1   COMPONENTS OF HYPERSPECTRAL IMAGING SYSTEM

The main components of hyperspectral imaging system (Figure 9.7) includes a light source used for illumination of objects, an area detector having camera or scanner used to take images of sample object, a transportation stage used for placing and movement of sample, a wavelength dispersion device usually a spectrograph used for acquiring spectral information of object, and a computer with software for processing and analysis of data [87, 121].

Generally, light source for spectral imaging techniques are divided into two groups [87]: one is for illumination purpose mainly utilizing the broadband lights and second is excitation source utilizing the narrowband lights.

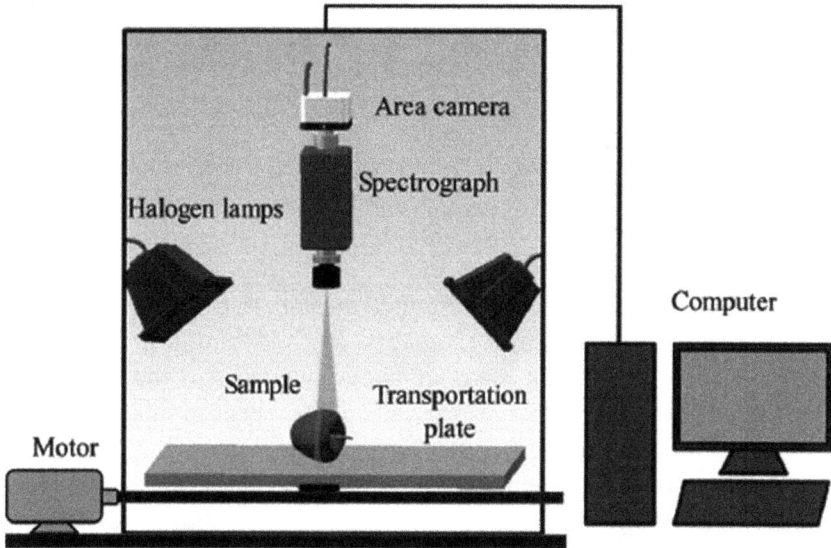

FIGURE 9.7 Configuration of a typical hyperspectral imaging system.
Source: Reproduced from Ref. [87] available through open access. https://creativecommons.org/licenses/by/3.0/

9.4.2.2 PRINCIPLE OF HYPERSPECTRAL IMAGING SYSTEM

In hyper-spectral imaging, mainly diffraction grating and prism are used as wavelength dispersion devices. These devices scatter the broadband light into various wavelengths and also project the light in the direction of the area detector. Mainly hyperspectral images are obtained using three approaches: whiskbroom (or point scanning), push broom (or line scanning) and tunable filter (or area scanning); and these three define basically the mechanism of measurement either taken at a single point, or taken of moving sample on conveyor or fixed scene of steady samples are taken [40]. There exist three modes of hyperspectral imaging (viz. reflectance, transmission, and interacting mode), which are used for spectral imaging of objects [147]:

- Reflectance mode is primarily used for study of external features of food samples;

- Transmittance mode is mainly used for study of internal constituents; and
- Interactance mode combines the both reflectance and transmittance modes.

Spectral and imaging data acquired through hyperspectral imaging needs pre-processing to extract useful information, and then, based on the extracted information, suitable calibration models are developed and validated for the intended purpose.

9.4.2.3 LIMITATIONS

Although extensive information obtained through hyper-spectral imaging can reveal external and internal quality of food samples, yet there is a need for a long time for spectral image acquisition, the complexity associated with spectral image processing and spectral analysis, and it involves high computational cost. Further, commercial implementation of hyperspectral imaging techniques is a costly; therefore, usage of this imaging technique is mostly limited to laboratory studies. Consequently, the hyperspectral-imaging technique is mostly used to acquire images of high spatial and spectral resolutions for fundamental researches that help in the selection of suitable wavelengths, which are further used for the development of customized multispectral imaging system for real-time inspection of food.

9.4.2.4 APPLICATIONS OF HYPERSPECTRAL IMAGING FOR QUALITY INVESTIGATION OF NUTS

Hyperspectral imaging techniques is being utilized in the food industry for a variety of applications, such as determination of chemical constituents, internal defect detection, prediction of maturity, estimation of external attributes (such as size, shape, color, texture, surface defects, etc.), for grading and sorting of food items. In recent years, hyperspectral imaging has emerged as a powerful technique for quality and safety assessment of spices and nuts. Hyperspectral imaging technique is mainly used for detection of contaminants in spices, internal defects in nuts, insect/pest infestation in nuts, and disease detection in spice plants, etc.

9.4.2.4.1   Detection of Contaminants

Aflatoxin is a toxic metabolite produced by *Aspergillus* molds, and it is linked with many chronic and acute ailments, such as liver cancer in human and immune-suppression [11]. A wide variety of nuts (such as pistachio nut, almond, hazelnut, and spices, e.g., chili) are prone to aflatoxin contamination and a maximum threshold limit of 10 ppb (total aflatoxin) is set for herb and spices by EC Commission Regulation of 2006 [38].

In a study conducted by Ataş et al. [11], Hyperspectral Imaging technique is used for detection of aflatoxin contamination (threshold 10 ppb) in ground red chili flakes (53 samples) procured from the market in Turkey. They used CCD camera with liquid crystal tunable filters for acquisition of images; and UV (365 nm) and Quartz-tungsten-halogen (100W) were used as illumination sources. Series of Hyperspectral images were obtained in the range of 400 nm to 720 nm with a spectral bandwidth of 10 nm at 3 locations of individual chili samples under separate illumination sources. Most informative bands were reported at 540, 550, 560, 590, 640, and 650 nm corresponding to Dataset-1, obtained under the illumination of halogen and classification accuracy of 83.26% was obtained for discrimination of contaminated (aflatoxin greater than 10 ppb) and uncontaminated chili (aflatoxin less than 10 ppb) with the MLP classifier. In addition, for dataset-2 obtained under UV illumination, classification accuracy of 87% was reported using MLP classifier, for discrimination of contaminated and uncontaminated chili samples. Mean aflatoxin level was 16.78 ppb with an average of 3.2 ppb in aflatoxin negative samples and an average of 33.3 ppb for contaminated samples.

9.4.2.4.2   Detection of Internal Damage

Nuts may have internal damage due to insect infestation, environmental changes, or during processing used for drying and grading. For instance, exposure of almond nuts with varied drying temperature and moisture may result in internal damages [117]. Nuts having internal damage are difficult to distinguish from the healthy nuts through visual inspection only. Internal damage in nuts may result in off-flavor while lowering nutrition value. In a research work by Pearson [117], hyperspectral data in the range of 700 to 1400 nm was used to differentiate internally damaged nuts with error rate of 12.4% (Figure 9.8).

Nakariyakul and Casasent [101] used Hyperspectral imaging in range of 700 to 1400 nm for discrimination of internally damaged almond nuts

(*Prunus dulcis*, 454 samples). Samples were illuminated using a quartz tungsten halogen (100 W) lamp; and two spectrometers based on silicon photodiode array sensors and InGaAs (indium gallium arsenide) photodiode array were used for acquiring the transmission spectra of almonds in range of 700–1000 nm (0.48 nm intervals) and 950–1390 nm (3.2 nm intervals), respectively. Difference in the spectra of normal and internally damaged nut was observed at 710 nm and 930 nm, and it was reported that due to oxidization process, internally damaged nuts may have lesser absorbance at the 930 nm waveband. Two best ratio features (850/1210 nm and 1160/1335 nm) were selected for classification, and higher classification rate with ratio features were obtained in comparison to that of other algorithms of feature extraction based on a subset of separate wavebands or using all the spectral responses from wavebands.

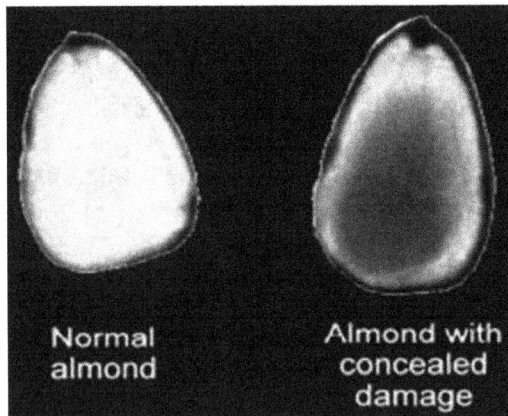

FIGURE 9.8   Hyperspectral images of almonds (normal and internally damaged).
Source: Reproduced from Ref. [117] and reprinted with permission from Elsevier.

9.4.2.4.3   *Grading and Sorting Operations*

HDR (high dynamic range) hyperspectral imaging is used to categorize the shelled hazelnut (*Corylus avellana L.*) kernels (raw) into different the quality classes [97]. A short-wave infrared (SWIR) camera with other basic components, linear translation stage (speed 3 mm.s⁻¹) for sample holding, LabVIEW software v8.5 for synchronization of camera with stage movement, and MATHLAB software for data analysis was used for the research work.

Different external features of hazelnut kernels (such as perimeter (mm), surface area (mm²), lengths (mm) of major and minor axis and kernel eccentricity) were extracted using Hyperspectral images (unprocessed) acquired in the range of 850–2500 nm. It was reported that changes in quality of hazelnut was predictable in the wavelength region 1934–2484 nm of acquired spectra that are linked with internal constituents of hazelnut, such as lipids, proteins, and carbohydrates [155]. PLS-DA models were developed using acquired data, and classification accuracy (>90%) for high quality of hazelnut was obtained, and misclassified kernels were categorized into lower-quality classes. Additionally, size, and shape parameters also showed significant variation among the kernels belonging to different quality classes.

9.4.2.4.4   Detection of Adulteration and Authentication

Various spices are frequently adulterated with different substances to gain economic profit that is a serious threat to consumer health and trust [81]. For instance, adulteration of powdered paprika with lead oxide lead to the death of many people in Hungary during 1994 [84]. In order to safeguard the communal health, it is imperative to detect various adulterants in nuts and spices.

Orrillo et al. [110] used NIR (near-infrared) hyperspectral imaging for detection of adulteration of papaya seeds in black pepper berries (*Piper nigrum*) in Peru. The spectra of pure and adulterated black pepper (0 to 30% w/w) mixed with papaya seed powder, as intact berries and papaya seeds were acquired in the wavelength range of 900–1710 nm (5 nm intervals). Although higher absorbance values were obtained for spectra of black pepper than that of papaya seeds, yet similar pattern of spectra was observed for both. Using statistical models (such as SIMCA (Soft independent modeling of class analogy) and PCA), the berry samples were classified with 100% accuracy and powdered samples were classified with sensitivity greater than 90%.

Although nuts are considered a great source of various macronutrients but some sensitive individuals may have allergic reaction towards consumption of specific nuts, such as peanuts. Mishra et al. [95] used hyperspectral and multispectral imaging for detection of traces of peanut and hazelnut in wheat flour. It was reported that using spectral index extracted from hyperspectral images, it was possible to detect traces (0.01% by weight) of peanut and other tree-nuts in wheat flour.

In a research work by Kiani et al. [70], the hyperspectral imaging technique in conjunction with chemometrics was explored for authentication of Nutmeg. Hyperspectral line-scan camera with push-broom technology was used to obtain spectral information in the wavelength range of 400–1000 nm of various samples, such as authentic nutmeg powder (51 samples), adulterant materials including pericarp (1), shell (1), and spent (5) samples, retail product (31 samples) and adulterated mixtures (5 to 60%).

Various statistical techniques (such as Partial Least Squares-Discriminant Analysis (PLS-DA), ANN and PCA) were used for data exploration and building the model for detection of adulteration. Although, PCA was able to distinguish authentic and adulterated samples, yet ANN models were able to detect low level (5% of own material of product) of adulteration and had more accuracy as compared to the PLS-DA model also (Figure 9.9). Results from Hyperspectral image analysis were confirmed with microscopic analysis of samples.

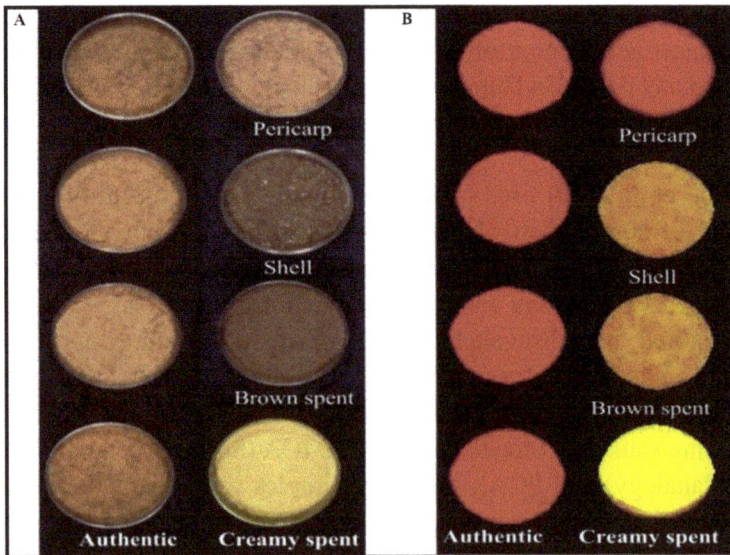

FIGURE 9.9   (A) RGB; and (B) PCA; score images of the typical authentic (nutmeg) and its adulterant samples on PC1 (90%) representing the distribution of particles between the samples.
Source: Reproduced from Ref. [70] and reprinted with permission from Elsevier.

Visible and near-infrared (VNIR) hyperspectral imaging was used by Khan et al. [65] for detection and classification of various adulterants (such as sawdust, wheat bran, and rice bran) in red chili samples collected from open markets of India and Pakistan. Spectra of pure chili samples, pure adulterants, and adulterated (0 to 30% by weight) mixtures were acquired in the range 395–1000 nm. Using pre-processing technique SVM and k-means clustering on spectral data, it was found that in wavelength range of 450–550 nm darkest red color was observed for chili and particularly at 500 nm bright pixels were related with adulterants. Using SVM classifier, pure chili samples were classified with overall accuracy of 99%; whereas, adulterated samples were classified with an accuracy of 85%.

9.4.2.4.5 Other Applications

Fluorescence-based hyperspectral-imaging system with independent component analysis (ACA) and k Nearest Neighbor classifier (kNN) approach was utilized to distinguish black walnut meat and its shells [162]. CCD camera with two-halogen lamp (150 W) setup was used for acquiring the images of cracked walnut.

They selected 4 to 10 optimal wavelengths in total wavelength region 425–775 nm to reduce data. Using only 10 optimal wavelengths, they achieved a detection rate of 90.6% for 5,496 samples analyzed in the study (Figure 9.10). In addition, they compared the performance of ICA-kNN method and kNN classifier alone. It was found that few wavelengths (optimal) as selected by ICA-kNN has same performance as that of total wavelength selection by kNN classifier indicating that reduced data by ICA-kNN is capable for discriminating the walnut shell and its meat [79].

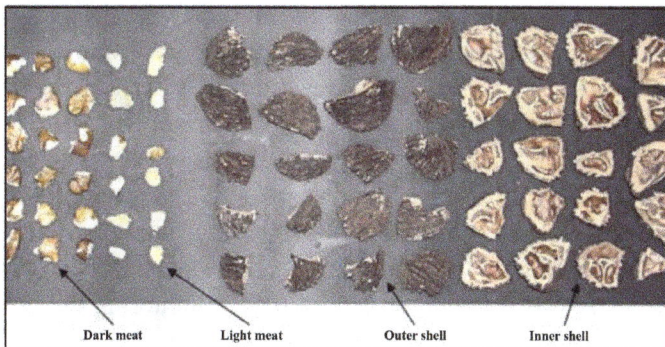

FIGURE 9.10 Four sample categories of walnut depending on shell/meat conditions in the dataset.
Source: Reproduced from Ref. [162] and reprinted with permission from Springer.

Moisture content of peanut kernels could be successfully predicted using hyperspectral imaging alone [58] and also coupled with chemometrics [140]. Fluorescence-based hyperspectral imaging has been used for differentiating the pulp and shells of walnuts [57]. For detection of fungal aflatoxins in pistachio, the long-wave NIR hyperspectral imaging has proven to be instrumental [66].

9.4.3   THERMAL IMAGING (TI)

Thermal imaging (also known as infrared (IR) imaging) records the temperature by measuring the amounts of infrared beams discharged from the body surface. TI technique utilizes these discharged radiations and creates pseudo-images pertaining to the surface temperature distribution (also known as thermogram) [9]. Sensitive infrared cameras are used to capture the thermal images and to record the temperatures between–10 and 1500°C owing to their high sensitivity for thermal infrared band (3–5 mm). TI is based on the fact that every material emits a certain degree of infrared radiation, falling in the invisible light band on electromagnetic spectrum having wavelengths in the range of 0.75–100 μm. The infrared radiation (IR) are classified as: short (1.4–3 μm), mid (3–8 μm), longwave (>8 μm) and extreme (15–100 μm). Among these five IR classifications emitted by material, radiation from short waves can be effectively detected by TI system [54, 91].

The TI system consists of a thermal camera, detectors, focusing, and collimating lenses, channels, filters, signal process and an image processing unit (Figure 9.11). Active or passive imaging modality is used to acquire the digital images. No external power/energy is required for the passive system for imaging, whereas some sort of heating and/or cooling system (such as laser beam, halogen lamp) is used for controlling the temperature of the system and observing heat processes. In active thermography, thermal waves are created by use of pulse thermography, Lock-in thermography, pulsed phase thermography, vibro thermography, etc. In food processing operation, surface temperature is mostly measured by passive thermography; while the surface and subsurface defects in products are detected through active thermography. Passive thermography is useful for monitoring the non-contact temperature of product during processing [114].

TI possesses a broad capability to store the temperatures of every pixel of an image and associated software produces better results regarding quality assessment of an object. This technique has the capability to acquire the temperature of a wide variety of agricultural produces under different working environments. One of the major causes of food safety concerns is the presence of extraneous matter (it could be alive or dead microbe, pest or insect, dirt, dust, etc.). Thus with the use of TI, it is possible to distinguish foreign bodies from a food sample by using thermographic images [26]. To categorize the thermal images, a principal elements analysis, neighbors clustering, k-nearest, and hierarchical clustering are applied in spatial domains. Partial least squares (PLS), fisher's discriminate analysis (FDA), linear discriminate analysis (LDA), and ANNs are also applied for classification and optimization of images [144].

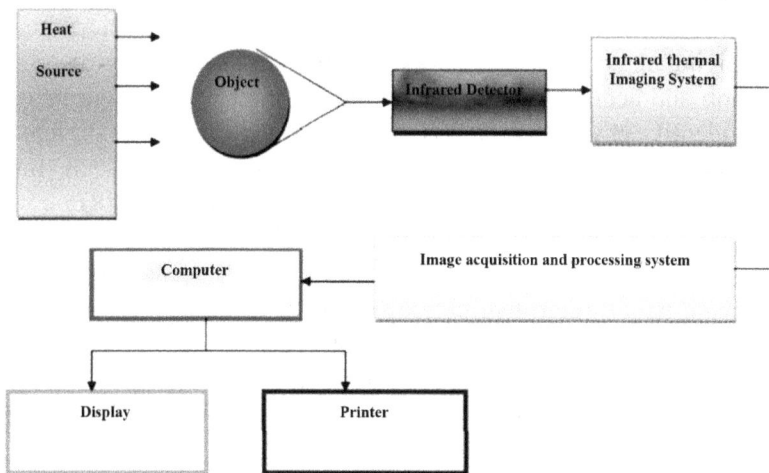

FIGURE 9.11 Component of thermal imaging system.
Source: Reproduced from Ref. [114] and reprinted with permission from Elsevier.

Warmann and Margner [152] successfully applied thermal image processing for quality control and detection of extraneous matter in hazelnuts. Foreign matters (such as stones, wooden sticks, cardboard, metal chips) in almond were detected using TI [44]. Further, Wang et al. [151] applied the TI technique for determination of time-temperature history and thermographs of radiofrequency treated in-shell walnuts. Kheiralipour et al. [67] investigated the total emissivity of pistachio kernel using TI technique for determining its surface temperature and managing some of agricultural engineering operations to study the effects of surface temperature and moisture content of pistachio kernel. They found the direct relation of temperature on total emissivity and inverse relation moisture content of pistachio kernel.

9.4.4   MAGNETIC RESONANCE IMAGING (MRI)

In this technique of MRI, sample images are acquired by recording the absorption of resonance from a particular nucleus on exposure to magnetic energy either from pulsed radiofrequency or by rotating magnetic field. The number of nuclei present in the food determines the amount of resonance energy absorbed. The 2D and 3D images of an object are determined by MRI using proton density. The generation of magnetic resonance images can be controlled by the radiofrequency pulse sequence used for excitation of the nuclear spins. The essential components of an MR imaging include: (a) a large magnet which is used to generate the uniform magnetic field; (b) smaller electromagnetic coils used to generate magnetic field gradients for imaging; (c) a radio transmitter and receiver and its associated transmitting and receiving antennae or coils; and (d) computer used to coordinate signal generation and acquisition and image formation. This technique is widely used to identify physical structural changes in food matrix (e.g., infection, aging, microbial defects) and chemical changes, etc., [20, 137]. Internal quality features of spices and nuts could also be assessed by MRI.

Lakshminarayana et al. [82] applied the NMR imaging technique for obtaining images of different transverse and vertical sections of groundnut for studying its oil/water distribution. The MRI technique was successfully employed for differentiating the fresh and stored hazelnuts through observing physical and chemical processes commonly taking place during storage. An increase in moisture and fat oxidation was used as indicators for storage and were assessed through MRI [15].

9.4.5   X-RAY IMAGING

X-rays is an electromagnetic wave, which travels in straight lines having wavelength in range between 0.01 and 10 nm of electromagnetic spectrum with photon energies of 120 eV–120 keV. They possess high penetrating power. The magnetic field or electric field does not affect the flow of X-rays. X-ray image is a transmittance image, while the visible light (VL) picture is a reflectance image.

Hard X-rays are avoided for inspection and quality assessment of agricultural produce [5]. It can contaminate and damage agricultural produce due to its high radiation activity with greater penetration. Therefore, mostly soft x-rays are used for quality evaluation of spices and nuts as they can provide details on changes in the internal density. Soft X-rays take only 3–5 seconds for providing the X-ray image of a sample [105].

Soft X-ray inspection technique works on the differentiating the samples due to the presence of contaminants and its density (Figure 9.12(top)). The X-ray loses some of its energy when it penetrates through a food product. The presence of contaminants in food creates denser area, further reducing the energy of X-rays. When the X-rays travel through the food sample, depending upon the density of food material, some energy is absorbed, and some is emitted out as signal and reaches a sensor, which is later converted into an interior image of the food. After analyzing the converted processed image, the contaminant/foreign matter in the food seems to be darker gray which helps in identifying foreign contaminants [26].

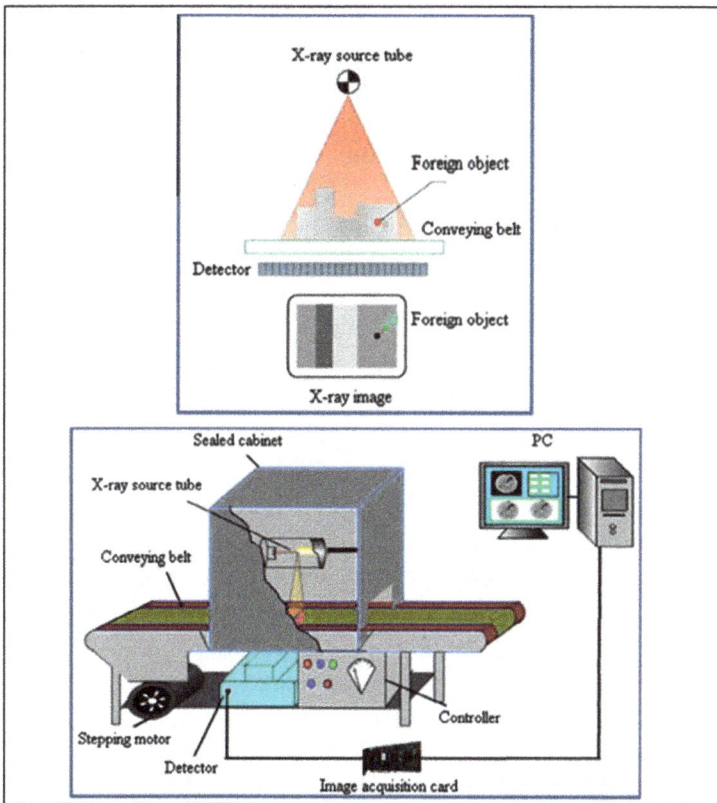

FIGURE 9.12   Principle of soft x-ray imaging (top); the soft X-ray inspection system (bottom).
Source: Reproduced from Ref. [26] and reprinted with permission from Elsevier.

The soft X-ray inspection system (Figure 9.12(bottom)) mainly consists of an X-ray tube (i.e., computer-controlled X-ray generator), conveying belt, a line-scanning sensor for X-ray detection, stepping motor, computer, and image-acquisition card. Soft X-ray imaging is cheaper, simple to use, accessibility, and material restrictions as compared to MRI. [17]. A cross-sectional view of an object could be obtained by a typical X-ray image, thereby making them ideal for internal quality assessment of nuts and spices [96].

Apart from its medical usage of X-ray, there are a number of utilities of X-rays imaging, such as detection of metal and other harmful extraneous matter in food and dairy industries, etc., [76]. The application of this technique for non-destructive quality assessment of spices and nuts are still in the dormant stage, and there is great scope for exploration of this method. The X-ray imaging technique are mainly employed for inspecting the quality of spices and nuts to detect product density, damages, internal defects, presence of foreign materials or contaminants, maturity, and infestations [64].

Keagy et al. [64] used 12-bit digital images of pistachio nuts by X-ray imaging at a resolution of $(0.125 \text{ mm})^2$/pixel for detecting insect infestation. Further, navel orange worm damage in pistachio nuts and other quality attributes of pistachio kernels were detected by X-ray images [23]. X-ray imaging method has been successfully used for the quality inspection of apricots [163]. X-ray line scan images and scanned films were used for the detection of navel orange worm in almonds using the Algorithms [71].

Digital radiography technique was used by Kotwaliwale et al. [75, 78] for obtaining apparent linear attenuation coefficients of pecan nutmeat and shell. Kotwaliwale et al. [77] estimated weight of Pecan nutmeat with less than 10% error using soft X-ray digital imaging technique using different combinations of voltage and current. They could successfully distinguish between defects and insects by contrast stretching or high-frequency emphasis modifications of X-ray images (Figure 9.13). Kotwaliwale et al. [77] applied soft X-rays imaging on unshelled pecans at a voltage and current combination of 15 to 50 kVp and 0.1 to 1 mA, respectively, and good contrast x-ray image of a pecan was captured for defects and insect damage. Further, Kotwaliwale et al. [79] successfully assessed physiological and insect damage in pecans at 40 kVp voltage and 0.5 mA current of X-ray having an integration time of 460 ms. Also, the error in determining nutmeat weight was reduced to only ±5.8%.

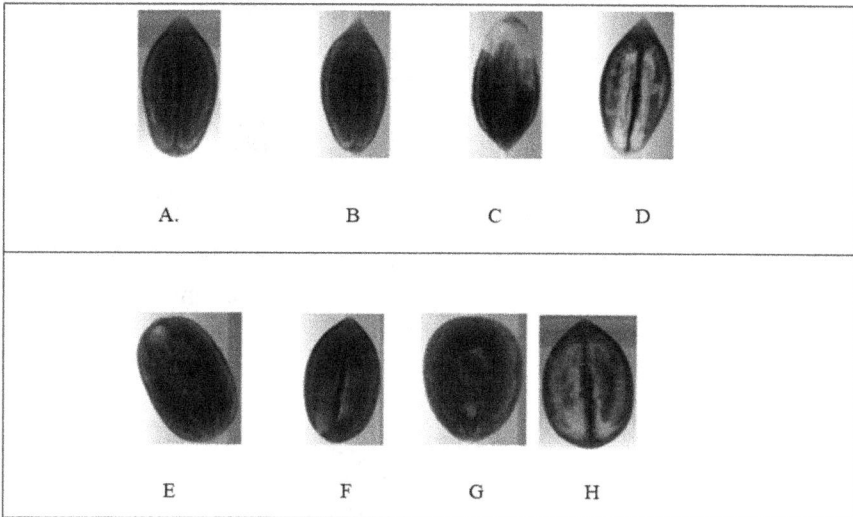

FIGURE 9.13 X-ray images of pecans with different visible attributes: (A) good nut; (B) good nut with shucktight ; (C) nut with mechanical damage; (D) nut with shriveled nutmeat; (E) insect damage to one cotyledon from inside; (F) visible insect hole; (G) insect damage and insect; (H) hollow nut.
Source: Reproduced from Ref. [79] with permission from Elsevier.

Yanniotis et al. [159] applied the X-ray imaging on pistachio nuts for assessment of fungal necrotic spot. Khosa and Pasero [69] showed the application of X-ray imaging in developing method for automatic real-time detection, segmentation, and classification of hazelnuts.

9.4.6 COMPUTED TOMOGRAPHY (CT)

Compared to conventional X-ray radiography, higher-energy X-ray photons are employed to generate CT. The main difference between image acquired by X-ray and CT is that x-ray images are in 2D while CT scan images are 3D. It basically generates 2-dimensional and 3-dimensional images by accumulating X-ray images in the form of thin projected slices of a sample [109] (Figure 9.14). The CT technique has been recently used for non-destructive internal quality inspection (pore topology, internal corruption, tiller numbers, moisture, grain hardness, volumes, densities, temperature, and mechanical damage) of various food commodities, when quality features are not visible on the surface of the products [8, 27, 31, 145, 157]. The 2-dimensional slices extracted from CT imaging show better contrast among the constituents of the respective slice. However, more time is needed to generate 2-D and 3-D slices compared to transmission radiography.

According to the decay characteristics of X-ray on the material, the basic working principle of CT technology is based on focusing the X-ray beam on food sample in different directions and recording a shadow image reflecting X-ray radiation attenuation along the beam path (Figure 9.14). The rotation of the sample generates successive images that are stored and subsequently analyzed by computer-assisted tomography (CAT) scanning. The decay characteristics of different materials represent different selected X-ray energies [37].

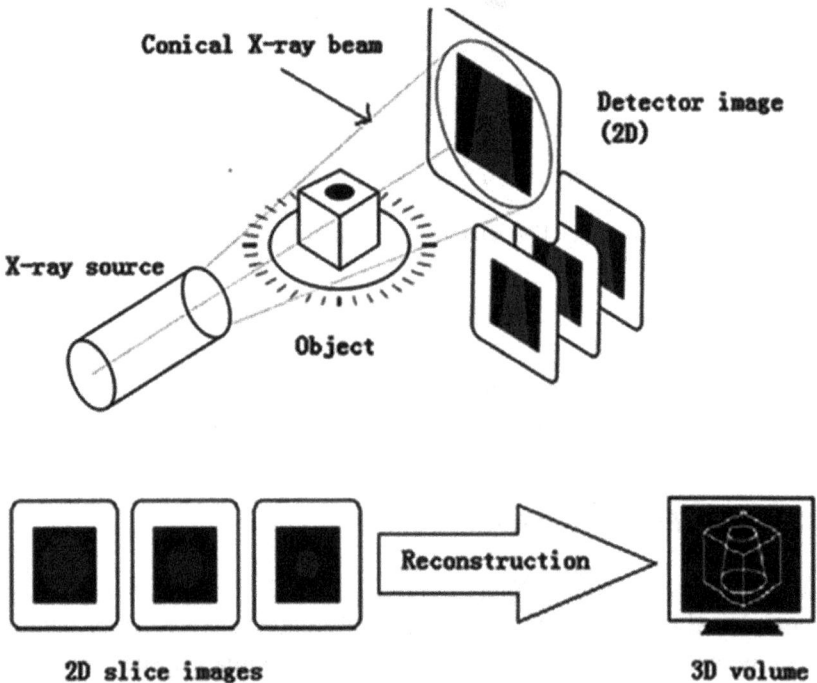

FIGURE 9.14   Working principle of x-ray CT technology. Some radial projections are captured at different angles on a sample by the x-ray to obtain 2D slice images. Besides, a 3D image could be produced by the restructure of scanned numerous slice images of the sample. *Source:* Reproduced from open-access source [37]. https://creativecommons.org/licenses/by/4.0/

The X-ray energy of medical radiation is generally 140 keV, while that of industrial CT is 50 keV to 16 MeV. In the inspection of agricultural products, the selected energy is usually as variable as variety [37]. The potential applications of CT for the internal quality evaluation of agricultural products are reviewed by Kotwaliwale et al. [76]. A number of reports are available for the application of CT in non-destructive quality assessment of spices and nuts. Kim and Schatzki [71] applied the CT for detection of pinhole damage in almonds. Donis-Gonzalez et al. [33] applied CT to investigate the internal properties (such as air, healthy tissue, decayed tissue, and void space) of fresh chestnuts.

Haug [50] used it to detect internal defects in pistachio nutshell. Bernard et al. [16] used the X-ray CT to investigate 3D characterization of walnut morphological traits. Ghodki et al. [43] investigated the effect of cryogenic treatment on microstructural changes (viz., volumetric size distribution of pores, total pore volume, total volume, total solid volume and pore-pore distance,) within cinnamon, black pepper, fenugreek, and chili, using X-ray micro-computed tomography coupled with image analysis.

9.4.7 MULTISPECTRAL IMAGING

Multispectral computer vision system captures only a few monochromatic images in the spectral domain, whereas a large number of images can be taken from hyperspectral imaging system. In Multi-spectral imaging, monochromatic images are taken at few selected wavebands in the spectrum. Different from traditional digital cameras (which capture the light of a particular frequency falling on it for imitating human color perception), the multispectral imaging system generates two or more definite single-band monochromatic images with the use of specific narrowband filters. The image acquisition takes place in a quick time followed by rapid image processing through simple algorithms resulting in on-site instant and precise decision-making.

The important features in designing a multispectral imaging system is that there must be less data storage space for both spectral and spatial coordinates, thereby resulting in rapid processing of images. This is attained by acquiring images having low spatial resolutions at some important wavelengths. These images behave as fundamental datasets helpful in establishing optimal wavebands. These wavebands can then be applied for an application by the multispectral imaging solution [59].

Kalkan et al. [62] applied the multispectral imaging in detection of aflatoxin in red chili pepper flakes. One of the common adulterants in turmeric powder is tartrazine colored rice flour. It was detected by Bandara et al. [13] using a multispectral imaging system developed by utilizing nine spectral bands having peak wavelengths varying from 405 nm to 950 nm. The multispectral technique was evaluated by Kalkan et al. [62] for determining low concentrations of aflatoxin in hazelnuts and chili pepper using an algorithm. They concluded that the developed method had 95.6% accuracy for classifying the fungal contaminated and uncontaminated hazelnut kernels. Damiani et al. [30] successfully detected cistus, olive leaves, and myrtle as contaminants in Mediterranean oregano using NIR hyperspectral imaging in combination with chemometrics.

The technique has certain drawbacks, such as it works only for an inline assessment; there is frequent distortion of images, requirement of regular calibrations and debugging for smooth operation of high precision systems, thereby making the whole system quite complex.

9.4.8 NEAR INFRARED (NIR) IMAGING

NIR imaging is a useful nondestructive food quality evaluation technique, with two major wavebands being used for the purpose of imaging: (a) 700–1100 nm; and (b) greater than 1100 nm. The lower wavebands could be successfully used for studying the compositional changes such as moisture and protein content of food products. They could also be useful in the assessment of various defects [111]. It is well known that the conventional NIR spectroscopic techniques are also being used for food quality evaluation, but they are unable to give spatial information as provided by NIR imaging techniques. To understand this difference better, it's vital to mention that the NIR spectroscopy provides the quantitative amount of food constituents viz. fat, protein, and moisture, while NIR imaging is instrumental in representing the distribution pattern of these food constituents inside the food matrix.

The NIR spectra of some adulterated spices (viz.: cinnamon, garlic powder, and black pepper) having pre-determined quantity of talc, corn starch and millet was measured and the results revealed that when this spectra is coupled with Fourier transform spectrometry (PerkinElmer Frontier™ FT-NIR spectrometer), it can predict the quantity of various adulterants [83].

The NIR imaging technique has been used by several researchers [118, 143, 150] for estimation of different food components, such as unsaturated fatty acids, amino acids, etc., in peanuts. Further, this technique can be used for detection of *Aspergillus flavus* contamination in peanut [88], mycotoxins in red paprika [53]. Other than safety aspects, NIR has been used for detecting various physicochemical changes in hazelnuts due to storage [14].

9.4.9 OTHER POTENTIAL IMAGING TECHNIQUES

Although there exists a range of imaging techniques suitable for external and internal quality evaluation of nuts and spice, yet each technique has certain advantages associated with some limitations. For instance, some techniques are suitable for surface inspection (such as camera-based vision), whereas others are suitable mainly for internal structure (such as X-ray); and hyperspectral imaging is suitable for inspection of both external and internal quality but it is expensive. Therefore, Researchers are trying to explore other novel techniques having potential for quality evaluation of different food items.

9.4.9.1 FLUORESCENCE IMAGING (FI)

It is based on the luminescence of the pigments/compounds present in the sample [26]. Fluorescence is the process of emission of light by a material or substance which has absorbed light or other electromagnetic radiation. Mainly fluorescence imaging (FI) technique uses laser confocal scanning microscopy (LCSM) associated with other visualization components and is utilized for food safety assessment such as microbial detection in food samples [61]. FI technique is employed by researchers for understanding and visualizing the distributions of fat in batter coating for deep-fat fried chicken-nugget [4] and distribution of gluten, starch, and air bubbles in dough [73, 74]. In a research work by Byoung-Kwan et al. [22], laser-induced FI system (LIFIS) is utilized for detection of feces-contaminated poultry carcasses in the presence of ambient light and also fecal contamination on Golden Delicious apples is detected by FI [156]. This technique is also utilized for the detection of microbial biofilms developed on food-contact surfaces of processing equipment made of stainless steel [60] and for detection of organic residues present on the surface of equipment used in poultry-processing plants [120].

Potential of FI techniques may be explored for the detection of certain adulterants, contaminants, microbial load, and fecal materials in nuts and spices by utilizing the fluorescent properties of these materials. The main limitation of this technique is that phenomenon of fluorescence is highly dependent on the presence of materials or compounds having properties to emit light or fluorescence; therefore, this technique may be used in combination with hyperspectral imaging to widen its application domain in the food industry.

9.4.9.2   RAMAN CHEMICAL IMAGING (RCI)

Raman spectroscopy is a renowned technique in the area of food quality and food safety assessment. Raman chemical imaging (RCI) or Raman imaging is based on the integration of Raman spectroscopy with imaging technique so that spectral and spatial information can be obtained to determine the chemical composition and understanding their distribution in food items [148]. As compared to the hyperspectral imaging technique, Raman imaging provides information of micro or nano-scale level of food products [52, 121]. Additionally, RCI gives information on vibrations of covalent bonds and therefore have superiority in analyzing of contaminants and adulterants present in food materials. Using RCI technique, high-throughput detection can be achieved for a large number of samples.

Main components of RCI are: source of light (785-nm line laser) used for excitation, beam splitter used to reflect the laser on surface of the sample, motorized stage for positioning of sample, spectrograph used for dispersion of light into different wavelengths, charged-coupled device (CCD) detector for detection of signals, and the computer for analysis of data. Data of Raman imaging termed as Raman hypercube may be thought as an image I (x, y) at each individual wavelength λ or Raman spectrum I (λ) at each individual pixel (x, y) and unique spectral fingerprint of every pixel can be extracted to characterize the chemical component at that particular pixel.

RCI techniques has been explored for evaluation of quality of dairy, meat, cereals, nut, fruit, and vegetables, etc. Eksi-Kocak et al. [39] explored Raman hyperspectral imaging in combination with PCA and partial least squares regression (PLSR) for detection of adulteration of green peas in pistachio nut granules. A 1064 nm laser was used for excitation to acquire FT-Raman spectral (wavenumber range 200 to 3700 cm^{-1}) images of pistachio nuts adulterated with green peas (20–80% w/w). Difference in the spectra at band position of 1655 and 1441 cm^{-1} was observed with increasing adulteration of green peas with pure pistachio nut granules. Using statistical technique PCA, variation in pistachio nuts and green pea samples were observed, and using PLSR model level of adulteration were quantified with a good prediction results with $R^2 = 0.99$ and RMSEP (Root mean square error of prediction) $= 0.048$.

Main limitation associated with RCI is the weak Raman scattering signals due to low probability of Raman scattering event that demands high laser power with long acquisition time that may lead to destruction of thermally sensitive samples.

9.4.9.3 TERAHERTZ (THZ) SPECTROSCOPIC IMAGING

Terahertz (THz) spectroscopic imaging technique [149] utilizes the less explored range of electromagnetic spectrum starting from 0.1 to 10 THz [42]. It is alleged that THz radiations can be used for identification of dense and low-density materials as they can penetrate through nonpolar dielectric materials with lesser attenuation. Thus, THz imaging can be used for the detection of minute foreign bodies, and owing to the use of non-ionizing radiation, no harmful effect is observed in foodstuff leading to safer detection. Another important advantage is that THz technique is able to detect the compound of interest even in packaged food items.

Mainly THz time-domain spectroscopy (THz-TDS) is the most commonly used technique for food analysis and an additional imaging setup is included with THz-TDS setup to develop THz imaging technique. A good overview on the use of terahertz spectroscopic imaging for a variety of agri-food applications, such as detection of mycotoxin and microorganisms, detection of adulteration and antibiotic residues in food items, identification of genetically modified (GM) organisms, etc., is provided in the review by Wang et al. [149].

Lee et al. [86] explored the CW-THz (continuous wave-terahertz) imaging at 0.2 THz for detection of high-density metallic objects viz. aluminum and granite pieces as well as other low-density foreign bodies (such as maggots and crickets) embedded in powdered instant noodle. They compared the results of CW-THz imaging with X-ray imaging and reported that the X-ray imaging was unable to detect low-density foreign bodies but CW-THz imaging was able to detect both low-density and high-density foreign bodies hidden in food matrices.

Main limitation associated with THz imaging is that THz signals suffer high attenuation in presence of water that makes it unsuitable for food sample having moisture and thus detection by THz imaging is limited to dry foodstuff only. Another issue is the high cost and large-scale set up requirement by THz imaging. Although THz imaging has some limitations and most of the application of THz imaging is in the infancy stage, yet it has enormous potential for quality and safety assessment of foodstuffs. It may be explored in combination with chemometric methods for detection of adulteration and contamination in spices, and insect infestation in nuts, etc.

9.5   CHALLENGES

9.5.1   PHYSIOLOGICAL VARIABILITY AND HIGH VOLUME OF THE PRODUCT

Since, size of various types of nuts and many of whole spices (such as black pepper) are really small, therefore positioning of such samples for image capturing is a tedious task as the samples may overlap and hard to distinguish during image analysis [116]. Implementation of imaging techniques at an industrial scale requires sorting of the high volume of items at a faster speed, which is difficult to achieve with high accuracy. Further, some of the nuts possess complex morphological structures, such as walnut, and surface defects may remain hidden in images due to structural complexity. In addition, some variety of nuts needs pre-processing before sorting. For instance, Pistachio nuts are usually placed in water to separate floating nuts from those of sink ones (heavier nuts). Wet nuts may produce specular reflections that are problematic for imaging sensors and image recognition algorithms [32, 127].

Some of the spices and nut may get damaged during processing operations that may increase the complications during image analysis as it would be difficult to judge a natural defect versus damage caused during processing. Variation in physiological properties of the product and related variation in relevant quality attributes makes it difficult to achieve any single blanket strategy suitable for all nuts or all spices. Research efforts needs to be focused on development of customized imaging system suitable for specific product for faster analysis. Besides, some additional mechanism or instrumentation may be explored for better placement of samples prior to image acquisition process.

9.5.2   COST

Cost is an important factor that is hampering the wide-scale adoption of imaging techniques. Most of the imaging techniques, particularly hyperspectral imaging, multispectral imaging, MRI, etc., are quite expensive, and requiring highly sophisticated instrumentation. There is a need to develop user-friendly low-cost imaging systems suitable for speedy analysis of nuts and spices. It is expected that with advancement in research on semiconductor devices will help in lowering down the cost of basic imaging components like CMOS or CCD cameras that will pave the way for futuristic development of low-cost imaging devices.

9.5.3   VOLUMINOUS DATA AND HIGH COMPUTATION TIME

Imaging techniques generates a large volume of data, which needs processing to extract useful information. In the case of spectral imaging technique generated data is much higher as compared to other imaging techniques, as it involves both spectral information and spatial information. Storage and processing of such a voluminous data is a cumbersome process. A number of statistical methods and techniques are used alone or in combination to extract relevant information for analysis that increases the computation time for specific analysis. Many of existing software are capable of processing and analysis of images, but several times, specific algorithms or programs are required to be developed for data analysis of images. More research is required for the development of statistical procedures, algorithms, and software for faster processing and analysis of image data while reducing overall computation time for analysis.

9.6   SUMMARY AND FUTURE TRENDS

Imaging techniques are gaining popularity in the food industry as a rapid tool for non-destructive quality assessment of agri-food products. Many of the imaging techniques have been widely explored for quality evaluation of fruits and vegetables, but limited exploration is carried out for quality investigation of nuts and spices. Quality analysis of nuts and spice using imaging techniques is much more complex due to their peculiar shape and physiological variability. In addition, it is easier to detect the quality of whole spices using images, but many of the spices are used in powdered form, and it's really tough to draw inferences about the quality of powdered spice using imaging techniques. Efforts have been made by the researchers for quality assessment of nuts and spice using imaging technique that mainly includes classification based on quality, detection of contaminants, adulterants, defects, and damages by insect infestation, etc. Some other potential techniques such as RCI and terahertz spectroscopic imaging need extensive investigation by researchers for quality assessment of nuts and spices.

Research efforts need to be focused on improvisation of imaging technique, for instance, UV lights possess risk to humans involved in online inspection that needs to be replaced by suitable LED lights consuming less power with low noise. Spectral imaging techniques may be explored for the detection of toxicogenic fungal strains or microbial load in nuts and spices. Most of the research work is limited to laboratories only that needs to be transformed into portable automated imaging system suitable for onsite analysis. Imaging techniques may be coupled with other techniques, specially sensors that will enable the evaluation of the holistic quality of the product. Although there are some challenges like cost of imaging system, handling of large data from images and high computation time reducing speed of analysis but with advancement in fundamental electronics and artificial intelligence the cost of basic components will be lower down and it will help in overcoming these limitations in future.

KEYWORDS

- adulteration
- artificial neural network
- browning determinant
- complementary metal-oxide-semiconductor
- computer-assisted tomography
- fluorescence imaging

REFERENCES

1. Abbott, J., (1999). Quality measurement of fruits and vegetables. *Postharvest Biological Technology, 15*, 207–225.
2. Abdollahnejad, B. A., Adelinia, M., & Mohamadi, M., (2016). Sorting of pistachio nuts using image processing techniques and an adaptive neural-fuzzy inference system. *Journal of Agricultural Machinery, 6*(1), 60–68.
3. Abdullah, M. Z., (2008). Image acquisition system. In: Sun, D. W., (ed.), *Computer Vision Technology for Food Quality Evaluation* (Vol. 1, pp. 3–35). Boston – MA: Elsevier Inc.

4. Adedeji, A. A., Liu, L., & Ngadi, M. O., (2011). Microstructural evaluation of deep-fat fried chicken nugget batter coating using confocal laser scanning microscopy. *Journal of Food Engineering, 102*, 49–57.

5. Ahmed, M. R., Yasmin, J., Lee, W. H., Mo, C., & Cho, B. K., (2017). Imaging technologies for nondestructive measurement of internal properties of agricultural products: A review. *Journal of Biosystems Engineering, 42*(3), 199–216.

6. Akpo-Djènontin, D. O. O., Anihouvi, V. B., Vissoh, V. P., Gbaguidi, F., & Soumanou, M., (2016). Processing, storage methods and quality attributes of spices and aromatic herbs in the local merchandising chain in Benin. *African Journal of Agricultural Research, 11*(37), 3537–3547.

7. Alfatni, M. S., Shariff, A. R. M., Abdullah, M. Z., Ben, S. O. M., & Ceesay, O. M., (2011). Recent methods and techniques of external grading systems for agricultural crops quality inspection - review. *International Journal of Food Engineering, 7*(3).

8. Arendse, E., Fawole, O. A., Magwaza, L. S., & Opara, U. L., (2016). Non-destructive characterization and volume estimation of pomegranate fruit external and internal morphological fractions using x-ray computed tomography. *Journal of Food Engineering, 186*(4), 42–49.

9. Arora, N., Martins, D., Ruggerio, D., Tousimis, E., Swistel, A. J., Osborne, M. P., & Simmons, R. M., (2008). Effectiveness of a noninvasive digital infrared thermal imaging system in the detection of breast cancer. *The American Journal of Surgery, 196*(4), 523–526.

10. Ataş, M., Temizel, A., & Yardımcı, Y., (2010). Classification of aflatoxin contaminated chili pepper using hyperspectral imaging and artificial neural networks. In: *IEEE 18th Signal Processing and Communications Applications Conference* (pp. 9–12). Diyarbakır, Turkey.

11. Ataş, M., Yardimci, Y., & Temizel, A., (2012). New approach to aflatoxin detection in chili pepper by machine vision. *Computers and Electronics in Agriculture, 87*, 129–141.

12. Badhe, V. T., Singh, P., & Bhatt, Y. C., (2011). Development and evaluation of mango grader. *Journal of Agricultural Engineering, 48*(2), 43–47.

13. Bandara, W. G. C., Prabhath, G. W. K., Dissanayake, D. W. S. C. B., Herath, V. R., Godaliyadda, G. M. R. I., Ekanayake, M. P. B., Demini, D., & Madhujith, T., (2020). Validation of multispectral imaging for the detection of selected adulterants in turmeric samples. *Journal of Food Engineering, 266*, 8. Article ID: 109700, https://doi.org/10.1016/j.jfoodeng.2019.109700.

14. Bellincontro, A., Fracas, A., DiNatale, C., Esposito, G., Anelli, G., & Mencarelli, F., (2005). Use of NIR technique to measure the acidity and water content. *Acta Horticulturae, 686*, 499–504.

15. Bellincontro, A., Mencarelli, F., Forniti, R., & Valentini, M., (2008). Use of NIR-AOTF spectroscopy and MRI for quality detection of whole hazelnuts. In: *VII International Congress on Hazelnut* (Vol. 845, pp. 593–598). Viterbo, Italy.

16. Bernard, A., Hamdy, S., Le Corre, L., Dirlewanger, E., & Lheureux, F., (2020). 3D characterization of walnut morphological traits using x-ray computed tomography. *Plant Methods, 2020*, 8. doi: 10.21203/rs.3.rs-24599/v1.

17. Bischof, J. C., Mahr, B., Choi, J. H., Behling, M., & Mewes, D., (2007). Use of x-ray tomography to map crystalline and amorphous phases in frozen biomaterials. *Annals of Biomedical Engineering, 35*, 292–304.

18. Brosnan, T., & Sun, D. W., (2004). Improving quality inspection of food products by computer vision: A review. *Journal of Food Engineering, 61*(1), 3–16.
19. Brosnan, T., & Sun, D. W., (2002). Inspection and grading of agricultural and food products by computer vision systems: A review. *Computers and Electronics in Agriculture, 36*(2), 193–213.
20. Bushong, S. C., & Clarke, G., (2014). Magnetic resonance imaging. In: *Physical and Biological Principles* (4th edn., p. 658). San Diego, California, USA: Elsevier.
21. Butz, P., Hofmann, C., & Tauscher, B., (2005). Recent developments in noninvasive techniques for fresh fruit and vegetable internal quality analysis. *Journal of Food Science, 70*(9), 131–141.
22. Byoung-Kwan, C., Kim, M. S., Chao, K., Lefcourt, A. M., Lawrence, K., & Park, B., (2009). Detection of fecal residue on poultry carcasses by laser-induced fluorescence imaging. *Journal of Food Science, 74*, 154–159.
23. Casasent, D. A., Sipe, M. A., Schatzki, T. F., Keagy, P. M., & Lee, L. C., (1998). Neural net classification of x-ray pistachio nut data. *LWT-Food Science and Technology, 31*(2), 122–128.
24. Chang, C. I., (2003). *Hyperspectral Imaging: Techniques for Spectral Detection and Classification* (2nd edn., Vol. 1, p. 220). New York: Springer Science & Business Media.
25. Chen, P., & Sun, Z., (1991). Review of non-destructive methods for quality evaluation and sorting of agricultural products. *Journal of Agriculture Engineering and Research, 49*, 85–98.
26. Chen, Q., Zhang, C., Zhao, J., & Ouyang, Q., (2013). Recent advances in emerging imaging techniques for non-destructive detection of food quality and safety. *Trends in Analytical Chemistry, 52*, 261–274.
27. Chen, S. R., Xu, L., Yin, J. J., & Tang, M. M., (2017). Quantitative characterization of grain internal damage and 3D reconstruction based on micro-CT image processing. *Transactions of the Chinese Society of Agricultural Engineering, 33*(2), 144–151.
28. Cortés, V., Rodríguez, A., Blasco, J., Rey, B., Besada, C., Cubero, S., & Aleixos, N., (2017). Prediction of the level of astringency in persimmon using visible and near-infrared spectroscopy. *Journal of Food Engineering, 204*(2), 27–37.
29. Cubero, S., Aleixos, N., Molto, E., Gomez-Sanchis, J., & Blasco, J., (2011). Advances in machine vision applications for automatic inspection and quality evaluation of fruits and vegetables. *Food and Bioprocess Technology, 4*(5), 829–830.
30. Damiani, T., Dall'Asta, C., Pierna, J. A. F., Fauhl-Hassek, C., Arnould, Q., Kayoka, N., & Baeten, V., (2018). Near infrared microscopy and hyperspectral imaging for spices adulteration: Feasibility study. *Conference on Food Integrity*. https://www.researchgate.net/publication/331962628_Near_infrared_microscopy_and_hyperspectral_imaging_for_spices_adulteration_a_feasibility_study (accessed on 27 January 2021).
31. Dhondt, S., Vanhaeren, H., Loo, D. V., Cnudde, V., & Inze, D., (2010). Plant structure visualization by high-resolution x-ray computed tomography. *Trends in Plant Science, 15*(1), 419–422.
32. Domenech-Asensi, G., (2004). Applying optical systems. In: Edwards, M., (ed.), *Detecting Foreign Bodies in Food* (pp. 119–131). Cambridge, UK: Woodhead Publishing Limited.
33. Donis-González, I. R., Guyer, D. E., Fulbright, D. W., & Pease, A., (2014). Postharvest noninvasive assessment of fresh chestnut (*Castanea* spp.) internal decay using computer tomography images. *Postharvest Biology and Technology, 94*, 14–25.

34. Donis-González, I. R., Guyer, D. E., Pease, A., & Fulbright, D. W., (2012). Relation of computerized tomography Hounsfield unit measurements and internal components of fresh chestnuts. *Postharvest Biology and Technology, 64*(1), 74–82.

35. Dreher, M. L., Maher, C. V., & Kearney, P., (1996). The traditional and emerging role of nuts in healthful diets. *Nutritional Revolution, 54*, 241–245.

36. Du, C. J., & Sun, D. W., (2006). Learning techniques used in computer vision for food quality evaluation: A review. *Journal of Food Engineering, 72*, 39–55.

37. Du, Z., Hu, Y., Ali, B. N., & Mahmood, A., (2019). X-ray computed tomography for quality inspection of agricultural products: A review. *Food Science and Nutrition, 7*(10), 3146–3160.

38. EC (European Commission) Regulation 1881/2006, (2006). Setting Maximum 523 Levels for Certain Contaminants in Foodstuffs. *Official Journal of the European Union*, 29.

39. Eksi-Kocak, H., Mentes-Yilmaz, O., & Boyaci, I. H., (2016). Detection of green pea adulteration in pistachio nut granules by using Raman hyperspectral imaging. *European Food Research and Technology, 242*(2), 271–277.

40. Elmasry, G., Kamruzzaman, M., Sun, D., & Allen, P., (2012). Principles and applications of hyperspectral imaging in quality evaluation of agro-food products: Review. *Critical Reviews in Food Science and Nutrition, 52*(11), 999–1023.

41. Ercisli, S., Sayinci, B., Kara, M., Yildiz, C., & Ozturk, I., (2012). Determination of size and shape features of walnut cultivars using image processing. *Scientia Horticulture, 133*, 47–55.

42. Ferguson, B., & Zhang, X. C., (2002). Materials for terahertz science and technology. *Nature Materials, 1*(1), 26–33.

43. Ghodki, B. M., Singh, S. S., Chakraborty, S., Jana, S., Ghodki, D. M., & Goswami, T. K., (2019). Influence of cryogenic treatment on micro-structural characteristics of some Indian spices: X-ray micro-tomography. *Journal of Food Engineering, 243*(3), 39–48.

44. Ginesu, G., Giusto, D. D., Margner, V., & Meinlschmidt, P., (2004). Detection of foreign bodies in food by thermal image processing. *IEEE Transactions on Industrial Electronics, 51*(2), 480–490.

45. Giraudo, A., Calvini, R., Orlandi, G., Ulrici, A., Geobaldo, F., & Savorani, F., (2018). Development of an automated method for the identification of defective hazelnuts based on RGB image analysis and color grams. *Food Control, 94*, 233–240.

46. Gonzalez, R. C., Woods, R. E., & Eddins, S. L., (2004). *Digital Image Processing Using MATLAB* (p. 98). New Jersey: Prentice Hall.

47. Gonzalez, R. C., & Woods, R. E., (2004). *Digital Image Processing* (2nd edn., pp. 119–124, 298–300, 523–525). Upper Saddle River, New Jersey, USA: Prentice-Hall, Inc.

48. Goswami, P. R., & Jain, K. R., (2013). *Non-Destructive Quality Evaluation in Spice Industry with Specific Reference to Black Pepper* (pp. 1–5). Nirma University International Conference on Engineering (NUiCONE), Ahmedabad.

49. Gunasekaran, S., Paulsen, M. R., & Shove, G. C., (1985). Optical methods for nondestructive quality evaluation of agricultural and biological materials. *Journal of Agricultural Engineering Research, 32*(9), 205–241.

50. Haug, R. H., (1989). Computed tomography of pistachio nutshell. *Oral Surgery, Oral Medicine, Oral Pathology and Oral Radiology, 68*(4), 493–502.

51. He, H. J., Zheng, C., & Sun, D. W., (2016). Image segmentation techniques. In: Sun, D. W., (ed.), *Computer Vision Technology for Food Quality Evaluation* (2nd edn., pp. 45–63). New York-USA: Academic Press.

52. He, H. J., Wu, D., & Sun, D. W., (2015). Nondestructive spectroscopic and imaging techniques for quality evaluation and assessment of fish and fish products. *Critical Reviews in Food Science and Nutrition, 55*(6), 864–886.

53. Hernández-Hierro, J. M., García-Villanova, R. J., & González-Martín, I., (2008). Potential of near-infrared spectroscopy for the analysis of mycotoxins applied to naturally contaminated red paprika found in the Spanish market. *Analytica Chimica Acta, 622*(2), 189–194.

54. Hu, W., Sun, D. W., Pu, H., & Pan, T., (2016). Recent developments in methods and techniques for rapid monitoring of sugar metabolism in fruits. *Comprehensive Reviews in Food Science and Food Safety, 15*, 1067–1079.

55. Husin, Z. B., Ali, A. H. B. A. A., Shakaff, Y. B. M., & Farook, R. B. S. M., (2012). Feasibility study on plant chili disease detection using image-processing techniques. In: *Third International Conference on Intelligent Systems Modeling and Simulation* (pp. 291–296). Kota Kinabalu.

56. Jain, K. R., Modi, C. K., & Pithadiya, K. J., (2011). Non-destructive quality evaluation in spice industry with specific reference to cumin seeds. *Journal of Food Engineering, 103*, 62–67.

57. Jiang, L., Zhu, B., Rao, X., Berney, G., & Tao, Y., (2007). Discrimination of black walnut shell and pulp in hyperspectral fluorescence imagery using gaussian kernel function approach. *Journal of Food Engineering, 81*(1), 108–117.

58. Jin, H., Li, L., & Cheng, J., (2015). Rapid and non-destructive determination of moisture content of peanut kernels using hyperspectral-imaging technique. *Food Analytical Methods, 8*(10), 2524–2532.

59. Jin, J. W., Chen, Z. P., Li, L. M., Steponavicius, R., Thennadil, S. N., Yang, J., & Yu, R. Q., (2011). Quantitative spectroscopic analysis of heterogeneous mixtures: The correction of multiplicative effects caused by variations in physical properties of samples. *Analytical Chemistry, 84*(1), 320–326.

60. Jun, W., Kim, M. S., Cho, B. K., Millner, P. D., Chao, K., & Chan, D. E., (2010). Microbial biofilm detection on food contact surfaces by macro-scale fluorescence imaging. *Journal of Food Engineering, 99*, 314–322.

61. Jun, W., Kim, M. S., Lee, K., & Millner, P., (2009). Assessment of bacterial biofilm on stainless steel by hyperspectral fluorescence imaging. *Sensing and Instrumentation for Food Quality and Safety, 3*, 41–48.

62. Kalkan, H., Beriata, P., Yardimcia, Y., & Pearson, T. C., (2011). Detection of contaminated hazelnuts and ground red chili pepper flakes by multispectral imaging. *Computers and Electronics in Agriculture, 77*(1), 28–34.

63. Kamruzzaman, M., ElMasry, G., Sun, D. W., & Allen, P., (2013). Non-destructive assessment of instrumental and sensory tenderness of lamb meat using NIR hyperspectral imaging. *Food Chemistry, 141*(1), 389–396.

64. Keagy, P. M., Parvin, B., & Schatzki, T. F., (1996). Machine recognition of navel orange worm damage in x-ray images of pistachio nuts. *LWT-Food Science and Technology, 29*(2), 140–145.

65. Khan, M. H., Saleem, Z., Ahmad, M., Sohaib, A., & Ayaz, H., (2019). *Unsupervised Adulterated Red-Chili Pepper Content Transformation for Hyperspectral Classification.* https://arxiv.org/abs/1911.03711 (accessed on 27 January 2021).

66. Kheiralipour, K., Ahmadi, H., Rajabipour, A., Rafiee, S., Javan-Nikkhah, M., Jayas, D. S., & Siliveru, K., (2016). Detection of fungal infection in pistachio kernel by long-wave near-infrared hyperspectral imaging technique. *Quality Assurance and Safety of Crops and Foods, 8*(1), 129–135.

67. Kheiralipour, K., Ahmadi, H., Rajabipour, A., Rafiee, S., & Javan-Nikkhah, M., (2012). Investigating on total emissivity of pistachio kernel using thermal imaging technique. *International Journal of Agriculture and Technology, 8,* 435–441.

68. Khojastehnazhand, M., Omid, M., & Tabatabaeefar, A., (2010). Development of a lemon sorting system based on color and size. *African Journal of Plant Science, 4*(4), 122–127.

69. Khosa, I., & Pasero, E., (2014). Feature extraction in x-ray images for hazelnuts classification. In: *International Joint Conference on Neural Networks (IJCNN)* (pp. 2354–2360).

70. Kiani, S., Saskia, M. V. R., Leo, W. D. V. R., & Minaei, S., (2019). Hyperspectral imaging as a novel system for the authentication of spices: A nutmeg case study. *LWT-Food Science and Technology, 104,* 61–69.

71. Kim, S., & Schatzki, T., (2001). Detection of pinholes in almonds through x-ray imaging. *Transactions of the ASAE American Society of Agricultural and Biological Engineers, 44*(4), 997–1003.

72. Kodagali, J., (2012). Computer vision and image analysis based techniques for automatic characterization of fruits: Review. *International Journal of Computer Applications, 50*(6), 6–12.

73. Kokawa, M., Fujita, K., Sugiyama, J., Tsuta, M., & Shibata, M., (2011). Visualization of gluten and starch distributions in dough by fluorescence fingerprint imaging. *Bioscience, Biotechnology and Biochemistry, 75,* 2112–2118.

74. Kokawa, M., Fujita, K., Sugiyama, J., Tsuta, M., Shibata, M., Araki, T., & Nabetani, H., (2012). Quantification of the distributions of gluten, starch, and air bubbles in dough at different mixing stages by fluorescence fingerprint imaging. *Journal of Cereal Science, 55,* 15–21.

75. Kotwaliwale, N., (2003). *Feasibility of Physical Properties and Soft X-Ray Attenuation Properties for Nondestructive Determination of Quality of Nutmeat in In-Shell Pecans* (p. 217). Unpublished PhD Thesis, Oklahoma State University, Stillwater, USA.

76. Kotwaliwale, N., Singh, K., Kalne, A., Jha, S. N., Seth, N., & Kar, A., (2014). X-ray imaging methods for internal quality evaluation of agricultural produce. *Journal of Food Science and Technology, 51*(1), 1–15.

77. Kotwaliwale, N., Subbiah, J., Weckler, P. R., Brusewitz, G. H., & Kranzler, G. A., (2007). Calibration of a soft x-ray digital imaging system for biological materials. *Trans ASABE, 50*(2), 661–666.

78. Kotwaliwale, N., Weckler, P. R., & Brusewitz, G. H., (2006). X-ray attenuation coefficients using polychromatic x-ray imaging of pecan components. *Biosystem Engineering, 94*(2), 199–206.

79. Kotwaliwale, N., Weckler, P. R., Brusewitz, G. H., & Kranzler, G. A., (2007b). Non-destructive quality determination of pecans using soft x-rays. *Postharvest Biology and Technology, 45,* 372–380.

80. Kurtulmuş, F., Alibas, I., & Kavdır, I., (2015). Classification of pepper seeds using machine vision based on neural network. *International Journal of Agricultural and Biological Engineering, 9*(1), 51–62.
81. Lakshmi, V., (2012). Food adulteration. *International Journal of Science Inventions Today, 1*(2), 106–113.
82. Lakshminarayana, M. R., Joshi, S., Gowda, G. N., & Khetrapal, C. L., (1992). Spatial distribution of oil in groundnut and sunflower seeds by nuclear magnetic resonance imaging. *Journal of Biosciences, 17*(1), 87–93.
83. Lang, J., McNitt, L., & Robertson, I., (2020). *Application Note-FT-NIR Spectrometry.* https://www.perkinelmer.com/lab-solutions/resources/docs/APP_Determination_of_ Levels_of_Spice_Adulteration_using_Near-Infrared-Spectroscopy.pdf (accessed on 27 January 2021).
84. Lead Action News, (1995). *Adulteration of Paprika in Hungary.* https://www.lead.org. au/lanv3n3/lanv3n3-6.html (accessed on 27 January 2021).
85. Lee, D. J., Archibald, J. K., Chang, Y. C., & Greco, C. R., (2008). Robust color space conversion and color distribution analysis techniques for date maturity evaluation. *Journal of Food Engineering, 88*(3), 364–372.
86. Lee, Y. K., Choi, S. W., Han, S. T., Woo, D. H., & Chun, H. S., (2012). Detection of foreign bodies in foods using continuous wave terahertz imaging. *Journal of Food Protection, 75*(1), 179–183.
87. Li, X., Li, R., Wang, M., Liu, Y., Zhang, B., & Zhou, J., (2018). Hyperspectral imaging and their applications in the nondestructive quality assessment of fruits and vegetables: Chapter 3. In: *Hyperspectral Imaging in Agriculture, Food, and Environment* (pp. 27–63). London, UK; Intech Open Limited.
88. Li, Z., Tang, X., Shen, Z., Yang, K., Zhao, L., & Li, Y., (2019). Comprehensive comparison of multiple quantitative near-infrared spectroscopy models for *Aspergillus flavus* contamination detection in peanut. *Journal of Food Science and Agriculture, 99*(13), 5671–5679.
89. Lim, K. H., & Gopalai, A. A., (2013). Robotic vision system design for black pepper harvesting. *IEEE International Conference of IEEE Region 10(TENCON 2013)-Xi'an* (pp. 1–5). doi: 10.1109/TENCON.2013.6718825.
90. Liu, T., Xie, J., He, Y., Xu, M., & Qin, C., (2009). An automatic classification method for betel nut based on computer vision. *Proceedings of the IEEE International Conference on Robotics and Biomimetics* (pp. 1264–1267). Guilin, China.
91. Lloyd, J. M., (2013). *Thermal Imaging Systems* (p. 453). New York, USA: Springer.
92. Lohumi, S., Lee, H., Kim, M. S., Qin, J., & Cho, B. K., (2018). Raman imaging for the detection of adulterants in paprika powder: A comparison of data analysis methods. *Applied Sciences, 8*(4), 485–499.
93. Lorente, D., Aleixos, N., Gomez-Sanchis, J., Cubero, S., Garcia-Navarrete, O. L., & Blasco, J., (2012). Recent advances and applications of hyperspectral imaging for fruit and vegetable quality assessment. *Food and Bioprocess Technology, 5*(4), 1121–1142.
94. Mahendran, R., Jayashree, G. C., & Alagusundaram, K., (2012). Application of computer vision technique on sorting and grading of fruits and vegetables. *Journal of Food Processing Technology, 8.* doi: 10.4172/2157-7110.S1-001.
95. Mishra, P., Herrero-Langreo, A., Barreiro, P., Roger, J. M., Diezma, B., Gorretta, N., & Lleóa, L., (2015). Hyperspectral to multispectral imaging for detection of tree nuts and peanut traces in wheat flour. *Journal of Spectral Imaging, 4*(1), 1–9.

96. Morita, K., Ogawa, Y., Thai, C. H. I. N., & Tanaka, F., (2003). Soft x-ray image analysis to detect foreign materials in foods. *Food Science and Technology Research, 9*(2), 137–141.

97. Moscetti, R., Saeys, W., Keresztes, J. C., Goodarzi, M., Cecchini, M., Danilo, M., & Massantini, R., (2015). Hazelnut quality sorting using high dynamic range short-wave infrared hyperspectral imaging. *Food and Bioprocess Technology, 8*, 1593–1604.

98. Mythri, K. J., Divya, R. N., Poojitha, Prajna, M., & Kumar, M. R. P., (2020). Android based ripening stage identification for peppercorns. In: Smys, S., Senjyu, T., & Lafata, P., (eds.), *Second International Conference on Computer Networks and Communication Technologies (ICCNCT)* (pp. 20–44). 2019-Lecture Notes on Data Engineering and Communications Technologies, Springer, Cham.

99. Naik, S., & Patel, B., (2014). CIELab based color feature extraction for maturity level grading of mango. *National Journal of System and Information Technology, 7*(1), 9. Article ID: 0974-3308.

100. Naik, S., & Patel, B., (2017). Machine vision based fruit classification and grading: A review. *International Journal of Computer Applications, 170*(9), Article ID: 0975–8887.

101. Nakariyakul, S., & Casasent, D. P., (2011). Classification of internally damaged almond nuts using hyperspectral imagery. *Journal of Food Engineering, 103*, 62–67.

102. Narendra, V. G., & Hareesh, K. S., (2010). Quality inspection and grading of agricultural and food products by computer vision: A review. *International Journal of Computer Applications, 2*(1), 43–65.

103. Narendra, V. G., & Hareesha, K. S., (2011). Cashew kernels classification using color features. *International Journal of Machine Intelligence, 3*(2), 52–57.

104. Nasution, I. S., & Gusriyan, K., (2019). Nutmeg grading system using computer vision techniques. *IOP Conf. Series: Earth and Environmental Science (ICATES), 365*, 012003.

105. Neethirajan, S., Jayas, D. S., White, N. D. G., (2007). Detection of sprouted wheat kernel using soft x-ray image analysis. *Journal of Food Engineering, 81*(3), 509–513.

106. Nekvapil, F., Brezestean, I., Barchewitz, D., Glamuzina, B., Chiş, V., & Pinzaru, S. C., (2018). Citrus fruits freshness assessment using Raman spectroscopy. *Food Chemistry, 24*(2), 560–567.

107. NHB, (2020). *Black Pepper*. http://nhb.gov.in/NHBDPR/Black_Pepper_DPR.pdf (accessed on 27 January 2021).

108. Novini, A. R., (1995). The latest in vision technology in today's food and beverage container manufacturing industry. *Society of Manufacturing Engineers News, 1995*, 1–10.

109. Okochi, T., Hoshino, Y., Fujii, H., & Mitsutani, T., (2007). Nondestructive tree-ring measurements for Japanese oak and Japanese beech using micro-focus x-ray computed Tomography. *Dendrochronologia, 24*(2/3), 155–164.

110. Orrillo, I., Cruz-Tirado, J. P., Cardenasa, A., Orunaa, M., Carnero, A., Barbin, D. F., & Sichea, R., (2019). Hyperspectral imaging as a powerful tool for identification of papaya seeds in black pepper. *Food Control, 101*, 45–52.

111. Panigrahi, S., & Gunasekaran, S., (2001). Computer vision: Chapter 7. In: Gunasekaran, S., (ed.), *Nondestructive Food Evaluation Techniques to Analyze Properties and Quality* (pp. 39–98). New York: Marcel Dekker, Inc.

112. Park, B., & Lu, R., (2015). Hyperspectral imaging technology in food and agriculture. *Food Engineering Series* (p. 312). New York: Springer-Verlag.

113. Patel, J. J., Modi, C. K., & Jain, K. R., (2011). Quality evaluation of *Foeniculum vulgare* (fennel) seeds using colorization. *International Conference on Image Information Processing* (pp. 1–6). Shimla, India. doi: 10.1109/ICIIP.2011.6108884.

114. Pathmanaban, P., Gnanavel, B. K., & Anandan, S. S., (2019). Recent application of imaging techniques for fruit quality assessment. *Trends in Food Science and Technology, 94*, 32–42.

115. Paulus, E., & Suryani, M., (2019). Image analysis for smart machine of nutmeg sorting. *Journal of Physics: Conf. Series, 1196*, 8. Article ID: 012059.

116. Pearson, T. C., (1996). Machine vision system for automated detection of stained pistachio nuts. *Lebensmittel-Wissenschaft und Technologie (Food Science and Technology), 29*(3), 203–209.

117. Pearson, T. C., (1999). Spectral properties and effect of drying temperature on almonds with concealed damage. *LWT-Food Science and Technology, 32*(2), 67–72.

118. Phan-Thien, K. Y., Golic, M., Wright, G. C., & Lee, N. A., (2011). Feasibility of estimating peanut essential minerals by near-infrared reflectance spectroscopy. *Sensing and Instrumentation for Food Quality and Safety, 5*(1), 43–49.

119. Qiao, S., Tian, Y., Song, P., He, K., & Song, S., (2019). Analysis and detection of decayed blueberry by low field nuclear magnetic resonance and imaging. *Postharvest Biology and Technology, 156*, 7. Article ID: 110951.

120. Qin, J., Chao, K., Kim, M., Kang, S., Cho, B. K., & Jun, W., (2011). Detection of organic residues on poultry processing equipment surfaces by LED-induced fluorescence imaging. *Applied Engineering in Agriculture, 27*, 153–161.

121. Qin, J., Chao, K., & Kim, M., (2010). Raman chemical imaging system for food safety and quality inspection. *Transactions of the ASABE, 53*(6), 1873–1882.

122. Qin, J., Chao, K., Kim, M. S., Lu, R., & Burks, T. F., (2013). Hyperspectral and multispectral imaging for evaluating food safety and quality. *Journal of Food Engineering, 118*(2), 157–171.

123. Rao, P. S., Gopal, A., Revathy, R., & Meenakshi, K., (2009). Color analysis of fruits using machine vision system for automatic sorting and grading. *Journal of the Instrument Society of India, 34*, 284–291.

124. Razavi, S. M., Mazaherinasab, M., Nikfar, F., & Sanaeifard, H., (2008). Physical properties and image analysis of wild pistachio nut. *Iranian Food Science and Technology Research Journal, 3*(2), 61–70.

125. Riyadi, S., Rahni, A. A. A., Mustafa, M. M., & Hussain, A., (2007). Shape characteristics analysis for papaya size classification. In: *The 5th Student Conference on Research and Development* (pp. 1–5). SCOReD, Malaysia.

126. Ruiz-Altisent, M. F., & Moreda, G. P., (2011). Encyclopedia of agrophysics: Chapter 5. In: Glinski, J., Horabik, J., & Lipiec, J., (eds.), *Fruits, Mechanical Properties and Bruise Susceptibility* (pp. 318–321). Netherlands: Springer.

127. Ruiz-Altisent, M., Ruiz-Garcia, L., Moreda, G. P., Lu, R., & Hernandez-Sanchez, N., (2010). Sensors for product characterization and quality of specialty crops: A review. *Computers and Electronics in Agriculture, 74*(2), 176–194.

128. Rungpichayapichet, P., Nagle, M., & Yuwanbun, P., (2017). Mapping of physicochemical properties in mango by hyperspectral imaging. *Biosystems Engineering, 159*(3), 109–120.

129. Sajjan, M., Kulkarni, L., Anami, B. S., & Gaddagimath, N. G., (2016). Comparative analysis of color features for classification of bulk chili. In: *2nd International Conference on Contemporary Computing and Informatics (IC3I)* (pp. 427–432). Noida, India.

130. Salzer, R., & Siesler, H. W., (2014). *Infrared and Raman Spectroscopic Imaging* (2nd edn., p. 460). New York, USA: John Wiley & Sons.

131. Sandoval, J. R., Sandoval, E. M., Rosas, M. E., & Velasco, M. M., (2018). Color analysis and image processing applied in agriculture. In: Travieso-Gonzalez, C. M., (ed.), *Colorimetry and Image Processing* (pp. 71–78). IntechOpen. doi: 10.5772/intechopen.71539.

132. Saremi, H., Okhovvat, M., & Saremi, H., (2007). Control managements of *Aspergillus flavus* a main aflatoxin producers and soil borne fungi on pistachio in Kerman. *Iranian Food Science and Technology Research Journal, 3,* 27–31.

133. Sharma, S., Dhalsamant, K., & Tripathy, P. P., (2019). Application of computer vision technique for physical quality monitoring of turmeric slices during direct solar drying. *Journal of Food Measurement and Characterization, 13,* 545–558.

134. Shashua, A., & Riplin-Raviv, T., (2001). The quotient image: Class-based re-rendering and recognition with varying illuminations. In: *IEEE Transactions on Pattern Analysis and Machine Intelligence* (Vol. 23, pp. 129–139).

135. Shirai, Y., (1987). *Three Dimensional Computer Vision* (pp. 1–297). Berlin: Springer-Verlag.

136. Sonka, M., Hlavac, V., & Boyle, R., (1999). *Image Processing, Analysis, and Machine Vision* (p. 542). San Diego-California, USA: Springer.

137. Suchanek, M., Kordulska, M., Olejniczak, Z., Figiel, H., & Turek, K., (2017). Application of low-field MRI for quality assessment of conference pears stored under controlled atmosphere conditions. *Postharvest Biology and Technology, 124,* 100–106.

138. Suktanarak, S., & Teerachaichayut, S., (2017). Non-destructive quality assessment of hens' eggs using hyperspectral images. *Journal of Food Engineering, 215,* 97–103.

139. Sun, D. W., (2000). Inspecting pizza topping percentage and distribution by a computer vision method. *Journal of Food Engineering, 44,* 245–249.

140. Sun, J., Shi, X., Zhang, H., & Xia, L., (2019). Detection of moisture content in peanut kernels using hyperspectral imaging technology coupled with chemometrics. *Journal of Food Process Engineering, 42*(7), 8. Article ID: 13263.

141. Sun, M., Zhang, D., Liu, L., & Wang, Z., (2017). How to predict the sugariness and hardness of melons: A near-infrared hyperspectral imaging method. *Food Chemistry, 218*(4), 413–421.

142. Thanushree, M. P., Sailendri, D., Yoha, K. S., Moses, J. A., & Anandharamakrishnan, C., (2019). Mycotoxin contamination in food: An exposition on spices. *Trends in Food Science and Technology, 93,* 69–80.

143. Tillman, L. B., Gorbet, W. D., & Person, G., (2006). Predicting oleic and linoleic acid content of single peanut seeds using near-infrared reflectance spectroscopy. *Crop Science, 46,* 2121–2126.

144. Vadivambal, R., & Jayas, D. S., (2011). Applications of thermal imaging in agriculture and food industry: A review. *Food and Bioprocess Technology, 4*(2), 186–199.

145. Verboven, P., Kerckhofs, G., Mebatsion, H. K., Ho, Q. T., Temst, K., Wevers, M., & Nicolaï, B. M., (2008). Three-dimensional gas exchange pathways in pome fruit characterized by synchrotron x-ray computed tomography. *Plant Physiology, 147*(2), 518–527.

146. Wang, H. H., & Sun, D. W., (2002). Correlation between cheese meltability determined with a computer vision method and with Arnott and Schreiber tests. *Journal of Food Science, 67*(2), 745–749.

147. Wang, H., Peng, J., Xie, C., Bao, Y., & Yong, H., (2015). Fruit quality evaluation using spectroscopy technology: Review. *Sensors, 15*(5), 11889–11927.

148. Wang, K., & Sun, D. W., (2018). Imaging spectroscopic technique: Raman chemical imaging. In: Sun, D. W., (ed.), *Modern Techniques for Food Authentication*; (2nd edn., pp. 287–319). Boston-MA: Academic Press.

149. Wang, K., Sun, D. W., & Pu, H., (2017). Emerging non-destructive terahertz spectroscopic imaging technique: Principle and applications in the agri-food industry. *Trends in Food Science and Technology, 67,* 93–105.

150. Wang, L., Wang, Q., Liu, H., Liu, L., & Du, Y., (2013). Determining the contents of protein and amino acids in peanuts using near-infrared reflectance spectroscopy. *Journal of the Science of Food and Agriculture, 93*(1), 118–124.

151. Wang, S., Tang, J., Sun, T., Mitcham, E. J., Koral, T., & Birla, S. L., (2006). Considerations in design of commercial radio frequency treatments for postharvest pest control in in-shell walnuts. *Journal of Food Engineering, 77*(2), 304–312.

152. Warmann, C., & Märgner, V., (2005). Quality control of hazel nuts using thermographic image processing. In: *MVA2005 IAPR Conference on Machine Vision Applications* (pp. 3–17). Tsukuba Science City, Japan.

153. Wasnik, P. G., Menon, R. R., Sivaram, M., Nath, B. S., Balasubramanyam, B. V., & Manjunatha, M., (2019). Development of mathematical model for prediction of adulteration levels of cow ghee with vegetable fat using image analysis. *Journal of Food Science and Technology, 56*(4), 2320–2325.

154. Wikipedia, (2020). https://en.wikipedia.org/wiki/RGB_color_model#/media/File:RGB_color_solid_cube.png (accessed on 27 January 2021).

155. Workman, J., & Weyer, L., (2008). *Practical Guide to Interpretive Near Infrared Spectroscopy* (p. 344). Boca Raton - FL: CRC Press.

156. Yang, C. C., Kim, M. S., Kang, S., Cho, B. K., Chao, K., Lefcourt, A. M., & Chan, D. E., (2012). Red to far-red multispectral fluorescence image fusion for detection of fecal contamination on apples. *Journal of Food Engineering, 108,* 312–319.

157. Yang, W., Xu, X., Duan, L., Luo, Q. M., Chen, S. B., Zeng, S. Q., & Liu, Q., (2011). High-throughput measurement of rice tillers using a conveyor equipped with x-ray computed tomography. *Review of Scientific Instruments, 82,* 761–793.

158. Yang, X., Fu, J., Lou, Z., Wang, L., Li, G., & Freeman, W., (2006). Tea classification based on artificial olfaction using bionic olfactory neural network. *Lecture Notes in Computer Science, 39*(2), 343–348.

159. Yanniotis, S., Proshlyakov, A., Revithi, A., Georgiadou, M., & Blahovec, J., (2011). X-ray imaging for fungal necrotic spot detection in pistachio nuts. *Procedia Food Science, 1,* 379–384.

160. Yaseen, T., Sun, D. W., & Cheng, J. H., (2017). Raman imaging for food quality and safety evaluation: Fundamentals and applications. *Trends in Food Science and Technology, 62*(2), 177–189.

161. Zhang, B., Huang, W., Li, J., Zhao, C., Fan, S., Wu, J., & Liu, C., (2014). Principles, developments, and applications of computer vision for external quality inspection of fruits and vegetables: A review. *Food Research International, 62,* 326–343.

162. Zhu, B., Jiang, L., Jin, F., Qin, L., Vogel, A., & Tao, Y., (2007). Walnut shell and meat differentiation using fluorescence hyperspectral imagery with ICA-kNN optimal wavelength optimal wavelength selection. *Sensory Instrumental Food Quality and Safety, 1*, 123–131.
163. Zwiggelaar, R., Bull, C. R., Mooney, M. J., & Czarnes, S., (1997). Detection of soft materials by selective energy x-ray transmission imaging and computer tomography. *Journal of Agricultural Engineering Research, 66*(3), 203–212.

CHAPTER 10

FOURIER TRANSFORM INFRARED (FTIR) SPECTROSCOPY WITH CHEMOMETRICS: EVALUATION OF FOOD QUALITY AND SAFETY

NEELAM UPADHYAY, C. G. HARSHITHA, NILESH KUMAR PATHAK, and RAJAN SHARMA

ABSTRACT

Fourier transform infrared (FTIR) spectroscopy is referred to as a finger print technique as no two different samples are known to yield an overlapping spectrum, and it can be used for solid, liquid, or gaseous samples, which give spectra based on the vibrational transitions taking place in the molecule. For the purpose of easy understanding and concept building of young and innovative food technologists, the chapter focuses on the application of FTIR spectroscopy in food quality and safety.

10.1 INTRODUCTION

Spectroscopy deals with the study of the interaction of electromagnetic (EM) radiations with matter. The history of spectroscopy goes back to the 17th century, when Sir Isaac Newton used the word 'spectrum' for describing the rainbow of colors, which combined to form the white light. Joseph von Fraunhofer (inventor of spectroscope) experimented with dispersive spectrometers in the early 1800s, which resulted in accurate instruments as well as quantitative scientific technique.

Spectroscopy is now the indispensable part of an analytical chemistry and offers several applications in almost all the basic and applied sciences. The term "electromagnetic (EM) radiation" represents the radiant energy emitted from any source in the form of light, heat, etc. The important characteristics of these EM radiations are [78]:

- They have dual character, i.e., particle character and wave character. For example, a beam of light is a stream of particles called photons moving through the space in the form of waves.
- These waves are associated with electric and magnetic fields oscillating perpendicular to each other and also perpendicular to the direction of propagation, hence are called electromagnetic waves.
- All EM radiations travel at the velocity of light, c (= 2.998 x 10^8 ms^{-1}).
- The wavelength, λ, of EM radiation is related to their frequency, υ, and velocity, c, as per the equation c = υ λ.
- EM radiations are made up of photons for which photon energy is given by Planck's equation: E = h υ where, h = Planck's constant (= 6.6262 × 10^{-32} Js): Eqn. (8).

The chapter gives an exhaustive list of the research studies that have been carried out to couple the FTIR spectroscopy and chemometrics for the purpose of evaluation for various analytes that influence quality and safety of food products.

10.2   REGIONS OF ELECTROMAGNETIC SPECTRUM AND INTRODUCTION TO SPECTROSCOPY

The complete arrangement of different types of EM radiations in order of their increasing wavelengths or decreasing frequencies is called electromagnetic spectrum. The regions into which the EM radiations have been classified are on the basis of their source and detectors required. The different radiations differ in their physical and chemical effects as they possess different energies, and based on this, they can be absorbed by atoms or molecules, resulting in atomic spectrum of molecular spectrum [78].

Atomic spectrum is obtained when an atom absorbs some energy (in the form of fixed photon or quantum specific for a specific atom), its electron jumps from one orbit to another higher orbit, and when the electron jumps back to the lower energy orbit, this results in emission of energy (which is equal to the difference between the energies of the two energy levels (ΔE)) leading to atomic spectrum. However, in the case of molecules, three effects may be observed, namely rotation, vibration, or electronic transitions, when energy is absorbed by a molecule. These effects depend upon the amount of energy being absorbed and the corresponding energy levels of molecules, i.e., rotational, vibrational, and electronic are collectively known as molecular energy levels.

The transition of energies taking place between these levels results in molecular spectrum. The energy required for electronic transition (i.e., between electronic energy level) is higher than that required for vibrational (i.e., from one vibrational level to another) followed by rotational (i.e., from one rotational level to another). Thus, electronic transitions are also accompanied by transition between vibrational and rotational levels, i.e., one electronic level, say, n = 1, has vibrational sub-levels which are represented by v = 0,1,2,3, so on; and each of these vibrational levels have a number of rotational sub-levels which are represented by J = 0, 1, 2, 3, etc., where v and J are called vibrational and rotational quantum numbers, respectively (Figure 10.1). This further indicates that since in molecular spectra, various transitions are possible in different energy levels, therefore it is complex than the atomic spectra. The energy is quantized in both, i.e., atomic, and molecular spectrum.

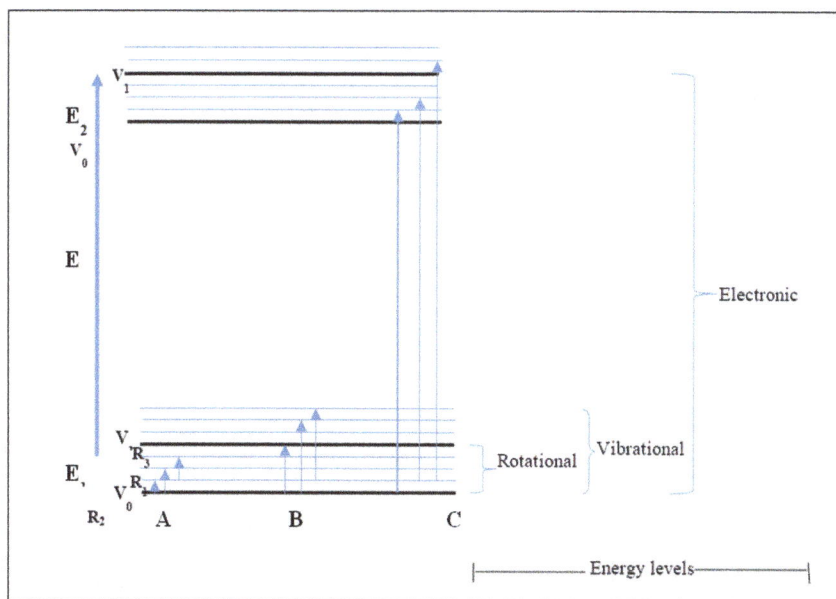

FIGURE 10.1   Rotational, vibrational, and electronic transitions between the first two electronic energy levels of a molecule (molecular energy levels).

The molecules are composed of atoms which are held together by electrical force. It is electrically neutral, and each molecule has its own structure and composition. The vibrations in molecules can be understood by classical and quantum mechanical model [1, 8, 40].

Spectrophotometers are the instruments that measure the intensity of light, i.e., EM radiations being absorbed or transmitted from the sample under study. As discussed earlier, these radiations vary based on the sample and hence the spectrophotometers required for a particular kind of analysis/ samples differ. The different types of spectroscopy are mentioned in Figure 10.2 based on their principle and application.

10.2.1 VIBRATIONAL SPECTROSCOPY: VIBRATION IN MOLECULE

10.2.1.1   CLASSICAL MODEL

In the classical model, it is assumed that the molecules behave like harmonic oscillators. Harmonic oscillator is a system, in which a mass is attached to the spring having fixed spring constant k as depicted in Figure 10.3(B). If we displace the particle of mass 'm' from its equilibrium position, a restoring force starts working on it, which tries to bring back the particle to the original position. Suppose, we displace the particle along x-axis, the restoring force acting on it is expressed as:

$$F = -kx \tag{1}$$

where; F is restoring force, k is spring constant and x is the extension length of spring along x-axis.

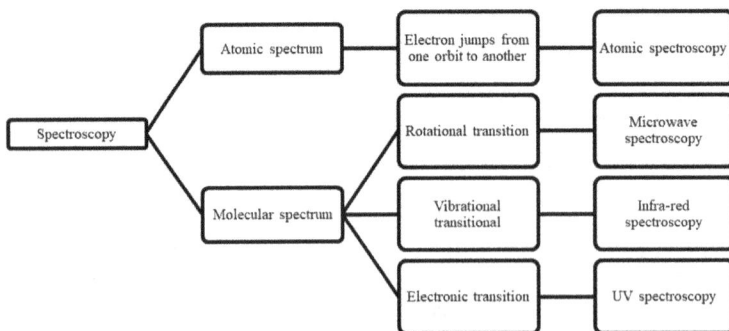

FIGURE 10.2   Classification of spectrophotometers for measuring different spectrum.

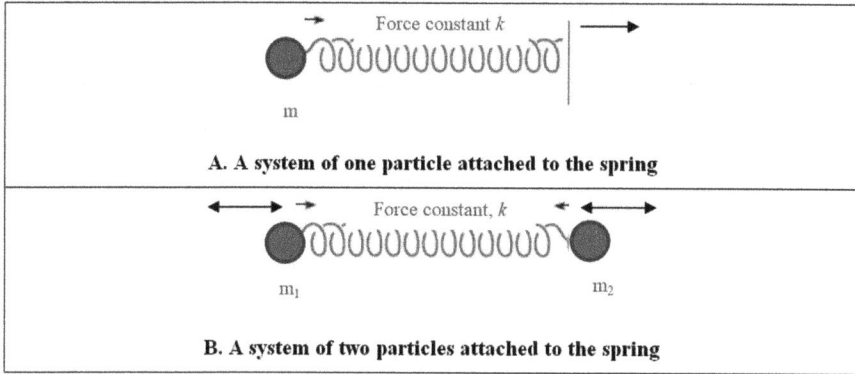

FIGURE 10.3 System of a particle attached to a spring. (A) A system of one particle attached to the spring; (B) a system of two particles attached to the spring.

From Newton's second law, the rate of change of momentum is as follows:

$$F = ma = m\frac{d^2x}{dt^2} \qquad (2)$$

Thus, from Eqns. (1) and (2), we have:

$$\frac{d^2x}{dt^2} + \omega^2 x = 0$$

which is a second-order linear differential equation having oscillatory solution as shown in Eqn. (3)

$$x = A\sin \omega t \qquad (3)$$

where; $\omega = \sqrt{k/m}$ is the frequency of vibration, k is spring constant and m mass of single-particle (Figure 10.3(A)).

If we take, a system of two particles attached to the spring (of spring constant, k: Figure 10.3(B)) as depicted above, then from above equations, we can find the frequency of vibration for this system as shown in Eqn. (4).

$$\omega = \sqrt{k/\mu} \qquad (4)$$

where; $\mu = \dfrac{m_1 m_2}{m_1 + m_2}$ = is reduced mass of system.

Similarly, for the system of more than two particles, we will have to solve more complex situation of differential equation to predict the behavior of a particle. However, for two-particle system, we have two different modes of vibrations, and these vibrations may be in phase or out of phase. The classical picture will tell us that the system oscillates with continuous energy. Classically, the total energy E is the sum of kinetic and potential energy as follows:

$$E = T + V \qquad (5)$$

where; T and V are kinetic and potential energies, respectively, as expressed below:

$$T = \frac{1}{2} m \left(\frac{dx}{dt} \right)^2 \qquad (6)$$

$$V = \frac{1}{2} kx^2 \qquad (7)$$

At equilibrium position, potential energy is zero and total energy of oscillator is purely kinetic; and at turning point kinetic energy is zero therefore, total energy is purely potential. In the classical model, the oscillator can have a continuous value of energy, while quantum mechanical model, according to Planck harmonic oscillator, has the discrete form of energy.

10.2.1.2   QUANTUM MECHANICAL ANALYSIS

In 1900, the German physicist Max Planck formulated that the oscillator in the cavity does not possess the continuous value of energy, but have certain discrete value of energies. According to him, when the transition occurs between two energy levels (E_1 and E_2), the energy changes and is expressed as follows:

$$\Delta E = E_2 - E_1 = hv \qquad (8)$$

where; h is the Planck's constant and v is frequency of radiation.

When EM radiations interact with the molecules, the radiation is absorbed by the molecule and transition occurs in the energy state E_1 and E_2. Since, large number of energy levels exist corresponding to which several possible transitions may occur, and the spectra of molecules becomes complex. The simple and complex spectra of molecules are analyzed in the vibrational spectroscopy, in which molecules interact with the EM radiation and start vibrating with different modes of vibration and these modes has discrete value of energy.

10.2.1.3 VIBRATIONAL SPECTRUM

In this section, the vibration of the molecule is discussed under the influence of EM radiation. When the molecule interacts with EM wave, it possesses vibrational energy, and possible energy of vibration of molecules is obtained by solving the Schrodinger equation to find the energy, wavenumber, and possible frequency mode of vibrating molecules. For analyzing the spectroscopy of a molecule, let us consider the simplest situation for the vibration of a diatomic molecule, which is similar to the vibration of spring attached with two masses m_1 and m_2 as was depicted in Figure 10.3(B). The Schrodinger equation for such a system is:

$$\nabla^2\psi + \frac{8\pi^2\mu}{h^2}(E-V)\psi = 0 \tag{9}$$

where; ψ is a wave function (which contains all the needed information about the system), μ is reduced mass, E is total energy, and V is the potential energy ($V = \frac{1}{2}kx^2$).

The compression or extension of bond is like a spring and it obeys the Hook's law. Equation (9) is in three dimensions. The extension or compression in the spring is in one dimension, therefore the Eqn. (9) in one dimension can be written as:

$$\frac{d^2\psi}{dx^2} + \frac{8\pi^2\mu}{h^2}(E - \frac{1}{2}kx^2)\psi = 0 \tag{10}$$

Solving Eqn. (10) to obtain the energy of a vibrating molecule, we get:

$$E_v = \frac{h}{2\pi}\sqrt{\frac{k}{\mu}}\left(v + \frac{1}{2}\right) = hv_{osc}\left(v + \frac{1}{2}\right), \quad v = 0,1,2,3,... \tag{11}$$

In wavenumber representation, we have Eqn. (12):

$$\varepsilon_v = \frac{E_v}{hc} = \frac{V_{osc}}{c}\left(v + \frac{1}{2}\right) = \frac{\omega}{2\pi c}\left(v + \frac{1}{2}\right) = \omega_e\left(v + \frac{1}{2}\right)$$

(12)

Subsequently for each vibrational quantum number, we get:

$$\varepsilon_v = \frac{1}{2}\omega_e, \frac{3}{2}\omega_e, \frac{5}{2}\omega_e, ... \quad for\ v = 0, 1, 2, ...$$

(13)

where; v is a vibrational quantum number.

This indicates that the vibrational levels are discrete, equispaced with common separation of ωcm^{-1}. We have found the energy of vibrating oscillator in terms of wavenumber. This quantized energy value of vibrating molecules is obtained by assuming that molecules are executing simple harmonic motion and corresponding potential profile is exactly parabolic in nature. However, in reality, molecules do not obey harmonic oscillation and Hook's law; because the real bonds between the molecules are not as homogeneous as assumed in the case of a spring-mass system. Therefore, we have to consider some new profile of potential, which contains the overall signature of molecules, and that potential profile is known as anharmonic potential and oscillator corresponding to this potential is known as anharmonic oscillator.

The potential of Anharmonic oscillator is:

$$V(r) = f(r - r_e)^2 - g(r - r_e)^3 = fx^2 - gx^3$$

(14)

The energy values of Anharmonic oscillator is:

$$E_v = hc\omega_e\left(v + \frac{1}{2}\right) - hc\omega_e x_e\left(v + \frac{1}{2}\right)^2$$

(15)

where; ω_e is the spacing of energy level when potential energy curve is a parabola, X_e is Anharmonic constant. The value of $\omega_e x_e < x_e$, which indicates that the energy levels of Anharmonic oscillator are not equidistant.

The transition between any two vibrational energy levels is governed by certain rules known as the selection rule. These selection rules (Eqn. (16)) are as follows:

$$\Delta v = \pm 1, \pm 2, \pm 3, ...,$$

$\Delta v = \pm 1$, $(v = 0 \rightarrow v = 1)$: corresponds to intense beam and gives fundamental band; and

$\Delta v = \pm 2, \pm 3, \ldots$: appears with decreasing intensity and gives overtones (16)

This is the general discussion about the interaction of EM radiations with the vibrating diatomic molecules, which can be applied to several other molecules. As food technologists, we want to employ this concept to food, which is a complex system involving several molecules and their bonds, for example, amide in proteins, carboxyl in lipids, etc. Therefore, transition between the energy levels of these molecules are also governed by the said selection rule and are exploited for several purposes that will be discussed in the latter part of this chapter.

10.3 FOURIER TRANSFORM INFRARED (FTIR) SPECTROSCOPY: INTRODUCTION

The infrared spectral region ranges from 0.8–100 μm wavelength of the EM spectrum, which has higher frequency and shorter wavelength than microwaves; and lower frequency and longer wavelength than visible waves. It is generally divided into near-infrared (NIR) region, mid-infrared region and far-infrared region, which range from 0.8–2.5 μm, 2.5–15 μm and 15–100 μm wavelengths, respectively.

Infrared (IR) Spectroscopy utilizes infrared light for acquiring information about the molecule by three different ways: determining the amount of absorption, emission or reflection of radiation by the molecule. Infrared absorption spectroscopy has been used to identify the functional group of the molecules based on the absorption of particular wavelength of radiations (considering vibration of molecule) [14].

IR spectroscopy is divided into two types: namely MIR and FIR; and infrared spectrum is referred to as molecular vibrational spectrum. The spectrum as obtained in MIR spectroscopy are based on fundamental vibration modes, which are related to specific functional groups existing in the sample and the spectra in this is obtained between range 400 to 4000 cm^{-1} wavenumber, while spectrum as obtained in NIR are due to molecular overtones and combination vibrations from the fundamental vibrational modes ranging from 4000 to 10000 cm^{-1} wavenumber [13, 25].

Overtones refer to the state when excitation in vibrational mode is from $\upsilon = 0$ to 2 (first overtone) or 0 to 3 (second overtone), i.e., when it is different from the fundamental transitions, which means $\upsilon = \pm 1$. However, the tendency of probability of occurrence of overtones sharply decreases when $\Delta\upsilon > \pm 1$ gets larger. As discussed earlier, after absorbing IR radiations, the spectrum obtained is due to the movement of electron from the ground state to excited state due to vibrational energy. The vibrational energy of the molecules determines the sample and vibrational freedom determines the number of absorption peaks and the transition energy levels determine the intensity of the absorption peaks. Therefore, by analyzing the infrared spectrum, one can readily obtain abundant information pertaining to the structure of a molecule [70].

The different kinds of IR spectrometers are divided into: first, second, and third-generation IR spectrometers. To understand the difference between these, the reader must know a simple ray diagram of the instrumentation of spectrophotometers as lamp/source (which produces light in the specific EM range):

- $\rightarrow$ slit 1 $\rightarrow$ monochromator (which is an optical device that transmits a narrow band of wavelength selected from polychromatic light coming from lamp);
- $\rightarrow$ slit 2 $\rightarrow$ sample holder $\rightarrow$ photocell (which converts light energy to electric energy) followed by amplifier and recorder.

The difference in the various generations of IR spectrophotometer lies in the monochromator, which may generally be of several types: absorption monochromator, prism, gratings/diffraction, and interference. In the year 1950s, the first-generation IR spectrometers were invented, in which prism made of NaCl was used for splitting EM radiations. These IR spectrometers had some disadvantages like narrow scan range and poor repeatability, and strict requirements of particle size and moisture content of sample to be analyzed.

However, to resolve the problems that occurred with first generation IR spectrophotometers, the second-generation IR spectrophotometer was introduced almost a decade later in which grating was used as a monochromator (dispersive instruments). The disadvantages associated with these were poor sensitivity, low accuracy, and scan speed.

Fourier transform infrared (FTIR) spectrometers are the latest and third-generation IR spectrometer, which collect all wavelengths simultaneously, referred to as Multiplex or Felgett Advantage.

In FTIR, interferogram is collected from the sample using an interferometer followed by performing Fourier Transform on the interferogram, which is then displayed as spectrum [31, 77]. An interferogram is obtained when light from two different sources is combined. For this, light from the source is first split into two beams by the beam splitter: (a) one is the reference beam; (b) the other one passes through the optical system (sample in our case) and is called a test beam. Reference beam provides the comparison of wavefront and the two beams when combined give interference pattern and yield an interferogram [28]. The instrument used for this purpose is interferometer. The prominent advantages of FTIR spectrometers that make it a robust technique over dispersive IR spectrometers are: increased signal-to-noise ratio of spectrum, improved accuracy of wavenumber, short scan time of all frequencies, high resolution, wide scan range, reduced stray light interference.

10.3.1 COMPONENTS OF FTIR SPECTROMETERS

The components of FTIR spectrometers include source $\rightarrow$ interferometer $\rightarrow$ sample compartment $\rightarrow$ detector $\rightarrow$ amplifier $\rightarrow$ A/D converter, i.e., the IR radiations are generated by source that is passed through the interferometer and sample followed by reaching the detector. The detector amplifies the signal and converts it to a readable signal, which is transported to a computer where Fourier transform takes place.

Fourier transformation is a mathematical tool applied to interferogram for converting it into a spectrum of intensity against frequency [74]. It is a mathematical transformation that converts a function of time, i.e., time-domain signal into a function of frequency, i.e., frequency domain. The presence of Michelson interferometer is integral part of FTIR and it majorly differentiates it from dispersive IR spectrometer [69]. It splits the beam of light/EM radiation into two and the paths of resulting beams are different [73]. Michelson interferometer (Figure 10.4) then recombines the two beams into the detector where the differences of their intensities are measured as a function of the difference of paths.

The instrumentation of the Michelson interferometer (Figure 10.4) involves two perpendicular mirrors (one being stationary and another being movable) and a beam splitter. The beam splitter is aligned so as to transmit half of the light (which subsequently hits the stationary mirror) and reflect half of the light (which hits movable mirror). The two beams recombine with each other at beam splitter after being reflected by the mirrors and pass through the sample followed by detection by the detector [75], where some radiation is absorbed, and the amount reaching the detector decreases. The radiation collected without a sample serves as background interferogram. The data transformed by the Fourier transformation algorithm forms a single beam of sample and is later divided into background single beam for finding out transmission spectrum, which reflects the change in intensity with respect to frequency or wavenumbers. Absorption is also affected by the number of bonds present in the molecule. The stronger the absorption, the larger are the bands in the spectrum.

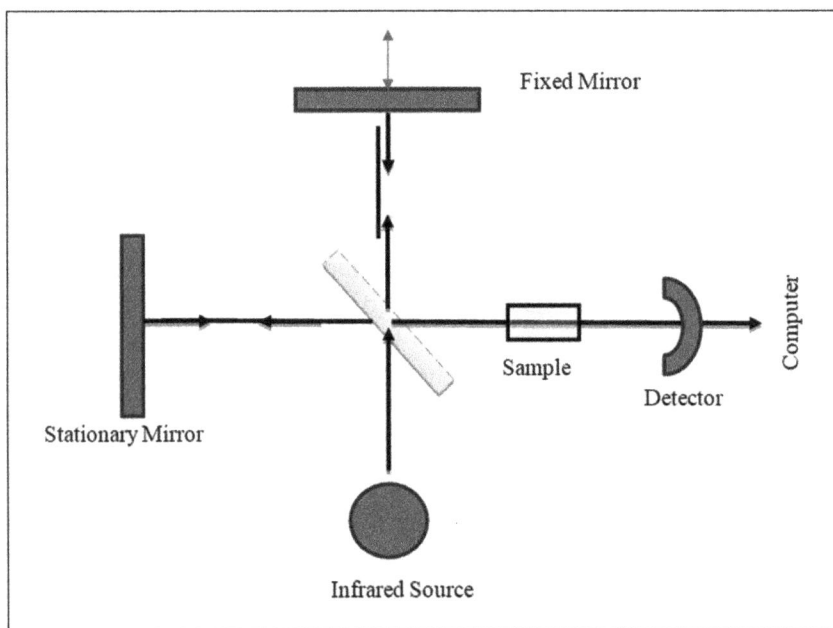

FIGURE 10.4 Schematic of the Michelson interferometer.

10.3.2 SAMPLING TECHNIQUES

FTIR utilizes the whole source range, in which all wavelengths are recorded instantaneously as opposed to covering the individual wavelengths produced by grating as well as prism system utilized in dispersive spectroscopy and numerous scans can be averaged in the same time as a single spectrum is taken via dispersive spectrometer [74]. As another option, the most seasoned and most clear transmission strategies, reflectance procedures have been favored for tests having complex structures, such as food matrices. In reflectance, the beam reflected back from the sample is estimated and as indicated by the kind of the reflectance, either from cell in contact with sample or straightforwardly from the sample, can be classified into two groups: internal or external reflectance measurements [74, 75].

10.3.2.1   DIFFUSE REFLECTANCE

In external reflectance, the radiation reflected from a sample surface is estimated. In the diffuse reflectance infrared technique (DRS), the infrared light is directed on a sample cup, including sample mixed with KBr, and the reflected light is collected by a mirror and is estimated by the detector. Pure KBr spectrum serves as the background. In this method, particle size heavily influences the spectral intensity [75]. Jaiswal and co-workers used diffuse reflectance sampling technique for predicting quality in banana [34] and sweetness in mango [38]. The authors acquired spectra at 299–1100 nm for 95 bananas using a portable NIR spectrometer and applied regression models, partial least square and multilinear regression methods on whole range of spectral data for obtaining information about TSS, dry matter, pH, and acid brix ratio [34, 38].

10.3.2.2   PHOTOACOUSTIC SPECTROSCOPY (PAS)

Photoacoustic spectroscopy (PAS) depends on the exchange of tweaked infrared radiation to mechanical vibration. In this strategy, the sample is set on a sample cup, which is loaded up with helium or air. At the point when the modulated infrared radiation is absorbed by a sample, it warms and cools in radiation. Accordingly, helium assimilates heat and expands, creating pressure waves. The sound of pressure waves is changed over into an electric sign utilizing a microphone. In this, carbon black is utilized as the background.

This system is appropriate for all phases of samples and is a non-invasive technique. PAS is useful, because the detected signal is proportional to the concentration of the sample and can be used with black or highly absorbing samples [72, 75]. FTIR-PAS consists of a portable mirror, whose speed can be changed; and it finds applications for depth profiling in an array of fields. Lv et al. [54] applied FTIR-PAS for the detection of fungicide in rice. Liu et al. [53] coupled PAS and artificial neural network (ANN) methods for non-destructive quantification of residues of dichlorvos on the apple cuticle.

10.3.2.3   ATTENUATED TOTAL REFLECTANCE

Attenuated total reflection (ATR) has been a mainstream technique because of fast strategy. It can analyze samples with a wide range of physical state or morphology. In this method, radiation is not transmitted through the sample; but is totally reflected, when infrared beam comes in contact with a sample. Thus, the thickness of the absorbing samples is not important, as it is in transmission radiation, as long as full surface contact between sample and crystal of sensing element is provided. ATR uses the theory of total internal reflectance, which indicates "when radiation passes through the boundaries of two mediums at an angle greater or equal to the critical angle" [24].

In ATR accessories, the infrared radiations are reflected through optically dense crystal while interacting with the sample, and absorbance information is collected while radiation penetrates only a short distance into the sample. Therefore, ATR is the preferred method for thick and strongly absorbing materials. It is an adaptable method for simple infrared examination and is helpful for testing outside of smooth materials that are either excessively thick or excessively murky for transmission estimations.

ATR is a non-destructive strategy, i.e., practically zero sample preparation is required, and it permits quick and straightforward examination of any kind of food (fluid, gel, powder, etc.). However, generally excellent contact must exist between sample and surface of crystal to ensure the assortment of top-notch spectral data [44]. The detailed applications of use of ATR sampling technique in food are discussed in this chapter.

10.3.2.4   SPECULAR REFLECTANCE

In specular reflectance, two mirrors are used. One mirror directs the light onto the sample and the reflected IR beam is directed by another mirror to the detector. Background is collected against gold reflective surface. It requires a smooth surface of the sample and has been applied efficiently in polymer science [44, 75]. Specular reflection estimations are appropriate for powdered samples, despite the fact that the sample surface should be smooth, requiring the utilization of a surface film in certain applications.

10.4   CHEMOMETRICS

Chemometrics (also referred to as chemoinformatics) is the interdisciplinary approach, which deals with extracting information pertaining to the chemical system by using mathematics (generally applied mathematics), statistics (such as multivariate method) and computer science. The data analyzed enables to select the optimal measurement procedures and experiments and thus, yields maximum appropriate chemical information [27]. It addresses the issues related to chemistry, biology, medicine, biochemistry, chemical engineering, etc.

The rapidly growing technological innovations have enabled the production of data having high dimensionality (HD) in a wide range of knowledge disciplines. HD data has multiple dimensions extending up to hundreds, thousands, or even millions, such as given in the form of absorbance/transmittance in FTIR spectral data in a wavenumber array. Since our brain processes in only three dimensions, thus making it very difficult for us to process and understand the visualization of data involving multiple dimensions. Therefore, it becomes very important to reduce the dimensionality of data or simplify the data. This can be achieved by combining the related data into groups using multidimensional scaling for identifying similarities in the data or clustering the data in-group items [33]. Further, since FTIR gives data at both informative and uninformative regions for a specific analyte; and if the whole region is selected for data analysis, this will not reveal the appropriate and relevant information. Therefore, before analysis of data, the selection of desired wavenumber range, data pre-processing (DP) or data pre-treatment (DPT) is carried out for removing/reducing the undesirable regions/signals/noise from HD data [51].

DP can be divided into unsupervised (applied frequently due to its efficiency and simplicity) and supervised methods. Unsupervised method can further be divided into transforming data column-wise or row-wise. In the former (i.e., variables column-wise), centering, and scaling are applied for transforming the dataset. However, even minor changes in dataset composition leads to changes/alterations in the transformed data matrix.

On the other hand, transformations on individual spectrum are carried out separately during row-wise operation method. Thus, the transformed data is not affected by changes in dataset composition. The example of reference dependent techniques (supervised data) is orthogonal signal correction (OSC) and involves complex mathematical calculations [9, 76]. Discrete wavelet transforms and wavelet packet transform algorithms can also be applied on the dataset, but are different from other DP method. The details of DP are mentioned in Table 10.1. Some definitions and overview related to each chemometric tool for its application in IR spectra are given below:

1. **Standard Normal Variate (SNV):** It is one of the normalization methods and is applied for correcting the differences in global intensity of signals which occur due to scattering of light, variations due to penetration of radiation, or due to size of sample. SNV is a normal variate having mean, i.e., $\mu = 0$ and standard deviation (σ) = 1. The interaction of IR radiations with sample particle often results in a shift of absorbance levels as scattering of light causes variations in path length which in turn lead to variations in the background signal levels with wavelength. This consequently produces baseline shift and curvature. This may vary within and/or among samples and causes difficulty in interpretation of spectra and also linear calibration of NIR diffuse reflectance spectra. Thus, Barnes et al. [7] introduced SNV transformation for reducing multiplicative effects caused due to scattering and particle size; and also, for reducing differences in global intensities of signals. This further improves PLS model prediction when applied on data of NIR with scattering effect. However, it may sometimes lead to the introduction of artifacts as the assumption of SNV is that multiplicative effects are uniform throughout the spectral range (which is not true always). In this, each spectrum is first centered followed by scaling via dividing by its σ [86].

2. **Multiplicative Scatter Correction (MSC):** It is a method of normalization. It focuses on issues related to particle size and light scattering. It is used for correcting both additive (like path length differences) and multiplicative (like particle size) scattering effects as these scatterings are not due to the chemical make-up of the sample, but due to the measurement geometry and morphology of the sample. Before carrying out MSC, two assumptions are made: firstly, the sample spectrum is a result of addition of two spectra, i.e., due to the absorption by chemical bonds (which we want to retain) and, due to diffusion/scattering of light (which we want to reduce by employing MSC); secondly, all the samples show same coefficient of diffusion spectrum at all the wavelengths which can be modeled via least-squares over the range of wavelength using reference spectrum which is free from the effect arising from

chemical bonds. The presence of reference spectrum is a must before going for MSC, which is not the case with SNV. Thus, from the discussion presented in points (a) and (b) above, it is clear that the shape of each spectrum could be due to three reasons: (a) absorption of different wavelengths of IR radiations by the sample, leading to a specific spectrum which is due to the chemical nature of sample and is of our interest; (b) difference in particle size of sample which causes radiations to deviate at different angle as per the wavelength (i.e., scattering effect); (c) difference in path length arising due to variation in position of sample and/or irregularities in surface of sample which leads to variations in spectra of same sample taken over and over again. To minimize (1) and (2), SNV and MSC are generally employed. In the presence of outliers, it is better to opt for SNV [29].

3. **Principal Component Analysis (PCA):** It is a descriptive statistical tool and adaptive exploratory method which is employed for reducing the dimensionality of the HD dataset, but also preserves the variability (which is reservoir of statistical information) as much as possible.

TABLE 10.1 Steps Involved in Spectral Data Analysis of FTIR via Chemometrics

SL. No.	ATR-FTIR Spectrum	Algorithm to be Applied
1.	**Data Acquisition:** It refers to obtaining organic profile of the sample in the form of absorbance or transmittance over wavenumber array (variables)	
2.	**Data Pre-Processing:** It is carried out so as to reduce/minimize/eliminate the following flaws and have the data which is suitable for modeling with respect to best fit. Conditions and obtaining optimum performance in subsequent steps	
2.1	Variations due to flaws/ limitations inherent to the instrument (e.g., Low signal intensity/scattering)	Scatter correction (corrects multiplicative effects like scaling and background effects): → Techniques includes: Standard Normal Variate (SNV), Multiplicative scatter correction (MSC)
2.2	Variations due to different means or SD	Simple algorithm of centering (mean centering and scaling (autoscaling)
		Benefits of these: ease the PCA calculation and interpretability of scores plots
2.3	Elevated baseline	Baseline corrections and derivative (or derivatization of data-first order, second order)

TABLE 10.1   *(Continued)*

SL. No.	ATR-FTIR Spectrum	Algorithm to be Applied	
2.3.1	Baseline lifted at constant value: → Offset; Generally, occurs in IR spectrum acquired using KBr disc (i.e., transmission sampling mode)	Purpose of baseline correction: reduces irrelevant variation due to elevated baseline. Purpose of derivatives: First-order: Removing offset baseline. Second-order: Removing offset baseline and overlapped peaks.	
2.3.2	Baseline lifted at inconsistent value: • Slope: Occurs in ATR-FTIR spectrum since as wavenumber decreases reflectance (concept behind ATR) also decreases • Curvature: It is similar to slope, but baseline resembles curve shape		
3.	Variable reduction: This means reducing the dimensionality of data		
3.1	Variable construction (VC): Involves re-constructing new variables by taking into account all raw wavenumbers)	Principal Component Analysis (PCA, Unsupervised): most preferred/chosen data exploratory technique Partial least square (PLS, supervised)	Classic variable construction methods
		Unsupervised: iPCA, sPCA, siPCA Supervised: sPLS	New variable construction method
3.2	Variable selection (VS): Involves sampling certain/ fixed number of raw wavenumber as input variable based on the pre-defined criteria. These are mostly based from machine learning.	Spectral window (semi-supervised); Genetic algorithm, successive projection algorithm (supervised).	

TABLE 10.1 *(Continued)*

SL. No.	ATR-FTIR Spectrum	Algorithm to be Applied	
4.	Modeling: Divided based on: • Goal of analysis; and • Nature of output variables required.	Classical Classification algorithm (categorical algorithm): e.g., Clustering algorithm (unsupervised); Linear discriminant analysis (LDA, supervised): dominant majority of IR-based modeling problem PCA and PLS coupled with LDA are used for constructing classification model.	
		Classical regression algorithm (numerical output), e.g., partial least square (PLS, classical)	
		Other algorithm emerging from other fields like pattern recognition or machine learning, e.g., classification and regression tree (CART) self-organizing maps (SOM)	Classification, supervised, non-linear algorithm
5.	Evaluation or validation of Model: Done to ascertain the risk of over-fitting (based on availability of testing set)	Internal validation • Cross-validation (CV) • Re-sampling: e.g., bootstrapping	
		External validation: • Sampling algorithm. • Random algorithm.	

PCA is the basis of multivariate data analysis. In this, new variables are constructed, called principal components (PC), which are the linear functions of the original dataset. These PCs successively maximize variance, but otherwise are uncorrelated with one another [42], which means that the first PC shows highest variance amongst all the linear combinations [20]. PCA also plays an important role in linear regression and simultaneous clustering of individuals and variables both (in the form of biplots). There are several adaptations of PCA that have been used in data analysis like incremental PCA (IPCA), sparse PCA (sPCA), etc. IPCA is applied when the dataset is large enough to fit in memory, and thus, low-rank approximation is built-in IPCA for input data [30]. sPCA is PCA-based model in which we induce sparsity on model parameters, i.e., scores, and/or loadings. In sPCA, non-correlation of PC and orthogonality of loading vectors is lost [68]. The advantages, disadvantages, and assumption of sPCA and PCA may be referred in Camacho et al. [12].

4. **Partial Least Square (PLS):** It is a statistical projection method which combines the features of PCA and multiple regressions. It is applied when a set of dependent variables is to be predicted from a large set of independent variables by finding out regression coefficient in linear model having large number of highly correlated x-variables. This algorithm utilizes the information from both matrices, i.e., spectroscopic data (X) and concentration (Y) during calibration and compresses data so that most of the variances due to X and Y are taken care of. As a result of this, the potential impact of variations in X while calibrating is reduced [20].

5. **Successive Projection Algorithm (SPA):** It is a forward variable selection strategy proposed by Araújo et al. [3] which means that one wavelength is selected first followed by incorporation of a new wavelength at fixed iteration, until a definite number of wavelengths, say N, are selected/reached. This is done for minimizing co-linearity problems (which means the case where some independent variables are highly correlated) in MLR (multiple linear regression, which uses explanatory variables for predicting outcome of response variable by explaining the relationship between a single continuous dependent variable and >1 independent variables which may be continuous or discrete) as wavelengths containing minimally redundant information are selected. SPA-MLR is reported to show better prediction ability than PLS/PCR models in several applications of NIR spectroscopy.

6. **Linear Discriminant Analysis (LDA):** It is supervised algorithm model for obtaining maximum class separability of data by reducing the dimensionality of data and classification of HD data into low-dimensional space [85].

7. **Self-Organizing Maps (SOM):** Teuvo Kohonen developed Kohonen maps, also called SOM [48, 49]. It is a non-linear method of representing HD data in 2-D plot (map), just like score plot of PCA. Since PCA is linear projection method, so when it fails then SOM are advantageous. The main purpose of SOM (which has rectangular, generally quadratic, array of fields varying from 5×5 to 100×100, making 25 to 10,000 fields, respectively, similar to chess-board) is cluster analysis (i.e., exploratory data analysis). In this map, the similar objects are assigned in neighboring squares.

8. **Cross-Validation (CV) and Bootstrap:** Both of these are internal validation technique and are used for splitting the data set into calibration sets and test sets. Calibration or training set is used for

creation and optimization of model. The test (or prediction) set is then tested on the model created by calibration set so as to have a realist estimation of performance via prediction. Resampling strategies is the term used for method of CV or bootstrap which are necessary for small data sets. In bootstraps, resampling with replacements is done and is considered to be better than a single CV.

9. **Sampling Algorithm and Random Algorithm:** Both of these are external validation methods and differ from internal validation technique as the latter only considers sampling variability, while the former addresses variations due to population or novel dataset [43]. In our case, suppose we first acquire spectra, apply required chemometrics techniques as per our requirement and finally establish an internal validation model by running the defined known samples (say, of milk) for the presence of some analyte (say, some antibiotic). For testing the efficiency of this developed model, we can procure milk samples from various manufacturers (making these independent set of samples) and run the algorithm so as to know whether the model developed can work in field space or not.

Root mean square error of calibration and root mean square error of prediction (RMSEP), residuals, and calibration variable factor selection are very important to be considered and have relevance in the NIR method with respect to accuracy and precision [10]. The applications of some of the chemometrics techniques are also given in the review by Granato et al. [23].

10.5 APPLICATIONS OF FTIR IN THE EVALUATION OF FOOD QUALITY AND SAFETY

Food is a nutritious solid or liquid commodity, which when ingested, is digested and assimilated for keeping the human body healthy and fit. The components of food can be grouped into proteins, carbohydrates, fat, vitamins, minerals, and water-based on not only the function they perform in the body, but also on the basis of the presence of a majority of chemical bonds (except in case of minerals) being present in these classes, such as [32]:

- **Alcohol:** O–H; C–O;
- **Amine:** N-H; C-N; N-H;
- **Aromatic:** C-H; C=C;
- **Carbohydrates:** C–H; C–O;

- **Carbonyl:** C=O;
- **Ether:** C–O;
- **Fats:** C–H; C=O; C=C–H;
- **Proteins:** Amide I, C=O; Amide II, N–H, C–N;
- **Water:** O–H; H–O–H.

It is very much clear from the discussions held till now that FTIR works on the vibrational transitions taking place in molecule, which are specific for a specific molecule. Thus, these vibrational transitions occur at specific wavenumber corresponding to a specific molecule/bond. Since food is a complex commodity having several macro and micro-molecules which have the potential to absorb IR radiations, therefore the spectra obtained for food is more complicated to understand. Therefore, the application of chemometrics becomes further useful in food matrices. However, it is very much interesting to note here that not any two different foods show an overlapping spectrum, making FTIR a fingerprint technique.

FTIR can be employed in food systems to know the composition of the food and for studying the characteristics of the macronutrients present in it [71]. FTIR coupled with chemometrics find several applications in food science and allied disciplines, like food microbiology, effect of processing, monitoring of oxidation processes in oil, etc., for gathering qualitative and/or quantitative information about it. It has been widely used for studying the quality and safety aspects of foods and the same are covered in this chapter.

Quality is generally defined as the degree of excellence. Quality of food refers to series of several attributes (like, flavor, texture, color, nutritive value of food, fat content, etc.), related to product that collectively influence the judgment of consumers towards buying the food. On the other hand, food safety focuses on the prevention of the presence of any chemical or biological toxicants/potential toxins and overdose of food additives in the food product [26]. The details of aim and outcomes of studies carried out on food quality and safety are mentioned in Tables 10.2 and 10.3, respectively. Few of these are described below for the food groups:

1. **Dairy Industry:** It is booming around the globe and FTIR has been widely used for the determination of milk composition. Besides this, it is also widely used for establishing the quality and safety of milk. Coitinho et al. [15] used FTIR for monitoring the adulterants like cornstarch, sodium bicarbonate, formaldehyde, sodium citrate, and saccharose, and addition of water or whey in milk. The researchers performed the analysis with the FTIR MilkoScan FT1 instrument as

it scans the whole of the middle infrared region having wavelengths ranging from 2.0 and 10.8 μm (5012–926 cm^{-1}). Leite et al. [52] used ATR-FTIR and PCA model for detection of soybean oil, which is now being used as an adulterant for replacing butter oil in the preparation of butter cheese. The authors reported that band at 3007 cm^{-1} evinced as the concentration of soybean oil increased for substituting content of butter oil.

2. **Meat and Fish Industry:** This industry serves as a source of some specific nutrients to population like omega-3 fatty acids of fish origin, but at the same time could also contain several hazardous chemicals in the form of antibiotics in meat or pollutants present in water that may come in fish. The studies related to these sectors are important, and considering the perishability of these commodities, FTIR, and chemometrics offer a stout solution on account of being quick and reliable techniques. Thus, researchers have assessed the potential of FTIR for authentication of meat and fish. Rohman et al. [64] used FTIR and partial least square calibration for the detection and quantification of pork in beef meatball formulation. The authors developed calibration model using PLS regression at the fingerprint region (1200–1000 cm−1). Pink et al. [58] studied frozen, minced hake samples using FT-MIR with an ATR cell together with FT-NIR (in transmission mode). PLS regression was applied for spectral regions of 1600–890 cm^{-1} range for FT-MIR and 1530–1866 nm range for FT-NIR for prediction of dimethylamine. Based on the results obtained with a narrow spectral range, the authors also suggested that a cheap filter instrument could be developed for its applications in industry. Papadopoulou et al. [56] applied the PLS model on the obtained FTIR data of minced pork meat with the aim of detecting spoilage in the minced pork meat stored under aerobic conditions. The authors reported that a good correlation existed between FTIR and microbial data in minced meat and concluded that FTIR coupled with chemometrics can be exploited for the quick detection of spoilage in meat.

3. **Oils and Fats:** These are of great concern because economic fraud is increasing in oil and fats and related industries by blending cheap/synthetic/residual oils in high quality oils with the aim of making profit. Upadhyay et al. [79, 80] used ATR-FTIR for detection of adulteration of pig body fat and goat body fat in ghee. PLS regression was applied for spectral window 1786–1680, 1490–919 and 1260–1040

cm^{-1} for developing a prediction model for the presence of goat body fat in ghee and 3030–2785, 1786–1680 and 1490–919 cm^{-1} spectral window for detection of pig body fat in ghee. The authors revealed that by this method, adulteration of pig and goat body fat in ghee could be detected at a level of 3% and 1%, respectively. Jiang et al. [41] developed a simple FTIR based method for detection of acid values (AVs) of edible oils using O-H stretching (3535 cm^{-1}) band of carboxyl group in free fatty acids along with peak valley (3508 cm^{-1}) and spectral data from 3340–3390 cm^{-1}.

4. **Cereals and Grains:** Moisture meters based on IR spectroscopy are commonly used for detection of moisture content in grains. Amir et al. [2] used FTIR for analyzing different quality parameters in wheat, i.e., the moisture, fat, protein, carbohydrates, ash, and hardness of grain on the basis of peaks obtained. The water peaks were observed in the range 1,640 cm^{-1} and 3,300 cm^{-1} on the basis of 'H and OH' functional group. Protein was observed based on amide I and amide II bands falling in the spectral range from 1,600 cm^{-1} to 1,700 cm^{-1} and 1,550 cm^{-1} to 1,570 cm^{-1}, respectively, while fat was perceived based on C-H bond. The authors reported that starch was perceived in spectral range from 2,800 and 3,000 cm^{-1}; and 3,000 and 3,600 cm^{-1} based on C-H and O-H stretch regions, respectively.

5. **Fruits and Vegetables:** Adulteration of fruits with sugar and other low-quality substances are common these days. Kelly and Downey [46] employed FT-MIR for detection of adulterants like sugars, beet sucrose (BS), partially inverted cane syrup (PICS), synthetic solutions of fructose: glucose: sucrose (FGS), high fructose corn syrup (HFCS) in single strength apple juice. FTIR has also been used as a non-destructive method for the prediction of sweetness in mango. Bureau et al. [11] developed a simple, rapid FTIR method for detection of both sugars and organic acids in apricot fruits. PLS was applied in the spectral region of 1500–900 cm^{-1} and good prediction performances were obtained.

6. **Wine:** Winery is a well-established field of food technology. The application of FTIR in wine is mainly restricted to large wineries. Patz et al. [57] used FTIR for analyzing compositional and physical parameters of wine such as relative density, alcohol, extract, sugar-free extract, conductivity, refraction, glycerol, total phenols, glucose, fructose, reducing sugars, sucrose, total acidity, pH, individual organic acids, and total SO$_2$.

TABLE 10.2 Evaluation of Quality Parameters of Various Food Commodities Using FTIR Coupled with Chemometrics

Food	Quality Parameter	Sampling Technique	Wavenumber (cm⁻¹)	Multivariate Model Used	Outcomes	References
Beef	Detection of chicken in beef meat	ATR and transmission FTIR	4000–500	PLS ANN	• When PCA was applied to the raw spectra of biological samples, the transmission FTIR technique (KBr plate) yielded better results than ATR-FTIR	[47]
Beef meat	Rat meat • Admixture of rat meat to beef meat is done in Indonesia for the formulation of meatballs for economic gain	ATR	4000–400	PLS, PCA	• Application of PCA and PLS at wavenumber 750–1000 cm⁻¹ was effectively used for quantification of rat meat in meatball formulation	[60]
Butter	Detection of Fat and Moisture content	ATR	3000–1000	MLR	• FTIR can be used for online monitoring of process • The technique required 20 s for analyzing samples	[81]
Cod liver oil	Canola, soybean, corn, and walnut • Admixture of above oils in cod liver as is difficult to differentiate by naked eyes	HATR	4000–400	PLS, PCR	• FTIR data coupled with PLS can be used for quantification of beef fat in cod liver oil in spectral range from 1200 to 1000 cm⁻¹	[65]
Cow-buffalo milk	Detection of admixture of soymilk in milk	ATR	4000–500	PCA, PLS	• FTIR with amalgamation of multivariate analysis demonstrated ability as a rapid method for monitoring the quality of milk	[35]

TABLE 10.2   *(Continued)*

Food	Quality Parameter	Sampling Technique	Wavenumber (cm⁻¹)	Multivariate Model Used	Outcomes	References
Ghee	Detection of pig body fat	ATR	4000–500	PCA, PLS	• Different absorption values obtained for pure ghee, pig body fat and pig body fat spiked in pure ghee which revealed that ATR-FTIR can be successfully used for detection of pig body fat in pure ghee	[79]
Ghee	Detection of goat body fat	ATR	4000–500	PCA, PLS	• The differences in the absorption at selected spectral regions of pure ghee, goat body fat and pure ghee containing goat body fat revealed that FTIR in combination with PLS can be used for detecting adulterants in ghee	[80]
Honey	Detection of sugar syrup in honey	ATR	4000–650	PCA, PLS	• FTIR spectra alone has great potential for determining purity/adulteration in honey, but when coupled with chemometrics, the results are more definite	[62]
Mango	Prediction of sweetness of intact mango	Diffuse reflectance	1200–2200	PLS	• Wavelength in the range of 1200–2200 nm could predict TSS and pH of mango using PLS and MLR regression model	[38]

TABLE 10.2 *(Continued)*

Food	Quality Parameter	Sampling Technique	Wavenumber (cm⁻¹)	Multivariate Model Used	Outcomes	References
Mango	Determination of maturity	Diffuse reflectance	1200–2200	PLS	• NIR in combination with PLS could be used as a rapid method for prediction of maturity index in mango at 1200–2000 nm spectra range	[39]
Meat	Non-meat ingredients • Injecting solutions of maltodextrin salt, carrageenan, phosphates, for increasing water holding capacity of meat	ATR	4000–525	PLS-DA	• Adulterated meat samples showed characteristics IR bands at specific wavenumbers. • Highest VIP scores was obtained at 1690 cm⁻¹ associated with adulteration using NaCl in meat which causes aggregation of β-sheets vibrations of proteins	[55]
Milk	Detection of corn starch, sodium bicarbonate, formaldehyde, saccharose, and sodium citrate	ATR	5000–900	PLS, PCA	• FTIR is an ideal substitute for detecting adulterations, as the equipment has been in use several industries for determining composition of milk	[15]
Olive oil	Detection of palm oil adulteration in extra virgin olive oil	ATR	4000–650	PLS, PCR	• PLS and PCR calibration models at spectral region 1500–1000 cm⁻¹ could successfully be used for quantification of palm oil in extra virgin olive oil	[66]

TABLE 10.2 *(Continued)*

Food	Quality Parameter	Sampling Technique	Wavenumber (cm⁻¹)	Multivariate Model Used	Outcomes	References
Olive oil	Detection of corn and sunflower oil in olive oil	ATR	4000–650	PLS, PCR	• Application of multivariate analysis to the spectral data can quantify the adulterants like corn oil and sunflower oil in extra virgin olive oil. • FTIR can be used as green analytical tool for rapid detection of adulterants in oil	[63]
Rapeseed oil	Adulterated with purification wastage cooking oil	ATR	4000–450	PLS, PCR	• FTIR coupled with PLS could distinguish pure rapeseed oil and rapeseed oil adulterated with wastage cooking oil	[83]
Sage	Detection of olive leaves, sumac, myrtle leaves, hazelnut leaves cistus and phlomis, sandalwood, and strawberry tree leaves	ATR	4000–400	PCA, PLS-DA	• FTIR in combination with multivariate analysis showed feasibility in detection of the admixture of olive leaves, sumac, myrtle leaves, hazelnut leaves cistus and phlomis, sandalwood, and strawberry tree leaves in sage	[21]
Wheat flour	Detection of barley flour in wheat flour	ATR	4000–450	PLS	• FTIR spectroscopy and multivariate analysis were feasible techniques for detection of barley flour adulteration in wheat flour	[4]
Wine	Prediction of total antioxidant capacity	ATR	4400–600	PLS PCA	• FTIR in combination with PLS provided rapid non-destructive technique for detection of total antioxidant capacity of red wines	[82]

TABLE 10.3 Detection of Toxic/Potentially Toxic Analytes in Food Commodities Using FTIR Coupled with Chemometrics

Food	Toxic/Potentially Toxic Analyte	Sampling Technique	Wavenumber cm⁻¹	Multivariate Model Used	Outcome	References
Cow milk	Formalin • Formalin is a throat and eye irritant and toxic when consumed	ATR	4000–400	PLS PCR	• ATR-FTIR method can detect 0.5% level of formalin in cow milk	[6]
Freshwater	Pyrethroid and organic chlorine • These cause feelings of numbness, burning, itching, tingling, stinging, or warmth that could last for a few hours.	ATR	4000–400	PLS	• ATR-FTIR could be used to classify and quantify pesticides in freshwater samples	[16]
Milk	Aflatoxin M1 • Aflatoxin M1 is a hydroxylated metabolite of aflatoxin B1, the presence of aflatoxin above MRL level in food causes hepato-toxicity in human	ATR	4000–500	PCA, PLS	• The variations in spectral regions appeared in the spectral window of 423–1123 and 3550–3499 cm⁻¹ between pure and spiked milk samples • Developed method could detect aflatoxin M1 in milk even at 0.02 µg/l	[36]
Milk	Detection of urea • Urea is added to milk for increasing non-protein nitrogen • The overdose of urea causes indigestion, diarrhea, acidity, malfunctioning of kidneys, ulcers, and impaired vision; and damage to the intestinal tract and digestive system	ATR	4000–700	PCA, PLS	• The well-defined differences appeared in 1670–1564 cm⁻¹ spectral region among pure and urea spiked samples of milk • ATR-FTIR can detect urea even at 100 ppm in milk	[37]

TABLE 10.3　*(Continued)*

Food	Toxic/Potentially Toxic Analyte	Sampling Technique	Wavenumber cm^{-1}	Multivariate Model Used	Outcome	References
Peanut	To discriminate moldy peanut with reference to aflatoxin • Aflatoxin B1 is a well-known group 1 carcinogen and causes hepatotoxicity	ATR	4000–625	PLS	• ATR-FTIR in combination with multivariate analysis has the potential to differentiate between moldy peanut and uninfected peanut • The limit of detection for the developed rapid method is 20 ppb	[45]
Poultry Egg Yolk	Fluoroquinolone • Ciprofloxacin (CIP) and norfloxacin (NOR) are used as a growth promoter and against bacterial infections, the residues cause allergic reactions and antibiotic resistance	DRS	4000–400	NA	• The FTIR spectra revealed the characteristic absorption peak for detection of CIP and NOR at 1627 and 1026 cm^{-1}, respectively • The limit of detection for CIP and NOR were 0.032 and 0.028 ng/mL, respectively	[50]
Rice	Tricyclazole • Tricyclazole is effective against the *Magnaporthe oryzae* (a fungus) associated with the blast disease. • The overuse of pesticide causes cancers, birth defects, neurological, and developmental toxicity, immunotoxicity, and disruption of the endocrine system.	PAS	4000–500	PCA	• The FTIR-PAS spectra revealed strong absorption at 1200 cm^{-1} wavenumber associated to C-N bond of tricyclazole.	[54]

TABLE 10.3 (*Continued*)

Food	Toxic/Potentially Toxic Analyte	Sampling Technique	Wavenumber cm⁻¹	Multivariate Model Used	Outcome	References
Turmeric	Metanil Yellow • Joint FAO/WHO Expert Committee on food additives classified this substance in CII category • Neurotoxic in nature	ATR FT-Raman	4000–650 3700–100	NA	• The characteristic peak for metanil yellow in ATR and FT-Raman were at 1140 and 1406 cm⁻¹, respectively • ATR can detect metanil yellow at concentration of 5% whereas FT-Raman can detect 1% of metanil yellow in turmeric	[17]
Vegetables	Flonicamid • The over usages of pesticide causes skin rashes, eye damage, cancer, vomiting, and diarrhea	DRS	4000–400	NA	• The limit of detection of flonicamid in vegetables using DRS was 0.007 µg mL⁻¹	[67]
Vine fruit	Ochratoxin A • Produced by *Aspergillus ochraceus* and toxic to liver	ATR	4000–600	PCA PLS	• ATR-FTIR with amalgamation of multivariate analysis can be used to detect and classify Ochratoxin A in dry sultanas	[22]

It is evident that both quantitative and qualitative information about the food sample can be obtained from the mid-infrared and near infrared region. FTIR has also been used for the detection of food fraud and adulterants in several foods because of its rapidity and non-destructive nature. Chemometrics is applied to FTIR data for obtaining the useful information. It has also showed up as a promising instrument and is applicable in juices, dairy, agricultural produce, oils, and fats. The number of reports on the utilization of FTIR and chemometric tools to food investigation has expanded significantly over the last two decades. Some workers have employed both MIR and NIR in the same food material for drawing the concrete inference regarding the advantages or disadvantage of one technique over another (Table 10.4).

TABLE 10.4   Comparative Analysis Of various Analytes/Food Commodities by Coupling Different Analytic Techniques (like NIR and MIR) with Chemometrics

Analyte	Technique Used	Statistical Tool	Outcome	References
Beans • Prediction of Phytochemical composition	NIR and MIR	PLS	• The performance of NIR spectroscopy was reported to be better than MIR for quantification of phytochemical composition, *in vitro* antioxidant activity and individual phenolic compounds	[13]
Bread • Staling of wheat bread	2D MIR-NIR	PCA	• NIR spectroscopy is equally efficient in detection of bread staling as MIR	[61]
Coffee • Authentication of coffee	NIR and MIR	Factorial discriminant analysis and PLS	• Both NIR and MIR spectroscopy have the potential to differentiate between *Arabica* and *Robusta* coffee.	[19]
Meczal • Quantification of adulterants like water, methanol, and ethanol in meczal	NIR and MIR	PLS	• Could detect adulterants at 5% • PLS2 showed the best predictive results for FT-MIR (R^2 = 0.9579–0.9895); while PLS1 for FT-NIR (R^2 = 0.9401–0.9665)	[59]

TABLE 10.4 *(Continued)*

Analyte	Technique Used	Statistical Tool	Outcome	References
Melamine • Melamine detection in liquid milk, milk powder and infant formula	MIR and NIR	PLS, ANN, SVR, LS-SVM, Poly-PLS	• Mid and NIR spectroscopy are effective tool for detection of melamine in dairy products • The LOD below 1 ppm can be achieved using precise algorithm	[5]
Vitamin C	NIR, FT-NIR, FTIR-ATR, diffuse reflectance (DRIFTS), FTIR-PAS, and FT-Raman	PLS	• The methods used have high predictive power, with overall prediction error of 0.2–3.0%. The time required for completing experiment ranged from 5 s to 3 min in NIR and FT-Raman, respectively • FTIR-ATR gave best prediction rate among all.	[84]
Wine • Monitor red wine fermentation	NIR and MIR	LDA	• NIR and MIR quantify the phenolic compounds in fermenting wines • NIR and MIR can be utilized to classify wines belonging to different fermentation steps	[18]

10.6 FUTURE TRENDS

FTIR is fingerprint technology as two different samples are reported to show different spectra. FTIR coupled with chemometrics helps in the extraction of useful information about the sample, making it a very powerful analytical technique. The future work pertaining to a combination of FTIR and chemometrics could be the development of comprehensive, accurate, and reliable models for each commodity of food so as to address the menace related to substandard or adulterated foods being sold to consumers. For this, external validation of the models is a must, i.e., to move from local to global model. In addition, both NIR and MIR can be compared for a range of food commodities for finding suitability of the same for drawing a model, which can directly correlate the spectra with the quality parameters of the commodity-like its proximate composition and several other functional properties.

10.7 SUMMARY

FTIR spectroscopy is a versatile tool for the identification of functional groups, the study of macronutrients as well as detection of various adulterants and contaminants in foods. It is less tedious and an effective technique that has the potential of eradicating the numerous problems of industries by coupling it with chemometrics. However, it is indeed a challenging task to choose the right kind of statistical tool as each method is aimed at addressing a specific issue pertaining to the spectra obtained via FTIR technique. This chapter has, however, tried to step-wise mention and discusses the sequence of chemometric methods that are required to be used and also the basic nature of each method and problem it targets to solve. Thus, if the best choice is made with respect to the kind of chemometric tool which is to be applied to the spectra, then this may resolve several confronting issues pertaining to the food industry and the researchers like identification of their raw samples with respect to quality and safety as FTIR coupled with chemometrics is a decently intelligent choice.

ACKNOWLEDGMENTS

The first author acknowledges the first-hand experience received at Dr. S.N. Jha's (presently, Assistant Director General-Process Engineering, Indian Council of Agricultural Research, New Delhi, India) laboratory at ICAR-CIPHET, Ludhiana, Punjab, India during Professional Attachment Training; and the book chapter written by Dr. Jha on NIR that resulted in inculcating interest in FTIR and learning the robustness of this technique, which has reaped the fruit in the form of this chapter.

KEYWORDS

- attenuated total reflection
- chemometrics
- food quality
- food safety
- Fourier transform infrared spectroscopy
- vibrational transitions

REFERENCES

1. Allen, H. C. Jr., & Cross, P. C., (1963). *Molecular Vib-rotors: The Theory and Interpretation of High Resolution Infrared Spectra* (p. 324). New York: John Wiley & Sons.
2. Amir, R. M., Anjum, F. M., Khan, M. I., Khan, M. R., Pasha, I., & Nadeem, M., (2013). Application of Fourier transform infrared (FTIR) spectroscopy for the identification of wheat varieties. *Journal of Food Science and Technology, 50*(5), 1018–1023.
3. Araújo, M. C. U., Saldanha, T. C. B., Galvao, R. K. H., Yoneyama, T., Chame, H. C., & Visani, V., (2001). The successive projections algorithm for variable selection in spectroscopic multicomponent analysis. *Chemometrics and Intelligent Laboratory Systems, 57*(2), 65–73.
4. Arslan, F. N., Akin, G., Elmas, Ş. N. K., Üner, B., Yilmaz, I., Janssen, H. G., & Kenar, A., (2020). FT-IR spectroscopy with chemometrics for rapid detection of wheat flour adulteration with barley flour. *Journal of Consumer Protection and Food Safety, 15*, 245–261.
5. Balabin, R. M., & Smirnov, S. V., (2011). Melamine detection by mid-and near-infrared (MIR/NIR) spectroscopy: A quick and sensitive method for dairy products analysis including liquid milk, infant formula, and milk powder. *Talanta, 85*(1), 562–568.
6. Balan, B., Dhaulaniya, A. S., Jamwal, R., Sodhi, K. K., Kelly, S., Cannavan, A., & Singh, D. K., (2020). Application of attenuated total reflectance-Fourier transform infrared (ATR-FTIR) spectroscopy coupled with chemometrics for detection and quantification of formalin in cow milk. *Vibrational Spectroscopy, 107*(103033), 1–7.
7. Barnes, R. J., Dhanoa, M. S., & Lister, S. J., (1993). Correction of the description of standard normal variate (SNV) and de-trend transformations in practical spectroscopy with applications in food and beverage analysis. *Journal of Near Infrared Spectroscopy, 1*, 185–186.
8. Barrow, G. M., (1962). *Introduction to Molecular Spectroscopy* (p. 318). New York: McGraw-Hill.
9. Blanco, M., Coello, J., Montoliu, I., & Romero, M. A., (2001). Orthogonal signal correction in near-infrared calibration. *Analytica Chimica Acta, 434*(1), 125–132.
10. Broad, N., Graham, P., Hailey, P., Hardy, A., Holland, S., Hughes, S., Lee, D., Prebble, K., Salton, N., & Warren, P., (2002). Guidelines for the development and validation of

near-infrared spectroscopic methods in the pharmaceutical industry. Reproduced from: *Handbook of Vibrational Spectroscopy* (p. 21). Chichester: John Wiley & Sons Ltd. https://www.farm.ucl.ac.be/tpao/portail_stat/cours_stat/des_indu/qualite/documents_qualite/hvs_nirpharma.pdf (accessed on 27 January 2021).

11. Buellens, K., Kirsanov, D., Irudayaraj, J., Rudnitskaya, A., Legin, A., & Nicolaï, B., (2006). The electronic tongue and ATR-FTIR for rapid detection of sugars and acids in tomatoes. *Sensors and Actuators B, 116*, 107–111.

12. Bureau, S., Ruiz, D., Reich, M., Gouble, B., Bertrand, D., Audergon, J. M., & Renard, C. M., (2009). Application of ATR-FTIR for a rapid and simultaneous determination of sugars and organic acids in apricot fruit. *Food Chemistry, 115*(3), 1133–1140.

13. Camacho, J., SMilde, A. K., Saccenti, E., & Westerhuis, J. A., (2019). All sparse PCA models are wrong, but some are useful. *Part I: Computation of Scores, Residuals, and Explained Variance.* https://arxiv.org/pdf/1907.03989.pdf (accessed on 27 January 2021).

14. Carbas, B., Machado, N., Oppolzer, D., Queiroz, M., Brites, C., Rosa, E. A., & Barros, A. I., (2020). Prediction of phytochemical composition, *in vitro* antioxidant activity, and individual phenolic compounds of common beans using MIR and NIR spectroscopy. *Food and Bioprocess Technology, 13*, 962–977. https://doi.org/10.1007/s11947-020-02457-2.

15. Cheng, I. F., Dharavath, S., Mustapha, A., Pal, P., & Kight, M. L., (1983). Fourier transform infrared spectrometry. *Science, 222*(4621), 297–302.

16. Coitinho, T. B., Cassoli, L. D., Cerqueira, P. H. R., Da Silva, H. K., Coitinho, J. B., & Machado, P. F., (2017). Adulteration identification in raw milk using Fourier transform infrared spectroscopy. *Journal of Food Science and Technology, 54*(8), 2394–2402.

17. Colume, A., Diewok, J., & Lendl, B., (2004). Assessment of FTIR spectrometry for pesticide screening of aqueous samples. *International Journal of Environmental Analytical Chemistry, 84*(11), 835–844.

18. Dhakal, S., Chao, K., Schmidt, W., Qin, J., Kim, M., & Chan, D., (2016). Evaluation of adulterated turmeric powder using FT-Raman and FT-IR spectroscopy. *Foods, 5*(2), 36.

19. Di-Egidio, V., Sinelli, N., Giovanelli, G., Moles, A., & Casiraghi, E., (2010). NIR and MIR spectroscopy as rapid methods to monitor red wine fermentation. *European Food Research and Technology, 230*(6), 947–955.

20. Downey, G., Briandet, R., Wilson, R. H., & Kemsley, E. K., (1997). Near-and mid-infrared spectroscopies in food authentication: Coffee varietal identification. *Journal of Agricultural and Food Chemistry, 45*(11), 4357–4361.

21. Espitia, V. C., Romía, M. B., & Bernardez, M. A., (2012). *Use of NIR Spectroscopy and Multivariate Process Spectra Calibration Methodology for Pharmaceutical Solid Samples Analysis.* PhD Thesis; Universitat Autònoma de Barcelona; https://www.recercat.cat/bitstream/handle/2072/218064/TFM_VanessaCardenasEspitia.pdf?sequence=1 (accessed on 27 January 2021).

22. Galvin-King, P., Haughey, S. A., Montgomery, H., & Elliott, C. T., (2019). The rapid detection of sage adulteration using Fourier transform infrared (FTIR) spectroscopy and chemometrics. *Journal of AOAC International, 102*(2), 354–362.

23. Galvis-Sánchez, A. C., Barros, A., & Delgadillo, I., (2007). FTIR-ATR infrared spectroscopy for the detection of ochratoxin-A in dried vine fruit. *Food Additives and Contaminants, 24*(11), 1299–1305.

24. Granato, D., Putnik, P., Kovačević, D. B., Santos, J. S., Calado, V., Rocha, R. S., Cruz, A. G. D., et al., (2018). Trends in chemometrics: Food authentication, microbiology, and effects of processing. *Comprehensive Reviews in Food Science and Food Safety, 17*(3), 663–677.
25. Griffiths, P. R., & De Haseth, J. A., (2007). *Fourier Transform Infrared Spectrometry* (p. 171). New York: John Wiley & Sons Ltd.
26. Hacisalihoglu, G., Larbi, B., & Mark, S. A., (2010). Near-infrared reflectance spectroscopy predicts protein, starch, and seed weight in intact seeds of common bean (*Phaseolus Vulgaris* L.). *Journal of Agricultural and Food Chemistry, 58*(2), 702–706. https://doi.org/ 10.1021/jf9019294.
27. Haghiri, M., (2016). Consumer choice between food safety and food quality: The case of farm-raised Atlantic salmon. *Foods, 5*(2), 22–29.
28. Héberger, K., (2008). Chemoinformatics: Multivariate mathematical-statistical methods for data evaluation: Chapter 7. In: Vékey, K., Telekes, A., & Vertes, A., (eds.), *Medical Applications of Mass Spectrometry* (pp. 141–169). London - UK: Elsevier.
29. https://diffractionlimited.com/help/quickfringe/QUICK_FRINGEWhat_is_an_ interferogram.htm (accessed on 27 January 2021).
30. https://nirpyresearch.com/two-scatter-correction-techniques-nir-spectroscopy-python/ (accessed on 27 January 2021).
31. https://scikit-learn.org/stable/auto_examples/decomposition/plot_incremental_pca. html (accessed on 27 January 2021).
32. https://www.newport.com/n/introduction-to-ftir-spectroscopy (accessed on 27 January 2021).
33. https://www.spectroscopyonline.com/ (accessed on 27 January 2021).
34. https://www.thinkingondata.com/5-basic-questions-and-answers-about-high-dimensional-data/ (accessed on 27 January 2021).
35. Jaiswal, P., Jha, S. N., & Bharadwaj, R., (2012). Non-destructive prediction of quality of intact banana using spectroscopy. *Scientia Horticulturae, 135*, 14–22.
36. Jaiswal, P., Jha, S. N., Borah, A., Gautam, A., Grewal, M. K., & Jindal, G., (2015). Detection and quantification of soymilk in cow-buffalo milk using attenuated total reflectance Fourier transform infrared spectroscopy (ATR-FTIR). *Food Chemistry, 168*, 41–47.
37. Jaiswal, P., Jha, S. N., Kaur, J., Borah, A., & Ramya, H. G., (2018). Detection of aflatoxin M1 in milk using spectroscopy and multivariate analyses. *Food Chemistry, 238*, 209–214.
38. Jha, S. N., Jaiswal, P., Borah, A., Gautam, A. K., & Srivastava, N., (2015). Detection and quantification of urea in milk using attenuated total reflectance-Fourier transform infrared spectroscopy. *Food and Bioprocess Technology, 8*(4), 926–933.
39. Jha, S. N., Jaiswal, P., Narsaiah, K., Gupta, M., Bhardwaj, R., & Singh, A. K., (2012). Non-destructive prediction of sweetness of intact mango using near-infrared spectroscopy. *Scientia Horticulturae, 138*, 171–175.
40. Jha, S. N., Narsaiah, K., Jaiswal, P., Bhardwaj, R., Gupta, M., Kumar, R., & Sharma, R., (2014). Non-destructive prediction of maturity of mango using near-infrared spectroscopy. *Journal of Food Engineering, 124*, 152–157.
41. Jha, S. N., (2010). Near-infrared spectroscopy: Chapter 6. In: Jha, S. N., (ed.), *Non-Destructive Evaluation of Food Quality: Theory and Practical* (pp. 141–212). Berlin, Germany: Springer.

42. Jiang, X., Li, S., Xiang, G., Li, Q., Fan, L., He, L., & Gu, K., (2016). Determination of the acid values of edible oils via FTIR spectroscopy based on the OH stretching band. *Food Chemistry, 212,* 585–589.
43. Jolliffe, I. T., & Cadima, J., (2016). Principal component analysis: A review and recent developments. *Philosophical Transactions of the Royal Society A: Mathematical, Physical and Engineering Sciences, 374*(2065), 20150202.
44. Justice, A. C., Covinsky, K. E., & Berlin, J. A., (1999). Assessing the generalizability of prognostic information. *Annals Internal Medicine, 130*(6), 515–524.
45. Karoui, R., Downey, G., & Blecker, C., (2010). Mid-infrared spectroscopy coupled with chemometrics: A tool for the analysis of intact food systems and the exploration of their molecular structure-quality relationships: A review. *Chemical Reviews, 110*(10), 6144–6168.
46. Kaya-Celiker, H., Mallikarjunan, P. K., Schmale, III. D., & Christie, M. E., (2014). Discrimination of moldy peanuts with reference to aflatoxin using FTIR-ATR system. *Food Control, 44,* 64–71.
47. Kelly, J. D., & Downey, G., (2005). Detection of sugar adulterants in apple juice using Fourier transform infrared spectroscopy and chemometrics. *Journal of Agricultural and Food Chemistry, 53*(9), 3281–3286.
48. Keshavarzi, Z., Banadkoki, S. B., Faizi, M., Zolghadri, Y., & Shirazi, F. H., (2020). Comparison of transmission FTIR and ATR Spectra for discrimination between beef and chicken meat and quantification of chicken in beef meat mixture using ATR-FTIR combined with chemometrics. *Journal of Food Science and Technology, 57*(4), 1430–1438.
49. Kohonen, T., (1995). *Self-Organizing Maps* (1st edn., p. 362). Berlin, Germany: Springer.
50. Kohonen, T., (2001). *Self-Organizing Maps* (3rd edn., p. 371). Berlin, Germany: Springer.
51. Kurrey, R., Mahilang, M., Deb, M. K., Nirmalkar, J., Shrivas, K., Pervez, S., & Rai, J., (2019). A direct DRS-FTIR probe for rapid detection and quantification of fluoroquinolone antibiotics in poultry egg-yolk. *Food Chemistry, 270,* 459–466.
52. Lee, L. C., Liong, C. Y., & Jemain, A. A., (2017). A contemporary review on data preprocessing (DP) practice strategy in ATR-FTIR spectrum. *Chemometrics and Intelligent Laboratory Systems, 163,* 64–75. http://dx.doi.org/10.1016/j.chemolab.2017.02.008.
53. Leite, A. I. N., Pereira, C. G., Andrade, J., Vicentini, N. M., Bell, M. J. V., & Anjos, V., (2019). FTIR-ATR spectroscopy as a tool for the rapid detection of adulterations in butter cheeses. *LWT-Food Science and Technology, 109,* 63–69.
54. Liu, L., Wang, Y., Gao, C., Huan, H., Zhao, B., & Yan, L., (2015). Photoacoustic spectroscopy as a non-destructive tool for quantification of pesticide residue in apple cuticle. *International Journal of Thermophysics, 36*(5–6), 868–872.
55. Lv, G., Du, C., Ma, F., Shen, Y., & Zhou, J., (2018). Rapid and non-destructive detection of pesticide residues by depth-profiling Fourier transform infrared photoacoustic spectroscopy. *ACS Omega, 3*(3), 3548–3553.
56. Nunes, K. M., Andrade, M. V. O., Santos, F. A. M., Lasmar, M. C., & Sena, M. M., (2016). Detection and characterization of frauds in bovine meat in natura by non-meat ingredient additions using data fusion of chemical parameters and ATR-FTIR spectroscopy. *Food Chemistry, 205,* 14–22.

57. Papadopoulou, O., Panagou, E. Z., Tassou, C. C., & Nychas, G. J., (2011). Contribution of Fourier transform infrared (FTIR) spectroscopy data on the quantitative determination of minced pork meat spoilage. *Food Research International, 44*(10), 3264–3271.

58. Patz, C. D., Blieke, A., Ristow, R., & Dietrich, H., (2004). Application of FT-MIR spectrometry in wine analysis. *Analytica Chimica Acta, 513*(1), 81–89.

59. Pink, J., Naczk, M., & Pink, D., (1999). Evaluation of the quality of frozen minced red hake: Use of Fourier transform near-infrared spectroscopy. *Journal of Agricultural and Food Chemistry, 47*(10), 4280–4284.

60. Quintero-Arenas, M. A., Meza-Márquez, O. G., Velázquez-Hernández, J. L., Gallardo-Velázquez, T., & Osorio-Revilla, G., (2020). Quantification of Adulterants in mezcal by means of FT-MIR and FT-NIR spectroscopy coupled to multivariate analysis. *Cy-TA-Journal of Food, 18*(1), 229–239.

61. Rahmania, H., & Rohman, A., (2015). The employment of FTIR spectroscopy in combination with chemometrics for analysis of rat meat in meatball formulation. *Meat Science, 100*, 301–305.

62. Ringsted, T., Siesler, H. W., & Engelsen, S. B., (2017). Monitoring the staling of wheat bread using 2D MIR-NIR correlation spectroscopy. *Journal of Cereal Science, 75*, 92–99.

63. Rios-Corripio, M. A., Rios-Leal, E., Rojas-Lopez, M., & Delgado-Macuil, R., (2011). FTIR characterization of Mexican honey and its adulteration with sugar syrups by using chemometric methods. *Journal of Physics-Conference Series, 274*(1), 8. Article ID: 012098.

64. Rohman, A., Che, M. Y. B., (2012). Quantification and classification of corn and sunflower oils as adulterants in olive oil using chemometrics and FTIR spectra. *The Scientific World Journal*, 12. https://hindawi.com/journals/tswj/2012/250795/ (accessed on 27 January 2021).

65. Rohman, A., Erwanto, Y., Man, Y. B. C., (2011). Analysis of pork adulteration in beef meatball using Fourier transform infrared (FTIR) spectroscopy. *Meat Science, 88*(1), 91–95.

66. Rohman, A., Man, Y. B. C., (2011). Application of Fourier transform infrared (FT-IR) spectroscopy combined with chemometrics for authentication of cod-liver oil. *Vibrational Spectroscopy, 55*(2), 141–145.

67. Rohman, A., & Man, Y. C., (2010). Fourier transform infrared (FTIR) spectroscopy for analysis of extra virgin olive oil adulterated with palm oil. *Food Research International, 43*(3), 886–892.

68. Sahu, D. K., Rai, J., Rai, M. K., Nirmal, M., Wani, K., Banjare, M. K., & Mundeja, P., (2020). Detection of flonicamid insecticide in vegetable samples by UV-visible spectrophotometer and FTIR. *Results in Chemistry*, 100059.

69. Santos, S. T., (2014). *Deep Exploration of the Benefits and Drawbacks of Sparse-based Models in NIR, Raman and Hyperspectral Imaging*. MSc Thesis; Tecnico Lisboa; https://fenix.tecnico.ulisboa.pt/downloadFile/844820067124097/dissertacao.pdf (accessed on 27 January 2021).

70. Saptari, V., (2003). *Fourier Transform Spectroscopy Instrumentation Engineering* (pp. 1–8). SPIE -Optical Engineering Press; Washington, DC.

71. Settle, F. A., (1997). *Handbook of Instrumental Techniques for Analytical Chemistry* (pp. 56, 57). Prentice-Hall, Upper Saddle River; New Jersey; USA.

72. Silverstein, R. M., Webster, F. X., Kiemle, D. J., & Bryce, D. L., (2014). *Spectrometric Identification of Organic Compounds* (8th edn., pp. 71–76). New Jersey, USA: Wiley.

73. Smith, B. C., (2011). *Fundamentals of Fourier Transform Infrared Spectroscopy* (pp. 15–45). Boca Raton, FL, USA: CRC Press.

74. Stuart, B. H., (2000). Infrared spectroscopy. In: Meyers, R. A., (ed.), *Encyclopedia of Analytical Chemistry* (pp. 529–559). Chichester, UK: John Wiley & Sons, Inc.

75. Stuart, B. H., (2004). *Infrared Spectroscopy: Fundamentals and Applications* (pp. 15–37). London-UK: John Wiley and Sons.

76. Sun, D. W., (2009). *Infrared Spectroscopy for Food Quality Analysis and Control* (pp. 8–11). Burlington, USA: Academic Press.

77. Svensson, O., Kourti, T., & MacGregor, J. F., (2002). An investigation of orthogonal signal correction algorithms and their characteristics. *A Journal of the Chemometrics Society, 16*(4), 176–188.

78. Thomas, (2020). https://chem.libretexts.org/Bookshelves/Analytical_Chemistry/ Supplemental_Modules_(Analytical_Chemistry)/Analytical_Sciences_Digital_ Library/Active_Learning/In_Class_Activities/Molecular_and_Atomic_ Spectroscopy/03_Text%3A_Molecular_and_Atomic_Spectroscopy/4%3A_Infrared_ Spectroscopy/4.3%3A_Fourier-Transform_Infrared_Spectroscopy_(FT-IR) (accessed on 27 January 2021).

79. Upadhyay, N., Goyal, A., & Rathod, G., (2011). Microwave spectroscopy and its applications in online processing. *Indian Food Industry, 30*(5/6), 63–73.

80. Upadhyay, N., Jaiswal, P., & Jha, S. N., (2018). Application of attenuated total reflectance Fourier transform infrared spectroscopy (ATR–FTIR) in MIR range coupled with chemometrics for detection of pig body fat in pure ghee (heat clarified milk fat). *Journal of Molecular Structure, 1153*, 275–281.

81. Upadhyay, N., Jaiswal, P., & Jha, S. N., (2016). Detection of goat body fat adulteration in pure ghee using ATR-FTIR spectroscopy coupled with chemometric strategy. *Journal of Food Science and Technology, 53*(10), 3752–3760.

82. Van, D. V. F. R., Sedman, J., Emo, G., & Ismail, A. A., (1992). A rapid FTIR quality control method for fat and moisture determination in butter. *Food Research International, 25*(3), 193–198.

83. Versari, A., Parpinello, G. P., Scazzina, F., & Del, R. D., (2010). Prediction of total antioxidant capacity of red wine by Fourier transform infrared spectroscopy. *Food Control, 21*(5), 786–789.

84. Wu, Z., Li, H., & Tu, D., (2015). Application of Fourier transform infrared (FT-IR) spectroscopy combined with chemometrics for analysis of rapeseed oil adulterated with refining and purificating waste cooking oil. *Food Analytical Methods, 8*(10), 2581–2587.

85. Yang, H., & Irudayaraj, J., (2002). Rapid determination of vitamin C by NIR, MIR, and FT-Raman techniques. *Journal of Pharmacy and Pharmacology, 54*(9), 1247–1255.

86. Ye, J., (2007). *Least Squares Linear Discriminant Analysis*, 7. http://staff.ustc.edu. cn/~zwp/teach/MVA/icml2007_Ye07.pdf (accessed on 27 January 2021).

87. Zeaiter, M., & Rutledge, D., (2009). Preprocessing methods. In: Tauler, R., Walczak, B., & Brown, S., (eds.), *Comprehensive Chemometrics-Clinical and Biochemical Data Analysis* (pp. 121–231). Roma London-UK: Elsevier. https://doi.org/10.1016/ B978-044452701-1.00074-0.

CHAPTER 11

ROBOTIC ENGINEERING: A TOOL FOR QUALITY AND SAFETY OF FOODS

RAVI PRAKASH and MENON REKHA RAVINDRA

ABSTRACT

Robots as quality and safety tools in food processing are basically reprogrammable devices intended for manipulating or moving parts, tools, or itemized industrial kits via adaptable-programed signals for performing commanded engineering tasks. Advances in food-robots since last few decades have open up numerous avenues for automation and process control in the food industry, which includes unit operations, such as sorting, chilling, freezing, packaging, labeling, serving, palletization, de-palletization, etc. The challenges in the widespread solicitation of food-robots are initial cost, need of sophisticated intelligent sensing elements to deal with varied size, shape, weight, and appearance of food materials, well-programmed end effecter to handle delicate food materials with safe lifting without zipping. The striking features in advances of food-robots could be no fatiguing, ease to work in tough and extreme physical situations (such as vacuum, life-threatening heat/cold/chemicals) and no emotional constraints (bored or drained by replicative work). The present chapter reviews the up-to-date interventions in food-robot as a tool for the quality and safety of foods.

11.1 INTRODUCTION

Robots being virtual and computational counterparts of farmworkers in food automation may be termed as *"food-robots,"* which are basically reprogrammable devices intended for manipulating or moving parts, tools, or itemized industrial kits via adaptable-programed signals coupled with feedback mechanisms for performing commanded engineering tasks. Robots could be one of the possible solutions to reduce human meddling in food processing to warrant quality and safety needs from the soil to fork along with the enhanced productivity, accuracy, and precision.

In general, it is appraised that processing of dairy and food commodities incur prodigious labor overheads up to about half of the total processing cost by employing humans, but it could be mitigated at large by deploying robots in long-run cost scheming by augmenting working hours, overall productivity, extra returns from the quality and safe products and ultimately the profitability. During food processing, rapidness, repetitiveness, and droning movement may demotivate farmworkers as time passes but do not impede robots, which will ultimately assist in moderating the handling losses and number of accidents, improve quality monitoring, control, and safety along with amplifying overall performance, exactness, and correctness.

This chapter reviews known interventions of food-robot as a tool for quality and safety during common unit operations related to foods with consequences and forthcoming drifts.

11.2 WHY FOOD-ROBOTS?

Food-robots can accomplish the number of particularly desired colossal tasks with augmented quality and safety unlike farmworkers, such as no tiring, resistant to tough physico-environmental situations, ability to work even in a vacuum, no tedious or fatigued by replicative jobs, zero chances of becoming out-of-focus from the assigned tasks, etc., [4, 33]. The outstanding features of a food-robot or a robot-assisted automated tool are briefly discussed in this section.

11.2.1 COMPACT FLOOR-SPACE DESIGN

The in-built design of a food-robot to be readily gesticulated with mountings, ceilings, walls, hangings, windows, and other accessorized paraphernalia could be accomplished in a confined floor-space to carry out manifold assignments with little movement of goods and services rendered apart from the supplementary output. It, therefore, contributes in maintaining quality and safety by minimizing the entrance of foreign elements into the food processing chain, which occasionally occur.

Moreover, a single robot sometimes may be programmed to carry out multiple assignments during time lag pre-set in the series of unit operations in food processing; therefore, flexibility in processing may also be achieved and improved. It has been reported that either deployment of full-fledged robots or semi-robotic systems accessorized with robotic tools in dairy and food processing could significantly enhance the overall productivity, quality, and safe performance in abridged floor area and total cost [40, 53].

11.2.2 ASEPTICISM

Since food-robots reduce or sometimes eliminate human interference, therefore ease of asepticism, one of the quality and safety implements, would be certainly up-surged as compared to direct human engagements. Moreover, there could be no chances of a robot getting infected by any of the diseases causing organisms or any disease transmission at any stage due to health issues occasionally seen in humans. In order to avoid batch to batch contamination or growth of spoilage/pathogenic organisms in robot grips used during the time lags of processing, the asepsis mechanisms and non-sticky hygienic protections has been recommended to be integrated with industrial food-robots.

11.2.3 EFFICIENCY AND EFFICACY

Of course, robots are basically machines, therefore can be engaged round the clock (24×7), unlike humans, without any limitations. Therefore, the overall output to input ratio per unit of labor overhead (efficiency) consumed could be maximized with improved efficacy.

11.2.4 LESS HANDLING, QUALITY, AND SAFETY LOSSES

A well-programmed robot with a peculiar design of grips and end-effectors for a food commodity can handle foods delicately, safely, and hygienically just enough to ensure little or no physical, chemical or microbiological injuries. Therefore, quality, safety, and handling losses occasionally arise due to these injuries; and these could be highly minimized by employing robotic tools, thereby overall qualitative and quantitative yields after processing could be enhanced.

11.2.5   ECOLOGICALLY RESILIENT

Robots as machine implements exhibit perfect resilience towards harsh or hostile surroundings with no physical, mental, or biological damages usually related with human interventions. It can be even engaged in severe hot/cold/ physically or chemically threatening atmospheres in food processing, such as inside cold storage, vacuum, steam chambers, dryers/evaporators, dark, intense light (such as inside UV chambers), no or less ventilated zones, etc.

The secondary benefits of robot designs include low energy consumption and total overheads per unit of human labor, which might have forgone for humans in fulfilling the said ecological facilities for obligatory comforts. Food-robots work well inside or near freezers, ovens, and furnaces more efficaciously during palletizing and de-palletizing.

11.2.6   RELIABILITY AND PRECISION

Since food-robots work inherently based on computational programmed artificial intelligence, therefore these remain reliable and precise even after repetitive, replicative, and gesture/posture catalyzed jobs, which generally affect physico-mental human health. Moreover, steadiness, and accuracy is also maintained; therefore food-robots exhibit marathon reliability and precision in operations.

11.3   COMPONENTS OF A FOOD-ROBOT

Based on engineering design requirements of a processing plant (such as layout, process flow, type of food materials being handled, sensor) and other artificial intelligence wants, extent of robotization (i.e., degree of human labor replacement), environmental, and other sophistication concerns, the structure of a food-robot or a robotic tool may vary from one assignment to another. In general, the chief constituents of a food-robot may include: a power source to stream energy needs, a control cum monitoring panel processor (such as human brain) to set, control, and monitor the process variable(s) along with the movement of a robot, a facility to set/manipulate (manipulator), the desired computational program, a feedback mechanism to receive signals of acts/deeds on assigned tasks, a commanding-cum-end-effecter manipulator for interacting with machine and job assigned.

The central processor resembles the human brain in a robot, and it enables various gesture, posture, and kinematics of the robot arms. A food-robot can work with an AC and/or DC supply [31].

Set and instantaneous indicators display the set and present values of process variables, actions taken and feedbacks recorded based on external and internal commands. The manipulator of the robotic tool is coupled with primary, secondary, and tertiary (or quaternary, if required) arms depending upon area coverage of robotization (Figure 11.1). The end effector interacts (i.e., senses, recognizes, grips, holds, carries, releases, and/or performs other likely tasks) bilaterally with the food-object being handled (according to the command from control panel) as well as the programming computational machine to simultaneously receive signals and re-directs feedback signals for further actions according to set and present process variables.

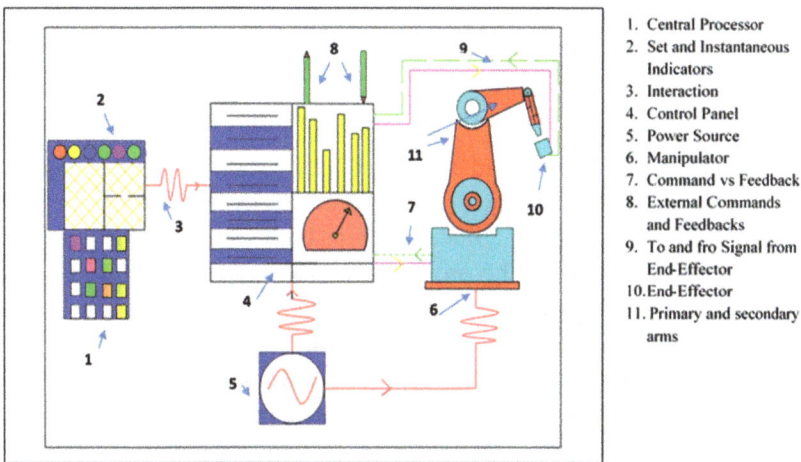

1. Central Processor
2. Set and Instantaneous Indicators
3. Interaction
4. Control Panel
5. Power Source
6. Manipulator
7. Command vs Feedback
8. External Commands and Feedbacks
9. To and fro Signal from End-Effector
10. End-Effector
11. Primary and secondary arms

FIGURE 11.1   Components of a food-robot.

11.4   TYPES OF FOOD-ROBOTS

Although broad categorization of food-robots based on few specific parameters appears to be hard-nosed, yet some of such classes are built on different definite benchmarks. The advances in food-robot took place leisurely only in few contemporary decades due to some of the inevitable challenges in food processing, which demand at least complex human interference basically to deal with a variety of complex handling and processing procedures for a range of food materials [9, 24].

The most prevalent food-robot with their peculiar specifications and handling ability [24] are presented in Table 11.1.

The various categories of food-robots based on engineering designs are being briefly discussed in this section.

11.4.1 BASED ON GESTURE, POSTURE, AND KINEMATICS

11.4.1.1 PORTAL FOOD-ROBOTS

These robots are basically straddling robotic systems spanning cubical control, supervision, monitoring, or handling zones by means of 3-rectilinear machetes parallel to the axial coordinates, whose kinematics are actually set along with the poignant constituents of the set-up situated overhead of the mounting.

11.4.1.2 ARTICULATED FOOD-ROBOTS

These are the most widespread industrial food-robots of higher degree of flexibility, kinematics, and ease of series connectivity, possessing compound interrelating modular arms facilitated with in-built gripping tools. The inter-relating modular arms may be programmed to move in the 3-dimensional space to carry out many-fold tasks with multiple (twice or four times) degree of freedom (DOF) to enable almost all combinations of movements. Since these robots possess the facility to be connected with interrelating modular arms to carry out fairly distant multi-fold assignments, therefore practical limitations are in bearing higher loads or heavy articles and had lower capacity after a circumscribed range.

11.4.1.3 SELECTIVE COMPLIANCE ARTICULATED ROBOT ARMS (SCARAS)

These robots are basically types of assembly/articulated robots, which consist of an assemblage of robots/robotic tools as series of arms, which are rigid along one axis (say Z-axis) and bendable along other two axes (i.e., respective XY-plane). They inherit foldable and retractable arms like humans that may be used for relocating food articles/packages from one cell to another or sometimes may be employed for loading and unloading purposes [8].

TABLE 11.1 Industrial Food-Robots and Their Specifications

Food-Robot	Model	Degrees of Freedom (DOF)	Workstation Covered (mm, dia.)/ Operating Ability (h)	Trifling Payload (kg)	Repeatability in Motion (mm)	Performance (cycles/min)	Make
Packaging robot	HD-RL3/4–1100	3	1100 mm	1	± 0.1	130	MAJAtronic, Germany
	KR16	6	3102 mm	35	± 0.05–0.08	>100	KUKA, Germany
	KR30/60						
Pick and place robot	M-430iA/2F	5	1130 mm	1–2	± 0.5	120–100	FANUC, USA
	KRAGILUS	6	1101 mm	6–10	±0.03	70–80	KUKA, Germany
	TS60SCARA	4	600 mm	2	±0.01	100	Stäubli, Switzerland
Palletizer robot	KRQUANTEC	5	3195 mm	120–240	±0.06	80–100	KUKA, Germany
	IRB660	4	3150 mm	180/250	±0.5	80	ABB, Switzerland
Inspect, test, and grader robot	FANUCM-1iA parallel link labeling robot	3/4/6	420 mm	1	±0.02	40	FANUC, Canada
	IRB360 Flex Picker robot	4	1600 mm	1–8	±0.1	variable	ABB Robotic
	IRB4600s	6+3	4600 mm	20–60	±0.05–0.06	variable	ABB Robotic
Serving robot	Pizza Hut pepper robot	17	12 h operation with motion up to 2 km/h	–	–	–	SoftBank Corp./ Aldebaran Robotics
	Connie	14	1–1.5 h	–	–	–	Hilton/IBM
	Nao	14	1–1.5 h	–	–	–	SoftBank Corp./ Aldebaran Robotics

11.4.1.4   DELTA FOOD-ROBOTS

Such robots are particular forms of parallel robots and resemble like spiders having a minimum of 3–4 articulated axes arms connected to a central base immobile actuator (sometimes multiple actuators) as universal joints. These actuators at the base allow limited inertia, high speeds, and accelerations to the articulated arms in the shape of a parallelogram to maintain the orientation of the end-effector (Figure 11.2). The features of such robots include: relatively compact in size, light in weight, expedient, profound, and flexible with higher degrees of freedom which facilitate better coverage in automation. The major applications of such robots include picking and placing, packaging, packing, etc. The various types of delta robots are: pocket delta, delta direct drive, delta cubical, linear delta.

11.4.2   BASED ON UNIT OPERATION

11.4.2.1   HARVESTING ROBOT

Harvesting agricultural produce is one of the most labor intense unit operations, own fabulous resource savings and quality prospects by deploying robotic tools. One of the earliest reports on the development of a laboratory-scale robot arm for harvesting mushrooms indicated a cycle period of 20s/3 mushrooms as compared with a classic human frequency of 12s [19]. The pick success rates were 69 and 92% in real-time and modeled results respectively, provided all other biological and environmental factors were controlled. In continuation of this work, the pick success rates without bruising damage were upgraded up to 85%, which comprised of an effective pick and scrap rates of 78.2 and 21.8% (5–10% for manual picking), respectively. The labor-intensive performance of the further developed robotic arms indicated the need of 31–34 robots to exchange contemporary 28 farmworkers [19].

Taqi et al. [49] indicated the development of a cherry-tomato harvesting robot based on image sensing mechanisms to differentiate between ripened and un-ripened tomato. This robot could harvest efficiently only the ripened and mature cherry tomato and leave unharvested the immature one without any damage to surrounding leaves or fruits. The major components of this robot were an arm base, pixy camera, linear actuator servomotor, vertical, and horizontal rod gears, grippers, infrared reflecting sensors, and load sensors. A typical food harvesting robot comprises of an end-effector equipped with a suitable recognition element to recognize a particular item to be harvested, a gripper to gently harvest and hold, a manipulator, and an integrated robotic arm(s) as schematically shown in Figure 11.3.

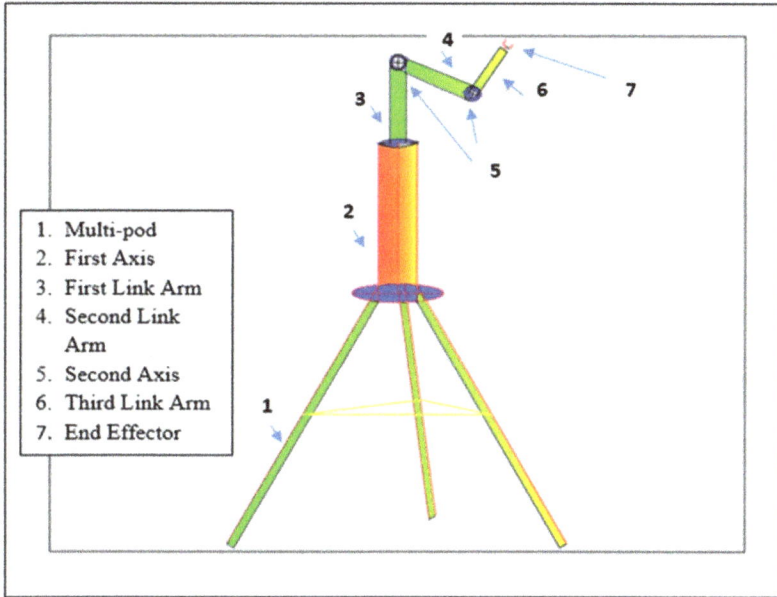

1. Multi-pod
2. First Axis
3. First Link Arm
4. Second Link Arm
5. Second Axis
6. Third Link Arm
7. End Effector

FIGURE 11.2   Schematic view of a delta-robot.

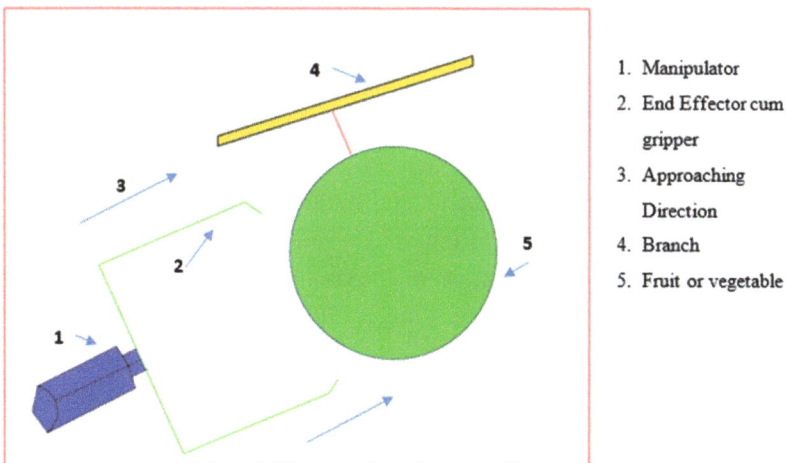

1. Manipulator
2. End Effector cum gripper
3. Approaching Direction
4. Branch
5. Fruit or vegetable

FIGURE 11.3   Schematic diagram of a harvesting gripper of a food-robot.

11.4.2.2   MILKING ROBOT

Automatic milking system (AMS) is a recent advancement in robotic milking, wherein a voluntary milking system is followed to facilitate a cow to set her own milking schedule as per wish of the animal. However, in the beginning, it needs training so as to make the dairy animal familiar with the system so that they can be milked with limited human interventions. Basically, each, and every dairy animal pertaining in a robotized milking set up is built-in with an automated electronic identifying label to assist a food-robot in identifying a particular milch animal during milking and milk production record keeping.

Technically, when a milch animal approaches a robot, the identifying tags are sensed and read by the computer programed robotic tools and thereafter, and the dairy animal gets feed recompense as made-to-order according to her milk production capacity. Then, the robot follows clean milk production procedures, such as cleaning of teats, attaching the milk cups of the automatic milking machine and performs gentle and rapid milking. At the end, the milking cups are automatically disconnected; as each quarter of the udder is milked sufficiently and then the dairy animal is permitted to exit from the robotized area. The various advantages of AMS include flexibility of animals as well as herd management, improved milking frequency and total yield per animal, better animal health and welfare (since case in point of mastitis and other udder infectious diseases could be reduced) [12, 21].

11.4.2.3   FOOD-ROBOTS IN DAIRY PROCESSING

Robotic tools are also engaged in the cheese manufacturing industry for packaging, picking, placing, piling, and re-pilling of heavy cheese blocks, hygienic curd slicing, stirring, and cooking of cheese curd, transferring sliced curds into cheese molds, turning, cutting, portioning, and so on.

These robots are integrated with artificially intelligent sensing elements to execute accurate and precise commands. The intelligent metering devices enable accurate filling and quantifying the bulk of processed food into a reliable package. The grippers integrated with robotic arms are used for heavy cheese blocks lifting, placing them on wooden platforms in the picking area, relocating them on a conveyor belt for transference [4].

11.4.2.4   PICKING AND PLACING ROBOT

The prime projection of deploying robots for picking and placing during various food handling operations helped in renovating the conventional practices in the food industry [31]. For instance, ABB IRB-660 and IRB-360 are nomenclatures for such robots. The first one is a serialized robot applicable in highly challenging consignment relocation while the second relied on parallel kinetic machine mechanism, which is specifically meant for heavy bulk gathering, picking, and placing in/onto platters, packets, boxes or sometimes feeding into other linked machines [1].

11.4.2.5   PACKING AND PALLETIZING ROBOT

Effective and efficient packing and palletizing by a robot in food processing rely on standardized verdicts pre-programmed to encounter the consignment load stipulations and the varieties of speeds obtainable. ABB is one of the latest technology leaders in such robots, which include a higher degree of flexibility, yield, and consistency to run into up-to-date packaging design, varied sizes of packs, product alternatives, and batch-to-batch patterns. Palletizing of cookies, drinks, pasta, confections, and related articles are currently loaded by such robots [1, 2].

11.4.2.6   SERVING ROBOTS

Serving-robots are one of the latest advances in food robotics to accept the order, deliver eatables/drinks and collect leftovers in a descending tray [5]. It can work round the clock and straightway pacts with the seller and customers. These robots are being deployed in real-time serving, hospitality at restaurants and receptions [23]. *Sushi* restaurants in Japan recently commenced execution of robotized food array for serving customers, whose ultimate aim is to have robotic servers and waiters in the future [24].

11.4.2.7   SCRUTINIZING, GRADING, AND TESTING ROBOT

Grading robot for fruits and vegetables has been in use for peaches, pears, and apples. These robots are programmed to inspect surfaces of fruits and vegetables from all sides, and pick and grade them based on shape, size, color, maturity, the extent of ripening and damage due to insects based on artificially intelligent sensors. Moreover, additional features of such robotization include: delicate handling using suitable grippers, improved degree of precision and season-wise historical database of products. A single robot with adequate design could be programmed to work for several types of fruits and vegetables.

Recent studies in these areas have reported reduced cost and improved performance robots having manifold imaging sensing, capturing, and recording means to inspect, grade, and scrutinize the food objects as per real-time quality parameters by use of deep learning algorithms [14, 35, 37].

11.4.2.8   COOKING ROBOTS

The major advantages of cooking robots are their compatibility with punitive and rugged surroundings occasionally faced near the cooking ovens (e.g., for pizza near 800°C), putative in baking and cooking in front of customers. In this robot, a vision relying on actuator regulator is integrated to facilitate hygienic and neat-conduct and also to shield human hands from hot surfaces [5, 24].

The developments in cooking robots are in infancy; therefore, these have tremendous potential to advance suitable robotized cooks to impart tireless cooking along with hygiene, quality, and safety. Moreover, it may reduce the total cost of cooking by deploying robot for round the clock with little physico-eco comforts.

11.4.3   BASED ON EXTENT OF ROBOTIZATION OF HUMAN REPLACEMENT

The extent of robotization in a process flow relies on multiple factors (such as initial cost, automation requisites, sequences of unit-operations, existing floor space available, etc.). Therefore, many companies robotize their processes step by step rather than substituting the entire work-power. Thus, food-robots are generally being deployed in a process industry either partially or semi-robotized or fully robotized.

11.4.4   BASED ON LEVEL OF AUTOMATION

The existing and wanted status of automation level in a processing plant (such as fully manual, semi-automated, fully automated) in reality will govern the upcoming robotization scheme. Sometimes, a robot-assisted tool or a robotic arm may be semi-automated along with partly human interference to carry out inhuman assignments.

11.4.5   BASED ON END-USER

Since food processing may capture almost all the segments like domestic, semi-industrial, and industrial sectors; therefore, it is remarkable to note that food-robots may be designed and engineered after specifying the end-user segmentations.

11.5   ADVANCES IN FOOD-ROBOTS

11.5.1   EARLY FOOD-ROBOTS

The food-robots were first brought into the bakery industry for direct handling of foods in the early 1990's. Such robotic tools were employed just to pick and place the food items at a set rational speed of 55–80 cycles per minute. Soon after the emergence of pick and place robot, the vision guidance technology was incorporated into it to locate precisely the picking site to facilitate efficient and rapid picking and placing operation even at highly repetitive rates. This integration of food-robots with vision technology resulted in the overall cost reduction per item [28, 39]. Thereafter, various packing and packaging setups installed such robots in wide ranges of applications, such as dairy, baking, confection, frozen, snack, and beverages, food storage houses.

Early days, setting out of food robotics were confined within the domain of few unit operations, such as packaging, palletizing, and de-palletizing in milk, beverages, chocolates, and foods.

At present, food robots cover a wide range of agricultural operations right from the field to the consumption, such as planting, spurting, spraying, picking, harvesting, cutting, transporting, handling, processing, quality detection, packaging, serving, etc. [45]. In beverages manufacturing, cleaning, counting, filling, ordering, sequencing, positioning, conveying, packing, lifting, and defect detections are being extensively carried out by robotic machines [43]. Moreover, sophisticated computer vision technologies based on multi-fold high definition (HD) cameras are being employed for identifying and inspecting defects via machine learning systems, which finally opened up numerous avenues for control monitoring and control in vegetables and fruits [42].

As far as productivity enhancements were concerned, an augmentation of +25% by engaging robotic tools has been reported as compared to actions by a human towline, which included varied rapidity of performance in diverse food segments [17]. Basically, the productivity enhancement depends on the number of factors, such as the extent of human interventions replaced by robots, the total number of robots engaged per operation, product, and process variation, speed, and accuracy demand and process-line specific requirements. A study carried out for the same derived from a pasta manufacturing Argentina based firm reported an enhancement of 10% in productivity by installing six robots [3].

At present, food robots cover a wide apart from unpacked products, food-robots have been reported to be successful in lifting, loading, and unloading (e.g., in retort processing), filling, picking, and arranging the packaged product (in market milk, ghee, etc.). Food robots in a form-fill-seal machine for packing liquid foods, at locations like crates, near the inlet of feeding section, the outlet of the bagging section, for orienting milk-bottles at the filling. Food robots are applicable for drinks, bags, stacks, vessels, carts, bulk amplues, cans and packets [32].

11.5.2 ADVANCES IN GRIPPER TECHNOLOGY FOR FOOD-ROBOTS

One of the most promising advances of food-robots include the design of suitable and hygienic grippers and integrate them with robotic holding elements to catch, grip, hold, carry, and properly release the food items (particularly soft and delicate objects) without injury or any physico-textural damage leaving no noticeable marks on their surfaces, which are yardsticks of quality, safety, physicochemical deteriorations and shelf-life, appearance, body, and texture and ultimately the customer's appeal. The fundamental design steps and principles underlying in good design of grippers comprise of the following elements in sequence:

- A means of imparting sufficient gripping force which could be either alone or in the combination of mechanical (mechanical holding by machine fingers mimicking human fingers), pneumatic (air pressure operated nozzle-diffuser arrangement to create holding force at the end of nozzle; Figure 11.4), hydraulic, pneumo-static/pneumo-dynamic (a kind of combined gripping mechanisms comprising the combination of air pressure and pseudo instant static/dynamic forces), electrical, vacuum (suck, hold, and release mechanism), froze forces (by instant freezing of surface and/or surrounding moisture of the food object being gripped), adhesive forces.
- A provision of sensing and recognizing element (s) to identify the specific object to be gripped among the group of similar/dissimilar objects (particularly in fruit and vegetable harvesting, sorting, etc.).
- A means of sensing exact closeness (locating sensor) and arrive appropriately secure position to commence gripping.
- A control mechanism to ensure adequate force imposition just enough to grip without textural damages coupled with command and feed-back mechanisms to orient the holding robotic arms, grip the object and hold enough to avoid zipping and finally gently releasing at the appropriate place.

The working steps of pinching grippers (typically possess at least two fingers) comprise of catching, enclosing the object into grippers (i.e., encloses the specimen partly or completely), holding, translating, and gently releasing. Out of these mentioned gripping forces obtainable in food-robotic grippers, the pneumatically operated grippers working as a vacuum suckers to hold food items are most widely used in the food industry.

In this field, recent advances include: the development of a universal hygienic food gripper by the German Institute of Food Technologies, Quakenbrück, working on negative/vacuum pressure existed between the gripper surface and food surface to generate sufficient holding force for gripping while released of the caught object was achieved by vacuum release mechanism. In this work, the constituents of the grippers were fabricated using food-grade stainless steel and plastics materials. Moreover, these grippers worked well for picking and crating of milk pouches in a form, fill, and seal machine also [18, 30].

The ease of designs, capital, and operation costs, maintenance cycles, accuracy, robustness, floor-area required and end-users performance play decisive role in calls for grippers for a particular food item. An appraisal-based links of grippers worked based on electric, pneumatic, and hydraulic forces are presented in Table 11.2.

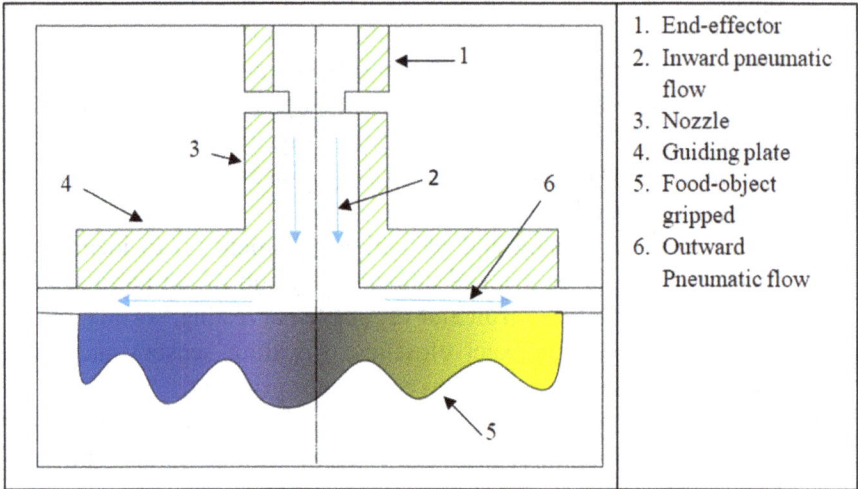

1. End-effector
2. Inward pneumatic flow
3. Nozzle
4. Guiding plate
5. Food-object gripped
6. Outward Pneumatic flow

FIGURE 11.4   Schematic diagram of a pneumatically operated food-gripper.

TABLE 11.2   Comparison of Grippers for Food-Robots

Parameters	Grippers for Food-Robots		
	Electrical	**Pneumatic**	**Hydraulic**
Accuracy and precision	High	High	Moderate
Strength and robustness	Moderate	Moderate	High
Operational speed	High	Moderate	High
Floor-space needed	Less	Less	Large
Cost	Low	High	High
Maintenance	Easy	Easy	Moderately difficult
Major drawbacks	Easily damaged, chances of electric shock and short-circuit, Ohmic heating due to electric current	Precise control needed, higher capital investment, good for low weight items	Good only for heavy payloads, higher capital investment, difficult to avoid surface injuries
End-user	Best for single-single fruit or vegetable	Best for bunch of fruits or vegetables	Best for single-single fruit or vegetable

The complications faced during designing of grippers for food-robots include: softness of foods (gripping forces sometimes may damage body and texture), irregular shapes and surfaces, hygiene requirements (e.g., microbial, toxic, chemical, or enzymatic deteriorations should be checked) during mechanical handling by grippers. Therefore, adequate calculations of gripping forces should be one of the pre-requisites of almost all designs of grippers to impart sufficient force to hold, carry, and gently place the objects.

In general, the gripping fingers should pass on a minimum amount of force (F), which must be sufficient to balance the sliding frictions (F_f) between the object's surface and the gripper which may be represented as follows [29]:

$$F_f = nF\mu \tag{1}$$

where; μ = the coefficient of friction between the food surface and the gripping fingers, n = total number of fingers.

To escape descending of a gripped object of mass 'm' at acceleration of 'a,' 'g' is acceleration due to gravity, F_f of a minimum two-finger gripper should be equilibrium in forces during vertical up, down, and horizontal movements.

For vertically upward motion:

$$F_f > m(g + a) \tag{2}$$

For vertically downward motion:

$$F_f > m(g - a) \tag{3}$$

In the case of horizontal carrying motion of the gripped object, following forces should be balanced to avoid slipping or zipping from grippers.

For horizontally parallel motion with finger faces:

$$F_f > m\sqrt{(g^2 + a^2)} \tag{4}$$

For horizontally normal motion with to finger faces:

$$F_f > mg \tag{5}$$

11.5.3 ADVANCES IN RECOGNITION DEVICES FOR FOOD-ROBOTS

11.5.3.1 MACHINE VISION-BASED RECOGNITION DEVICES

Recognition devices, basically working on the principle of computer vision technologies for food-robot, play a crucial role in distinguishing an actual object to be handled during harvesting, picking, plucking, gripping, placing, and many more unit operations. An efficient and apt recognition device will act as knowledge acquiring senses (particularly like 'eyes' in humans) in a food-robot, not only helps in good handling but also minimizes surface injuries, thereby maintains quality and safety.

The major challenges in designing and developing robotic recognition elements are: discerning schemes of fruit recognizers associated with devices to find out the location set out dimension(s) and identify the maturity of singular fruits. These stipulations should be backed up with suitable guiding mechanisms for machine-driven robotic arms towards the target object.

The general machine vision sensors for a fruit or vegetables robotic arm basically typify the target object based on color intensity, form, dimension, physique, and texture. These recognition elements are versatile in nature, and maybe employed for multiple applications (i.e., one element for multiple objects). Studies have been carried out to develop recognition elements for a particular fruit or vegetable, such as apple [46], cherry-fruit [48], cucumber [51], orange [20], tomato [28], strawberry [39], melon [15], and sweet pepper [7].

Apart from recognizing a single object, it is obligatory to ascertain cluster of similar objects (such as bunches of grapes) or a particular object from the cluster (i.e., fruit or vegetable from the cluster of leaves) or few particular objects from the cluster(s) of other objects. One of the early machine vision algorithms [27] for identifying locations of the number of tiny fruit items clustered in a 3-dimensional space reported a feat rate of around 70%. The further advances in these technologies included a vision recognition algorithm to contrast and treat the black and white images to identify fruits and vegetables guided by an intelligent control mechanism supported with independent sensing, planning, monitoring, and control units. Therefore, the feat rate of the food-robot was enhanced up to 85% [47].

This technology was further advanced by integrating a laser-ranging module [10] to ascertain three-dimensional locations of apples by a laser-guided apt and precise mechanism to calculate the real-time closeness of the end-effector with the fruit to be handled. This approach was found more accurate and realistic in locating an object as compared to the machine vision method which was computationally costly and delaying. In another study, a 3-dimensional stereoscopic machine vision was reported to recognize bunches of fruits based on dual (active as well as passive) reconstruction approach integrated with controlled illumination [50].

11.5.3.2   MULTI-SPECTRAL RECOGNITION DEVICES

Multi-spectral recognition devices for a food robot basically work on the principle of measuring absorbance/absorptivity, transmittance, scattering (or emissivity) of varied specific wavelengths (i.e., generally falls between Infrared (IR) and Near Infrared (NIR) regimes) of the incident and reflected (or refracted) light on the specimen (food objects); thereby characterize the physicochemical properties indirectly. The looked-for wavelengths for spectral study are achieved by a series of dichroic interference screens of pre-defined wavelengths and pass-through-band. The intensity of character-istic light vs. wavelength is generally evaluated by using suitable sensing elements fitted for every point of the captured scene. Uses of these methods may be extended to permit the supplementary data extraction, unlike ordi-nary human eyes, which miss the mark in capturing the receptors other than red, green, and blue lights.

In the recent past, NIR spectroscopic devices have non-destructive characterizations and online measurements by food-robots since they need negligible/no sample preparation. It has improved flexibility, rapidness and low cost. It is used for a variety of fruits and vegetables. For example, Chinese bayberry, citrus fruit, dates, dill, apple, avocado, banana, caraway, coriander, carrot, Japanese pear, kiwifruit, macadamia, mango, fennel, grape, green beans, guava, melon, mushroom, nectarine, peach, pepper, plum, tangerine, tomato, olive, onion, papaya, and many more in predicting chemical constituents, pH, total soluble solids, dry matter, protein, and sugar content, moisture level, starch, and essential oils compositions, interim injuries, maturity indices, nutrient levels (based on crop awning reflectance) based on surface characteristics (such as hardness, roughness, color, and appearance, etc.) [26, 34, 44].

The further advances in these areas included uses of hyperspectral recognition devices for detecting green un-mature apples in ordinary apple-tree foliage. This approach worked on sophisticated algorithms integrating methodologies like principal components analysis (PCA), extraction and classification of homogenous objects (ECHO) for scrutinizing spectra-line data and machine vision approaches, which reported a recognition rate of 88.1% (error = ± 14.1% due to overlying) [41].

In another study, a CCD camera-based recognition element integrated with 6 band pass-through screens to filter the appropriate wavelengths (600, 650 and 700 nm) of multispectral image analysis particularly recognition of citrus fruits [11]. The further advances in this work included the development of a multispectral recognition device of 780, 800, 900 and 960 nm NIR light screens to study the connection between spectrum data of green sweet peppers with the degree of identification, percent visibility, and ripening indices. It was reported that a 960 nm IR light was most realistic and operative to distinguish the sweet peppers [6].

11.6   CHALLENGES

Though the use of robots in automation has been well documented in all-encompassing industrialized unit operations related to various products [22, 52], yet the inevitable challenges accompanying food-robots limit their reputation at large [36]. These challenges may include the following arguments:

- One of the most prevalent challenges faced during robotizing the food industry was the higher initial cost (capital as well as operational costs) of robots and robotic tools, which are hardly affordable to most of the food processors (commonly belong into small to medium classes of firms).
- Needs of intelligent and sophisticated designs of sensing, gripping, carrying, holding, and locating elements in food processing chain to replace human labors. The various design and engineering challenges allied on the ways of widespread setting out of robots in food industries are widely varied ranges of sizes, shapes, weights, and positioning of raw as well as processed food commodities occasionally demanded and handled with inordinate care for quality and safety.
- Another noteworthy challenge in designing the end effecters, locators of food-robotic tools comprise of devising complicated holding gadgets to deal with occasionally delicate, greasy, icy, smooth, and sometimes slick food-objects during handling and processing at relatively higher rapidity, speed of response and accuracy along with safe lifting and no end tint on surfaces.

- Apart from designing and devising, other inevitable challenges include the issues related to sanitary and hygienic design and operational measures to be followed during food processing, which include the absence of dints, corners, etc., which are other requites of quality and safety.
- Moreover, robotization faced other obvious challenges while replacing humans were the absence of creativity, innovativeness, independent, and instant decision making (which are not pre-programmed), learning from mistakes and adaptability in a dissimilar pre-programed ecosystem.

11.7   FUTURE TRENDS

The unrivaled contributions of robots in carrying out tough and conscientious jobs in the food industry are advantageous part of the debate, but it faced numerous real world and procedural challenges as discussed in the preceding para, still remain furtive to the scientific world. The upcoming breakthrough in food-robots may include following investigates:

- Deployment of a cyber-physical system (CPS), a computational algorithm to tie the physical and virtual domain, which is a transdisciplinary approach rely on the idea of internet of things (IoT) to dig out the approaches to rationalize endwise production, processing, and supply links in food industry so as to attain the maximum level of safety confidence [25]. Moreover, CPS imparts sustenance in sanitary aids by self-directed gadgets and in precision agriculture by the use of drones. Presently, there have been trends of smart food labeling to trace the entire food chain to get deep dive into the exact trajectory food has followed till it reached the plate [38], therefore latest CPS based on cloud robotics are being deployed [13, 14] for these purposes. Another budding solicitation of CPS touching upon food quality includes the deterrence of contaminants from entering into food chain [16].
- Food industry holds marvelous possibilities for wireless sensor networks (WSN) for linking farm to the office (monitoring and managing of agricultural and animal farms), radio frequency identification (RFID) tools also for tracing, record keeping and documentation in upcoming future for consumer awareness and astringent standards to arrive across the demanded quality and safety norms by automated and integrated sensing elements to scan the defects, illness, and freshness.

- Apart from quality and productivity, food safety is another crucial factor linked with the end user's health and overall economy of a food business, which needs least human touch or intervention during any stage of processing, handling, transportation, and serving to nullify the entrance of pathogenic germs and other microorganisms in food chain. In order to fulfill the strict quality and safety standards, hygienic, and sanitary designs features of the food robotic components such as manipulators, end-effectors, monitoring, and vision tools, grippers, etc., need to be set and established.
- Moreover, cleaning and sanitization protocols for robotic tools using suitable cleaning agents, duration of application, concentrations, degrees of washing chemicals, their residual limit, and impact on human and eco-health need to be established.

11.8   SUMMARY

The present chapter covers an extensive review appraising the purview of food-robots along with their incredibility in augmenting quality and safety apart from precision, accuracy, speed, consistency, and overall productivity as matched with the traditional man-driven operations. Mostly food-robots have been traditionally deployed for heavy payload operations, such as lifting packages, palletizing, and de-palletizing of boxes; but emerging solicitations are more promising, which enables upcoming food-robots to be more delightful as well as fascinating.

KEYWORDS

- asepticism
- automation and process control
- food-robot
- fruits and vegetables
- gripper technology
- hygiene
- machine vision
- quality and safety
- robotic arms
- robotic milking

REFERENCES

1. ABB, (2007). ABB robotics launches simple programming tool. *Assembly Automation, 27*(1), 1–10.
2. ABB, (2016). *News: Picking.* São Paulo: ABB; http://new.abb.com/products/robotics/applications-by-industry/food-and-beverages/applications/picking (accessed on 27 January 2021).
3. ABB, (2015). *Power and Productivity for a Better World. São Paulo: ABB.* Available at: https://new.abb.com/betterworld (accessed on 27 January 2021).
4. Agrawal, A. K., Karthikeyan, S., Goel, B. K., Khare, A., Shrivastava, A. K., & Mishra, U. K., (2010). Robotization of Indian dairy industry: An indispensable step in futuristic processing plants. In: *Proceeding of National Seminar on Paradigm Shift in Indian Dairy Industry* (pp. 136–139). Karnal, India: NDRI.
5. Asif, M., Jan, S., UrRahman, M., & Khan, Z. H., (2015). Waiter robot: Solution to restaurant automation. In: *Proceedings of the 1ˢᵗ Student Multi-Disciplinary Research Conference (MDSRC)* (pp. 1–6). Wah, Pakistan.
6. Bachche, S., (2013). *Automatic Harvesting for Sweet Peppers in Greenhouse Horticulture* (pp. 1–20). PhD Dissertation; Kochi University of Technology; Kochi, Japan.
7. Bachche, S., Oka, K., & Ogawa, N., (2012). Distinction of green sweet pepper by using various color space models. In: *Proceedings of the Annual Conference of the Robotics Society of Japan* (pp. 1–10). Sapporo, Japan.
8. Brumson, B., (2011). *SCARA versus Cartesian Robots: Selecting the Right Type for Your Applications, Robotics.* http://www.robotics.org/content-detail.cfm (accessed on 27 January 2021).
9. Buckenhüskes, H. J., & Oppenhäuser, G., (2014). DLG-trend monitor: Roboter. *Robots in the Food and Beverage Industry, 9*(6), 16–17.
10. Bulanon, D. M., Burks, T. F., & Alchanatis, V., (2010). A multispectral imaging analysis for enhancing citrus fruit detection. *Environmental Control Biology, 48*, 81–91.
11. Bulanon, D. M., Kataoka, T., Okamoto, H., & Hata, S., (2004). Determining the 3D location of the apple fruit during harvest. In: *Proceedings of the Automation Technology for Off-Road Equipment* (pp. 91–97). Kyoto, Japan.
12. Butler, D., Holloway, L., & Bear, C., (2012). The impact of technological change in dairy farming: Robotic milking systems and the changing role of the stockperson. *Journal of Royal Agricultural Society and Engineering, 173*, 1–6.
13. Chaâri, R., Ellouze, F., Koubâa, A., Qureshi, B., Pereira, N., Youssef, H., & Tovar, E., (2016). Cyber-physical systems clouds: A survey. *Computer Networks, 108*, 260–278.
14. Chen, R. Y., (2017). An intelligent value stream-based approach to collaboration of food traceability cyber-physical system by fog computing. *Food Control, 71*, 124–136.
15. Dobrusin, Y., Edan, Y., Grinshpun, J., Peiper, U. M., & Hetzroni, A., (1992). Real-time image processing for robotic melon harvesting. *Trans. ASABE, 92*, 1–16.
16. Dong, Z., Li, F., Beheshti, B., Mickelson, A., Panero, M., & Anid, N. (2016). Autonomous real-time water quality sensing as an alternative to conventional monitoring to improve the detection of food, energy and water indicators. *Journal of Environmental Studies and Sciences, 6*(1), 200–207.
17. Gebbers, R., & Adamchuk, V. I., (2010). Precision agriculture and food security. *Science, 327*, 828–831.

18. Gjerstand, T. B., (2006). Handle of non-rigid products using a compact needle gripper. In: *Proceedings of the 39th CIRP International Seminar on Manufacturing Systems* (pp. 145–151). Ljubljana.

19. HDC, (1996). Effective mushroom harvesting, In: *Horticultural Development Company (HDC) Reports* (pp. 1–15). New York: USA.

20. Hannan, M. W., & Burks, T. F., (2004). Current developments in automated citrus harvesting. In: *Proceeding of ASAE Annual International Meeting* (pp. 1–10).

21. Higgs, D. J., & Vanderslice, J. T., (1987). Application and flexibility of robotics in automating extraction methods for food samples. *Journal of Chromatographic Science, 25*, 187–189.

22. Hurd, S. A., Carnegie, D. A., Brown, N. R., & Gaynor, P. T., (2005). Development of an intelligent robotic system for the automation of a meat-processing task. *International Journal of Intelligent System Technology Application, 11*, 32–48.

23. Jennings, C., Katchmar, M., Hickle, W., Michael, Z. A., & Traynor, C., (2005). *Robotic Beverage Server* (p. 9). Google Patents. U.S. Patent Application No. 11/210,244.

24. Khan, Z. H., Khalidb, A., & Iqbalc, J., (2018). Towards realizing robotic potential in future intelligent food manufacturing systems. *Innovative Food Science and Emerging Technologies, 48*, 11–24.

25. Khan, S. G., Herrmann, G., AlGrafi, M., Pipe, T., & Melhuish, C., (2014). Compliance Control and human-robot interaction, part I: Survey. *International Journal of Humanoid Robotics, 11*(3), 1–14.

26. Kim, Y. S., Reid, J. F., Hansen, A. C., & Zhang, Q., (2000). On-field crop stress detection system using multispectral imaging sensor. *Agricultural and Biosystems Engineering, 1*, 88–94.

27. Kondo, N., Nishitsuji, Y., Ling, P. P., & Ting, K. C., (1996). Visual feedback guided robotic cherry tomato harvesting. *Transactions of ASABE, 39*, 2331–2338.

28. Kondo, N., Yamamoto, K., Yata, K., & Kurita, M., (2008). A machine vision for tomato cluster harvesting robot. In: *Proceeding of ASABE Annual International Meeting* (Vol. 5, pp. 3111–3120).

29. Lien, T. K., (2013). Gripper technologies for food industry robots. In: Caldwell, D. G., (ed.), *Robotics and Automation in the Food Industry-Current and Future Technologies* (pp. 143–169). Cambridge, UK: Woodhead Publishing Ltd.

30. Lien, T. K., & Gjerstand, T. B., (2008). A new reversible thermal flow gripper for non-rigid products. *Transactions of the North American Manufacturing Research Institution of SME, 36*, 565–572.

31. Mahalik, N. P., & Nambiar, A. N., (2010). Trends in food packaging and manufacturing systems and technology. *Trends in Food Science and Technology, 21*(3), 117–128.

32. Naumann, M., (2010). Robots in food production. In: *DLG's Symposium* (pp. 1–10). Quakenbrück.

33. Nayik, G. A., Muzaffar, K., & Gull, A., (2015). Robotics and food technology: A mini-review. *Journal of Nutrition and Food Science, 5*, 380–384.

34. Nicolai, B. M., Beullens, K., Bobelyn, E., Peirs, A., Saeys, W., Theron, K. I., & Lammertyn, J., (2007). Non-destructive measurement of fruit and vegetable quality by means of NIR spectroscopy: A review. *Postharvest Biology and Technology, 46*, 99–118.

35. Pan, L., Pouyanfar, S., Chen, H., Qin, J., & Chen, S. C., (2017). Deep food: Automatic multi-class classification of food ingredients using deep learning. In: *IEEE 3rd International Conference on Collaboration and Internet Computing (CIC)* (pp. 181–189).

36. Peters, R., (2010). Robotization in food industry. In: *Proceeding of 5th International Conference on Food Factory for the Future 2010* (pp. 13–20). Gothenberg – Sweden.
37. Pierson, H. A., & Gashler, M. S., (2017). Deep learning in robotics: A review of recent research. *Advanced Robotics, 31,* 821–835.
38. Piramuthu, S., & Zhou, W., (2016). *RFID and Sensor Network Automation in the Food Industry: Ensuring Quality and Safety through Supply Chain Visibility* (pp. 53–87). New York, USA: John Wiley & Sons.
39. Rajendra, P., Kondo, N., Ninomoya, K., Kamata, J., Kurita, M., Shiigi, S., Hayashi, S., & Yoshida, H., (2009). Machine vision algorithm for robots to harvest strawberries in table-top culture greenhouse. *Engineering in Agriculture, Environment, and Food, 2,* 24–30.
40. Rene, J., Moreno, M., & Gray, J., (2010). Guidelines for the design of low-cost robots for food. *Industrial Robot: International Journal (Toronto, Ont.), 37*(6), 509–517.
41. Safren, O., Alchanatis, V., Ostrovsky, V., & Levi, O., (2007). Detection of green apples in hyperspectral images of apple-tree foliage using machine vision. *Transactions of ASABE, 50,* 2303–2313.
42. Saldaña, E., Siche, R., Luján, M., & Quevedo, R., (2013). Review: Computer vision applied to the inspection and quality control of fruits and vegetables. *Brazilian Journal of Food Technology, 16*(4), 254–272.
43. Saravacos, G., & Kostaropoulos, A. E., (2016). Equipment for novel food processes. *Handbook of Food Processing Equipment* (pp. 605–643). New York-USA: Springer.
44. Sui, R., Wilkerson, J. B., Hart, W. E., Wilhelm, L. R., & Howard, D. D., (2005). Multi-spectral Sensio for detection of nitrogen status in cotton. *Applied Engineering in Agriculture, 21,* 167–172.
45. Sun, D. W., (2016). *Computer Vision Technology for Food Quality Evaluation* (pp. 45–60). San Diego-CA, USA: Academic Press.
46. Tabb, A., Peterson, D., & Park, J., (2006). Segmentation of apple fruit from video via background modeling. *Transactions in ASABE, 2006,* 1–11.
47. Takahashi, T., Zhang, S., & Fukuchi, H., (2002). Measurement of 3D locations of fruit by binocular stereo vision for apple harvesting in an orchard. In: *Proceeding of ASAE Annual International Meeting ASABE* (pp. 1–10).
48. Tanagaki, K., Fujiura, T., Akase, A., & Imagawa, I., (2006). Cherry harvesting robot. In: *Proceedings of the International Workshop on Bio-Robotics, Information Technology, and Intelligent Control for Bio-Production Systems* (pp. 254–260). Sapporo, Japan.
49. Taqi, F., Al-Langawi, F., Abdulraheem, H., & El-Abd, M., (2017). A cherry-tomato harvesting robot. In: *Proceedings of the 18th International Conference on Advanced Robotics* (pp. 463–468). Hong Kong, China, IEEE.
50. Tarrio, P., Bernardos, A. M., Casar, J. R., & Besada, J. A., (2006). A harvesting robot for small fruit in bunches based on 3-D stereoscopic vision. In: *Proceedings of the World Congress Conference on Computers in Agriculture and Natural Resources* (pp. 270–275). Orlando, FL, USA.
51. VanHenten, E. J., Hemming, J., VanTuijl, B. A. J., Kornet, J. G., & Bontsema, J., (2003). Collision-free motion planning for a cucumber-picking robot. *Biosystem Engineering, 86,* 135–144.
52. Wallin, P., (1993). Advanced robotics in the food industry. *Industrial Robot, 20,* 12–13.
53. Zongwei, L., (2015). *Robotics, Automation, and Control in Industrial and Service Settings* (p. 150). Hershey, PA-USA: IGI Global.

INDEX

For Product Safety Concerns and Information please contact our EU
representative GPSR@taylorandfrancis.com
Taylor & Francis Verlag GmbH, Kaufingerstraße 24, 80331 München, Germany

www.ingramcontent.com/pod-product-compliance
Lightning Source LLC
Chambersburg PA
CBHW060757220326
41598CB00022B/2460